Generative AI-Powered Assistant for Developers

Accelerate software development with Amazon Q Developer

Behram Irani

Rahul Sonawane

Generative AI-Powered Assistant for Developers

Group Product Manager: Niranjan Naikwadi
Publishing Product Manager: Tejashwini R
Book Project Manager: Shambhavi Mishra
Senior Content Development Editor: Shreya Moharir
Technical Editor: Rahul Limbachiya
Copy Editor: Safis Editing
Proofreader: Shreya Moharir
Indexer: Hemangini Bari
Production Designer: Prashant Ghare
DevRel Marketing Executive: Vinishka Kalra

First published: August 2024
Production reference: 1230824

Published by Packt Publishing Ltd.
Grosvenor House
11 St Paul's Square
Birmingham
B3 1RB, UK.

ISBN 978-1-83508-914-9

www.packtpub.com

Contributors

About the author

Behram Irani is currently a technology leader at **Amazon Web Services** (**AWS**), specializing in data, analytics, and AI/ML. With over 20 years in the tech industry, he has helped organizations ranging from start-ups to large-scale enterprises modernize their data platforms. In the last 7 years at AWS, Behram has been a thought leader in the data, analytics, and AI/ML space. He has published a book and multiple papers, and has led digital transformation efforts for many companies worldwide. Behram holds a Bachelor of Engineering in Computer Science from the University of Pune and an MBA from the University of Florida. He is also the author of *Modern Data Architecture on AWS*.

Rahul Sonawane is currently a specialist solution architect at AWS specializing in data, analytics, and AI/ML. With over 24 years of technology experience, he has worked across multiple industries globally, including retail, consulting, and large-scale enterprises. Before joining AWS, Rahul worked on Google Cloud Platform for 2 years. For the last 7 years at AWS, he has been helping organizations modernize their data platforms. Rahul has published multiple public blog posts and articles focusing on data, analytics, and AI/ML. He holds a Master of Engineering in Computer Science from the University of Pune with a specialization in data warehouse technologies.

About the reviewers

Shantanu Kumar is an expert in designing and building large-scale data processing systems, AI infrastructure, and data platforms with over 8 years of engineering leadership experience at Amazon. He has designed and built many wide-ranging multi-tenanted systems for Amazon that support some of their key offerings such as 1-click product listing and prime video content recommendation. He led the Buy with Prime feature that elevated the customer experience and helped small and medium enterprises to offer Prime benefits to shoppers on their direct-to-customer (DTC) sites. He received several awards for his software engineering contributions and led a team of engineers to build scalable, reliable, and secure solutions in the cloud environment.

Bo Thomas leads the AWS Analytics Specialist Solutions Architect organization in the US for the financial services, retail, consumer packaged goods, automotive and manufacturing sectors. In this role, he advises some of the largest companies in the world on how best to design and manage their analytics and AI/ML platforms. He has more than 10 years of experience leading analytics, data engineering, and research science teams across Amazon's businesses. Prior to his current role, he led Amazon's enterprise people data warehouse and data lake platforms. He has a bachelor's degree in economics from West Point and an MBA from Duke University.

Table of Contents

Part 2: Generate Code Recommendations

3

Understanding Auto-Code Generation Techniques   51

4

Boost Coding Efficiency for Python and Java with Auto-Code Generation   83

5

Boost Coding Efficiency for C and C++ with Auto-Code Generation 109

6

Boost Coding Efficiency for JavaScript and PHP with Auto-Code Generation 125

7

Boost Coding Efficiency for SQL with Auto-Code Generation   143

8

Boost Coding Efficiency for Command-Line and Shell Script with Auto-Code Generation   159

9

Boost Coding Efficiency for JSON, YAML, and HCL with Auto-Code Generation   177

Part 3: Advanced Assistant Features

10

Customizing Code Recommendations   193

Part 4: Accelerate Build on AWS

14

15

Preface

In this fast-paced world of technology, developers face the challenge of managing repetitive tasks and maintaining productivity. This book provides a solution through Amazon Q Developer, a generative AI-powered assistant designed to optimize coding, streamline workflows and accelerate software development.

The journey of this book takes you through the setup and customization of Amazon Q Developer, demonstrating how to make use of its capabilities for auto-code generation, explanation, and transformation across multiple IDEs and programming languages. You'll learn how to use Amazon Q Developer to enhance coding experiences, generate accurate code references, and ensure security by scanning for vulnerabilities. Additionally, this book covers using Amazon Q Developer for **Amazon Web Services** (**AWS**)-related tasks, including solution building, applying architecture best practices, and troubleshooting errors. Each chapter provides practical insights and step-by-step guides to help you fully integrate this powerful tool into your development process.

By the end of this book, you will have mastered Amazon Q Developer, enabling you to accelerate your software development life cycle, improve code quality, and build applications faster and more efficiently.

Who this book is for

This book is for coders, software developers, application builders, data engineers, and technical resources using AWS services. You'll learn how to leverage Amazon Q Developer's features to enhance productivity and accelerate business outcomes. Basic coding skills are required. The book covers effortless code implementation, explanation, transformation, and documentation, helping you create applications faster and improve your development experience.

What this book covers

Chapter 1, *Introduction to Generative AI-Powered Assistants*, walks you through what generative AI-powered assistants are and how they work. We will look at some of the types of assistants for software developers and application builders and how they help boost productivity and improve user experiences.

Chapter 2, *Introducing and Setting Up Amazon Q Developer*, lays down the fundamentals of Amazon Q Developer and its features. We will look at how to enable the service so that it can work with multiple IDEs, command line, and also with various AWS services.

Chapter 3, Understanding Auto-Code Generation Techniques, provides insights into different auto-code generation techniques that Amazon Q Developer can assist with.

Chapter 4, Boost Coding Efficiency for Python and Java with Auto-Code Generation, looks at how you can use Amazon Q Developer to suggest code in the two most prominent programming languages used by developers: Python and Java. We walk you through the different auto-code generation techniques by building an example application using both of these programming languages.

Chapter 5, Boost Coding Efficiency for C and C++ with Auto-Code Generation, looks at how you can use Amazon Q Developer to suggest code in C and C++.

Chapter 6, Boost Coding Efficiency for JavaScript and PHP with Auto-Code Generation, looks at how you can use Amazon Q Developer to suggest code in the two important programming languages used by web developers: JavaScript and PHP.

Chapter 7, Boost Coding Efficiency for SQL with Auto-Code Generation, looks at how you can use Amazon Q Developer to suggest code in the most widely used database management and data manipulation language: SQL.

Chapter 8, Boost Coding Efficiency for Command-Line and Shell Script with Auto-Code Generation, looks at how you can use Amazon Q Developer to suggest code in command line and shell scripts.

Chapter 9, Boost Coding Efficiency for JSON, YAML, and HCL with Auto-Code Generation, looks at how you can use Amazon Q Developer to suggest code in JSON, YAML, and HCL formats that are used in prominent infrastructure as code services such as AWS CloudFormation and Terraform.

Chapter 10, Customizing Code Recommendations, looks at how you can use Amazon Q Developer's customization feature to allow code suggestions that align with the organization's internal libraries, proprietary algorithmic techniques, and enterprise code style.

Chapter 11, Understanding Code References, looks at how to use Amazon Q Developer to indicate the references of the generated code. We will also look at how you can turn it on/off and opt out of references.

Chapter 12, Simplifying Code Explanation, Optimization, Transformation, and Feature Development, looks at how Amazon Q Developer helps to explain, refactor, fix, and optimize code. We will also look at how it can upgrade projects by transforming the code to a newer version of the programming language. The concept of feature development will also be discussed with the help of an example.

Chapter 13, Simplifying Scanning and Fixing Security Vulnerabilities in Code, provides insights into how Amazon Q Developer scans for code vulnerabilities and also suggests how to fix security issues in code.

Chapter 14, Accelerate Data Engineering on AWS, covers how Amazon Q Developer assists data engineers and developers with coding in many of the services and tools provided by AWS.

Chapter 15, Accelerate Building Solutions on AWS, looks at how you can use Amazon Q Developer to get AWS-specific guidance and recommendations on a variety of topics, such as solution architecture, best practices, optimizing resources, and cost. It also helps with troubleshooting errors and support.

Chapter 16, Accelerate the DevOps Process on AWS, looks at how you can use Amazon Q Developer inside Amazon CodeCatalyst to speed up the application code building process.

To get the most out of this book

A basic understanding of programming languages such as Python, Java, C, C++, JavaScript, and PHP will help you build the example applications in this book faster with the help of Amazon Q Developer.

To start using Amazon Q Developer, you will either need an **AWS Builder ID**, which is free and easy to setup just by providing your email ID, or an **AWS Identity Center** login. For chapters that leverage Amazon Q Developer with other AWS services including Amazon SageMaker Studio, Amazon EMR Studio, Amazon Glue Studio, Amazon Redshift and AWS Lambda, you will need an **AWS account**. Many features of Amazon Q Developer are available for use in the **Free** tier, whereas certain features are available only in the **Pro** tier. Even though in this book we have highlighted features available only in **Pro** tier, it is advisable to always refer to the Amazon Q Developer pricing document for latest updates, the link to which is provided at the end of *Chapter 2*. Setup needed for multiple IDEs, command line, and also with various AWS services is also covered in that chapter.

Software/hardware requirements	Operating systems
VS Code, Visual Studio, or JetBrains IDE	Windows, macOS X, and Linux (any)

> **Note**
>
> This book does not have a GitHub repository since most of the code in the book is generated by Amazon Q and will vary with the readers' prompts. While the code examples are included to help illustrate the concepts discussed in the book and to ensure the continuity of the solution, they are intended for reference only. You are encouraged to experiment with your prompts and adapt the solutions to your specific needs. For those seeking additional resources, please refer to the official documentation and community forums for further guidance and support.

Note that *Chapters 4-9* demonstrate how Amazon Q Developer can enhance developer productivity by automatically generating code in numerous supported programming languages. Feel free to jump directly to the chapters that most interest you based on your expertise or preferences.

You may find typos in a few prompts and screenshots. We have purposely not corrected them to highlight that Amazon Q Developer understands the underlying meaning of what's being asked, even with incorrect grammar in the prompts.

Conventions used

There are a number of text conventions used throughout this book.

`Code in text`: Indicates code words in text, database table names, folder names, filenames, file extensions, pathnames, dummy URLs, user input, and Twitter handles. Here is an example: "Therefore, the code `compression:glue.Compression.SNAPPY` will fail to compile during the build stage."

A block of code is set as follows:

```
code to improve
def display_weather_table(temperature_data):
    df = pd.DataFrame(temperature_data, columns=['Date',
        'Temperature (°F)'])
    print(df)
```

Bold: Indicates a new term, an important word, or words that you see onscreen. For instance, words in menus or dialog boxes appear in **bold**. Here is an example: "The **Recommend Tasks** button is also presented on this screen, where Q can analyze the issue and assign more manageable tasks to users for faster actions."

> Use-cases or important notes
> Appear like this.

Get in touch

Feedback from our readers is always welcome.

General feedback: If you have questions about any aspect of this book, email us at `customercare@packtpub.com` and mention the book title in the subject of your message.

Errata: Although we have taken every care to ensure the accuracy of our content, mistakes do happen. If you have found a mistake in this book, we would be grateful if you would report this to us. Please visit `www.packtpub.com/support/errata` and fill in the form.

Piracy: If you come across any illegal copies of our works in any form on the internet, we would be grateful if you would provide us with the location address or website name. Please contact us at `copyright@packtpub.com` with a link to the material.

If you are interested in becoming an author: If there is a topic that you have expertise in and you are interested in either writing or contributing to a book, please visit `authors.packtpub.com`.

Share Your Thoughts

Once you've read *Generative AI-Powered Assistant for Developers*, we'd love to hear your thoughts! Scan the QR code below to go straight to the Amazon review page for this book and share your feedback.

https://packt.link/r/1-835-08914-3

Your review is important to us and the tech community and will help us make sure we're delivering excellent quality content.

Download a free PDF copy of this book

Thanks for purchasing this book!

Do you like to read on the go but are unable to carry your print books everywhere?

Is your eBook purchase not compatible with the device of your choice?

Don't worry, now with every Packt book you get a DRM-free PDF version of that book at no cost.

Read anywhere, any place, on any device. Search, copy, and paste code from your favorite technical books directly into your application.

The perks don't stop there, you can get exclusive access to discounts, newsletters, and great free content in your inbox daily

Follow these simple steps to get the benefits:

1. Scan the QR code or visit the link below

https://packt.link/free-ebook/978-1-83508-914-9

2. Submit your proof of purchase
3. That's it! We'll send your free PDF and other benefits to your email directly

Part 1: Generative AI-Powered Assistant

In this part, we will look at what generative AI-powered assistants are and how they work. We will then introduce Amazon Q Developer. We will also look at how to set it up so that developers and builders can start using it inside IDEs, the command line, and the AWS console and tools.

This part contains the following chapters:

- *Chapter 1, Introduction to Generative AI-Powered Assistants*
- *Chapter 2, Introducing and Setting Up Amazon Q Developer*

1

Introduction to Generative AI-Powered Assistants

In this chapter, we will look at the following key topics:

- What is generative AI?
- Common challenges faced by developers
- Generative AI-powered assistants for developers
- How developers benefit from assistants
- Types of assistants for developers

When ChatGPT came out, it revolutionized the way we interact with AI-based systems to seek the answers we need. ChatGPT opened the doors for **generative AI** (**GenAI**), a category of **artificial intelligence** (**AI**) that utilizes machine learning models to autonomously create new content, such as text, images, or other forms of data, based on patterns and information learned from existing datasets.

GenAI has disrupted multiple industries by transforming how they solve use cases. One of the areas that GenAI has revolutionized is the software development process. Instead of writing all the code by hand, GenAI can auto-generate a significant portion of the code, drastically improving the productivity of software engineers and developers who build applications.

GenAI not only auto-generates new code but also assists in many other tasks of the software development life cycle, including planning, analysis, design, testing, debugging, deployment, maintenance, and review. We will slowly unfold all aspects of it in this book.

This chapter will dive deep into how different GenAI-based assistants have emerged to help developers create software applications faster. But before we get into the details, we will explore the journey of generative AI, the challenges developers face on a daily basis, and how GenAI-based assistants can assist them. Let's begin by briefly understanding the power of generative AI.

What is generative AI?

Generative AI is a type of AI that can create new content, such as text, images, or even music, by learning from existing examples. It's like teaching a computer to understand patterns and styles from a huge amount of data, and then using that knowledge to produce something new and original, much like how a human might create a story or a painting based on their experiences and imagination. For instance, it can write articles, generate realistic pictures, or even help develop software code by predicting what comes next based on what it has learned.

Here are some key things to know about generative AI:

- **Generation versus analysis**: Generative AI can create brand new artifacts such as text, code, images, video, and so on, rather than just classify or extract insights from existing artifacts. For example, synthetic data generation is a brand-new generation of new artifacts.

- **Self-learning**: Many generative AI systems train themselves on large datasets to learn patterns and relationships. This allows them to generalize to new contexts.

- **Probabilistic models**: Generative AI build probabilistic models to generate new outputs that conform to patterns in the training data. Outputs are sampled from the learned probability distribution.

- **Varied applications**: Use cases include generating text (for example, articles, code, dialogue), images, audio, video game content, molecule designs, and much more.

- **Output diversity**: Generative models can produce a wide range of diverse, original outputs by capturing high-level patterns rather than copying verbatim.

- **Cutting-edge field** - Generative AI is an extremely active area of ML research, with innovations in models such as DALL-E, AlphaCode, and MuseNet demonstrating its rapid progress.

To put into perspective where in the AI stack generative AI fits, the following figure helps us understand the concept better.

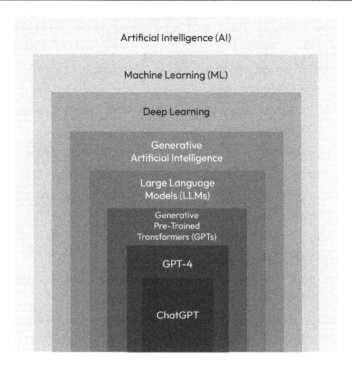

Figure 1.1 – AI stack

We want to keep the generative AI discussion brief as there is abundant material available to dive deeper into this field. The focus of this book is helping developers use GenAI to accelerate software development and improve their productivity. But before that, we also need to understand some of the challenges developers face.

Common challenges faced by developers/builders

If you look at what a typical day of a software developer looks like, you will realize that the bulk of the time is consumed by analyzing, creating, testing, and debugging code specific to a particular functionality required from the application. Many of the common challenges faced by developers are related to these themes:

- **Repetitive coding**: Manually implementing repetitive code or boilerplate without auto-complete or generation tools. This is not only time-consuming but also error-prone.

- **Understanding unfamiliar code**: Difficulty ramping up and understanding complex legacy code bases without AI explanations or summaries.

- **Finding code examples**: Tedious to manually search for and identify the right code examples to learn from or reuse without AI code search.

- **Diagnosing issues**: Debugging issues takes longer without AI assistance pinpointing potential causes and solutions for bugs.

- **Lack of standardization**: Code quality and consistency suffer without AI recommendations to standardize code patterns and styles.

- **Documentation**:– It is time consuming to manually write code documentation without AI automated documentation generation.

- **Reimplementing solutions**: Failing to discover and leverage existing solutions within a code base leads to redundant work.

- **Suboptimal efficiency**: Less guidance on improving code structure, performance, and efficiency without AI optimization.

- **Context switching**: Constant task switching disrupts developer flow without AI handling trivial tasks and lookups.

- **Knowledge gaps**: Beginners need guidance and intermediate developers have knowledge gaps without AI coding insights.

- **Creative limits**: Lack of idea stimulation and expanded solution search space without generative code suggestions.

- **Pace of technology changes**: New open source tools, libraries, and functionalities are getting added constantly so it's difficult to keep up to date with all new innovations and use them during application development.

Overall, generative coding AIs alleviate many pain points and augment human abilities at all skill levels. Developers stand to benefit greatly from adopting these rapidly emerging technologies.

Not everyone is tasked with coding applications. Many other technical personas exist in the organization who also assist in building applications by leveraging cloud-based services and tools. Typically, we refer to them as builders. Builders who solve use cases using AWS services are referred to as AWS builders. The kind of challenges these builders face also varies depending on their role and the tools they use. One of the most time-consuming challenges builders face is troubleshooting errors within a specific service or tool they use. Console errors and networking errors are the most prominent ones AWS builders encounter, for which they have to comb through log files to identify and fix the issues. AWS builders also need assistance with solution architecture, best practices, documentation, and support.

We will dive deep into solving each of these challenges in this book, but first, let's look at some areas of the software development process in which generative AI-powered assistants help developers and builders.

Generative AI-powered assistants for developers/builders

Generative AI-powered assistants work by leveraging advanced machine learning models, primarily trained on vast amounts of code and natural language data. Under the hood, these models analyze patterns and structures in existing code to predict and generate new code snippets. When a developer inputs a prompt or a partially written piece of code, the AI processes this input using deep learning techniques, understanding the context, syntax, and semantics. It then generates relevant code completions, suggestions, or even entire functions, mimicking the style and conventions of the existing code base.

Additionally, these assistants continuously learn and improve from user interactions, adapting to specific coding styles and preferences, thereby providing increasingly accurate and context-aware assistance over time. This intricate process of pattern recognition, contextual understanding, and continuous learning allows generative AI-powered coding assistants to significantly enhance developer productivity and efficiency.

Here are some examples of what they can do:

- **Code completion**: Predict and autocomplete code as the developer is typing based on the context. Similar to autocomplete but more powerful and contextual.

- **Code synthesis**: Generate entire code snippets or functions given a description of what the code should do. Saves developers time writing boilerplate or repetitive code.

- **Code explanation**: Provide plain language explanations of what a section of code is doing to help developers understand code bases.

- **Code summarization**: Summarize the overall purpose and flow of a code module at a high level. Useful for understanding legacy code.

- **Code error diagnosis**: Analyze error messages and stack traces to provide guidance on potential causes and fixes for bugs.

- **Code optimization**: Suggest improved ways to structure, consolidate, or streamline code to make it faster, more efficient, and so on.

- **Documentation generation**: Automatically generate code documentation and comments from code context.

- **Troubleshooting and issue resolution**: Understand the cause of the issue and provide possible solutions to builders.

- **Architecture and best practices**: Provide recommendations for builders on how to best architect a use case and also recommend the type of infrastructure to use to solve it.

The main value of these AI assistants is alleviating repetitive or rote aspects of coding to allow developers to focus on more high-value, creative parts of software development. They aim to increase developer productivity and software quality.

How developers/builders benefit from assistants

Generative AI assistants for developers are useful for several key reasons:

- **Improved productivity**: They automate repetitive coding tasks and workflows, allowing developers to get more done in less time. Things such as autocomplete, code generation, and debugging assistance directly save developers time and effort.

- **Reduce cognitive load**: By handling rote tasks and providing context-aware recommendations, AI assistants reduce the burden on a developer's working memory. This frees up mental bandwidth for more complex problem-solving.

- **Code discoverability**: Features such as natural language code search and summaries improve the discoverability of code bases. Developers can more easily find and understand relevant code examples.

- **Knowledge sharing**: AI models can encode programming best practices and patterns. This makes it easier to share knowledge across developer teams.

- **Consistency**: Code generated or optimized by AI tools adheres to consistent style and patterns. This improves code maintainability.

- **Beginner skill improvement**: Less experienced developers can leverage AI-powered completions, explanations, and recommendations to improve their skills faster.

- **Focus on creativity**: With rote coding work automated, developers can spend more time on creative problem-solving and optimizing algorithms.

- **Reduced errors**: Bugs and antipatterns can be automatically detected and fixed in real time as developers code. This improves software quality and reliability.

- **Latest technology integration**: Ease of usage of newer libraries, open source tools, and functionalities to improve the end-to-end code.

Overall, by augmenting human capabilities, generative AI enables developers to be more productive, write higher-quality code, discover new solutions, and focus their efforts on where humans add the most value. The potential for these tools to transform software development is very significant.

Types of assistants for developers/builders

Even though the list of generative AI-powered assistants keeps growing, here are some of the top generative AI-powered assistants for developers:

- **Amazon Q Developer**: A generative AI-powered assistant proficient in helping developers throughout the software development life cycle. When integrated into the IDE, Amazon Q offers comprehensive software development support, including code generation, explanation, optimization, and transformation, among many other automation features.

When used by AWS builders, it provides comprehension support for building faster solutions using various AWS services and assists with architecture, best practices, documentation, troubleshooting issues, and support.

- **GitHub Copilot**: A plugin from GitHub and OpenAI that provides context-aware code completions inside development environments.

- **DeepCode**: A code review assistant that identifies bugs, security issues, performance problems, and so on and suggests fixes.

- **Kite**: Autocompletion with documentation and code explainers to enhance code understanding.

- **Codex**: An API from OpenAI to generate code snippets from natural language descriptions.

- **TabNine**: A code completion tool that uses deep learning to suggest relevant code snippets in real time as developers type.

- **Pythia**: Facebook's IDE plugin that suggests code edits and transformations to fix issues.

- **Sourcery**: An AI tool that automatically refactors Python code, suggesting improvements and optimizations for cleaner and more efficient code.

The list keeps expanding as more start-ups, big tech companies, and open source projects integrate generative AI into the software development process. The goal is to augment productivity and software quality. Every GenAI assistant is different and assists developers in a variety of ways; however, in this book, we will go into detail about how developers and builders can use Amazon Q Developer to accelerate software development and boost their productivity on a daily basis.

Before we wrap up this chapter, here's a very important note about generative AI-powered assistants used by developers.

> **Always keep this in mind**
>
> Generative AI-powered assistants used for code generation exhibit a non-deterministic nature by producing different outputs for the same input under varying conditions. This variability arises from the underlying probabilistic models, which consider a range of possible solutions and select one based on factors such as context, learned patterns, and randomness. While this can introduce creativity and adaptability in code suggestions, it may also lead to inconsistent results, making it important for developers to review and validate the generated code to ensure it meets specific project requirements and standards.

Summary

In this chapter, we had a quick introduction to generative AI and how it has revolutionized many industries. Specifically in the software development industry, we examined some of the common challenges faced by developers. We then introduced various areas of the development process where generative AI-powered assistants can help developers and how they can benefit from such assistants.

Finally, we looked at some of the assistants available in the market today and how they help solve day-to-day challenges. We specifically highlighted Amazon Q Developer, a service provided by AWS that can help developers and builders. The rest of the book will revolve around Amazon Q Developer as the go-to assistant for software development.

In our next chapter, we will look at what Amazon Q Developer is and how it helps developers. We will also look at how to set it up in a variety of **integrated development environments** (**IDEs**) and other development tools typically used.

References

Generative AI: `https://en.wikipedia.org/wiki/Generative_artificial_intelligence`

2

Introducing and Setting Up Amazon Q Developer

In this chapter, we will look at the following key topics:

- Amazon Q nomenclature
- Amazon Q Developer basics
- Amazon Q Developer features
- Amazon Q Developer tiers
- Amazon Q Developer setup for third-party IDEs
- Amazon Q Developer setup for the command line
- Amazon Q Developer setup for AWS coding environments
- Build on AWS with support from Amazon Q Developer

In our previous chapter, we laid the foundation for how generative AI-powered assistants help developers improve their productivity, and we also looked at some of the assistants available in the market. In this chapter, we will focus on **Amazon Q Developer** – a developer tool that helps us understand and build applications using generative AI. It supports developers across the software development lifecycle. With Amazon Q Developer, employees receive timely, pertinent information and guidance, facilitating streamlined tasks, faster decision-making, and effective problem-solving, and fostering creativity and innovation in the workplace.

Let's get started with an important note on the service nomenclature.

Amazon Q nomenclature

Before we go any further, we want to clarify the nomenclature to prevent any misunderstandings for the remainder of this book.

Amazon Q is the flagship term used to refer to the generative AI-powered assistant from AWS. Under this term, there are multiple products and/or features of existing AWS services that specifically assist with certain types of technology domains and the personas that typically work in those domains. Let's quickly look at those:

- **Amazon Q Business**: This is a generative AI-powered assistant capable of answering questions, providing summaries, generating content, and securely completing tasks based on data from your enterprise systems. It empowers employees to be more creative, data-driven, efficient, prepared, and productive.

- **Amazon Q in QuickSight**: Amazon QuickSight is a **business intelligence** (**BI**) service from AWS. With Amazon Q in QuickSight, customers receive a generative BI assistant that enables business analysts to use natural language to build BI dashboards in minutes and effortlessly create visualizations and complex calculations.

- **Amazon Q in Connect**: Amazon Connect is a service from AWS that enables businesses to set up and manage a customer contact center with ease, offering various features to enhance customer service and support. Amazon Q in Connect leverages real-time conversations with customers and relevant company content to automatically suggest responses and actions for agents, enhancing customer assistance.

- **Amazon Q in AWS Supply Chain**: AWS Supply Chain is a service from AWS that unifies data and offers ML-powered actionable insights, built-in contextual collaboration, and demand planning. It seamlessly integrates with your existing **enterprise resource planning** (**ERP**) and supply chain management systems. With Amazon Q in AWS Supply Chain, inventory managers, supply and demand planners, and other stakeholders can ask questions and receive intelligent answers about their supply chain's status, underlying causes, and recommended actions. They can also explore "what-if" scenarios to evaluate the trade-offs of various supply chain decisions.

- **Amazon Q Developer**: This book focuses on this service. It assists developers and IT professionals with a wide range of tasks, including coding, testing, upgrading applications, diagnosing errors, performing security scans and fixes, and optimizing AWS resources.

> **Legacy name and nomenclature used in this book**
>
> The code assistant was previously called Amazon CodeWhisperer and has now been rebranded as part of Amazon Q Developer. You may see the legacy name CodeWhisperer in some places while using the tool with multiple AWS services, but it's all part of Amazon Q Developer now.
>
> Also, throughout the book, for brevity, we will sometimes use the terms Amazon Q, Q Developer, or just Q to refer to **Amazon Q Developer**.

Before we dive into how to set up Amazon Q Developer in different environments, let's first explore some of its basics.

Amazon Q Developer basics

After understanding the challenges developers face, AWS created a generative AI-powered assistant – Amazon Q Developer.

Developers typically follow the **software development lifecycle** (**SDLC**): plan, create, test, operate, and maintain. Each stage in this process is often repetitive and error-prone. As a result, the process takes significant time and effort, hampering developer productivity. The following figure shows the typical SDLC tasks on which developers spend time and effort.

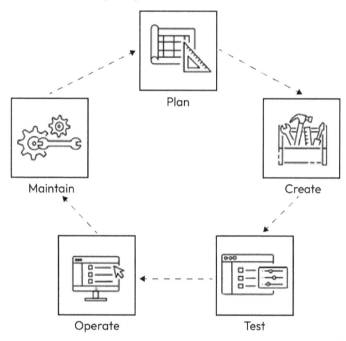

Figure 2.1 – The SDLC tasks that developers spend the most time on

Amazon Q Developer assists developers across the SDLC. Let's look at this at a high level before diving deep into each of these components throughout the book:

- **Plan**: Amazon Q assists during the planning phase by providing code explanations and helping with AWS best practices and recommendations.

- **Create**: Amazon Q enhances development productivity by offering in-line coding suggestions, generating new features using natural language, and allowing you to ask questions directly within the **integrated development environment** (**IDE**).

- **Test**: Amazon Q aids developers in verifying the functionality and security of their code. It assists in unit testing and identifies and resolves security vulnerabilities earlier in the development cycle.

- **Operate**: Amazon Q is equipped to troubleshoot errors, analyze VPC reachability, and provide enhanced debugging and optimization suggestions.

- **Maintain**: Amazon Q's Code Transformation feature assists in maintaining and modernizing code by upgrading projects to newer language versions.

Now, let's look at the features of Amazon Q Developer.

Amazon Q Developer features

We have dedicated chapters to dive deep into each of the features. This chapter will introduce the capabilities and help you complete the initial setup required for it to work with different tools. Let's begin with the most important feature, auto-code generation.

Auto-code generation

Amazon Q Developer's ability to generate substantial code accelerates application development, enabling developers to address previously unattended business-critical issues. This creates additional time for envisioning and crafting next-generation innovative experiences. Additionally, by conducting security scans within the **integrated development environment** (IDE), it identifies and rectifies potential vulnerabilities early in the application lifecycle, reducing costs, time, and risks associated with development.

Amazon Q Developer seamlessly integrates into the developer's IDE. By installing the Amazon Q IDE extension, developers can start coding immediately. As code is written, Amazon Q Developer autonomously assesses both the code and accompanying comments. Recognizing natural language comments (in English), Q provides multiple real-time code suggestions, even offering completion suggestions for comments as they are written.

Amazon Q goes beyond individual code snippets, suggesting entire functions and logical code blocks, often spanning 10–15 lines, directly within the IDE's code editor. The generated code mirrors the developer's writing style and adheres to their naming conventions. Developers can swiftly accept the top suggestion (using the *tab* key), explore additional suggestions (using arrow keys), or seamlessly continue with their own code creation process. A link to the complete list of user actions for Amazon Q Developer in different IDEs is provided in the *References* section of this chapter.

Amazon Q Developer supports many programming languages such as Python, Java, JavaScript, TypeScript, C#, Go, Rust, PHP, Ruby, Kotlin, C, C++, shell scripting, SQL, and Scala. Additionally, Q Developer is accessible as an extension in many IDEs such as Visual Studio, VS Code, and JetBrains IDEs, and is natively available in AWS Lambda, Amazon SageMaker Studio, Amazon EMR Studio, Amazon Redshift, Amazon CodeCatalyst, and AWS Glue Studio.

We have multiple chapters related to auto-code generation in *Parts 2, 3 and 4* of this book.

Code customizations

Amazon Q Developer enhances its suggestions by considering the nuances of internal codebases, which is crucial for organizations with extensive repositories, internal APIs, and unique coding practices. Developers often grapple with navigating large internal code repositories that lack comprehensive documentation. To address this, Amazon Q allows secure integration with an organization's private repositories. With just a few clicks, developers can tailor Amazon Q to provide real-time recommendations aligned with internal libraries, APIs, packages, classes, and methods.

This customization supports multiple data sources, enabling organizations to verify that recommendations align with coding standards, security protocols, and performance best practices. Administrators have granular control, selecting repositories for customization securely and implementing strict access controls. They decide which customizations to activate and manage access for developers within the organization. Each customization operates independently, preserving the integrity of the foundational model and safeguarding intellectual property. This ensures that only authorized members with specific access can view, access, and utilize these tailored recommendations.

We will cover this topic in depth in *Chapter 10* of the book.

Code transformations

Currently, Amazon Q lets you upgrade code written in Java 8 and Java 11 to Java 17. To assist with this feature, the Amazon Q Developer Agent for Code Transformation is available to generate a transformation plan used to upgrade your code. After transforming your code, it provides a transformation summary and a file difference, allowing you to review the changes before accepting them.

We will cover this topic in depth in *Chapter 12* of the book.

Code explanation, optimization, and update

Amazon Q Developer can explain, optimize, refactor, fix, and update specific lines of code within your IDE. To update your code, simply ask Amazon Q to modify a particular line or block of code. It will generate new code that incorporates the requested changes, which you can then insert directly into the original file.

We will cover this topic in depth in *Chapter 12* of the book.

Code feature development

The Amazon Q Developer Agent assists with developing code features or making changes to projects within your IDE. Describe the feature you want to create, and Amazon Q will use the context of your current project to generate an implementation plan and the necessary code to bring the feature to life.

We will cover this topic in depth in *Chapter 12* of the book.

Reference tracking

Trained on extensive datasets comprising billions of lines from both Amazon and open source code, Amazon Q Developer recognizes instances where a code suggestion resembles specific open source training data. It can annotate such suggestions with repository and licensing details. Additionally, it keeps a record of accepted suggestions that share similarities with the training data, facilitating the provision of proper attribution.

We will cover this topic in depth in *Chapter 11* of the book.

Security scanning

Amazon Q Developer also helps conduct scans on both generated and developer-written code to identify potential security vulnerabilities. It also provides recommendations for addressing identified vulnerabilities. The scanning process extends to detecting elusive security issues and is compatible with Python, Java, and JavaScript in VS Code and JetBrains IDEs.

Security scans cover various aspects, including compliance with **Open Worldwide Application Security Project (OWASP)** standards, enforcement of cryptographic library practices, adherence to AWS security standards, and other best practices.

We will cover this topic in depth in *Chapter 13* of the book.

Integration with AWS services

Amazon Q Developer also accelerates development on AWS by integrating with many services, allowing for rapid application building using AWS services.

Data engineers can expedite the creation of data pipelines by leveraging Q's integration with AWS Glue Studio Notebook and Amazon EMR Notebook. Creating SQL queries in Redshift becomes simple by articulating business outcomes using plain English sentences, as Q auto-generates SQL queries within the Redshift query editor. Data scientists and ML engineers can accelerate the ML development process by leveraging Q's integration with Amazon SageMaker Studio.

AWS builders can utilize Q Developer's integration with AWS Lambda to swiftly build event-driven logic. Q also supports the DevOps process through Amazon CodeCatalyst, assisting in many of its features such as pull requests and code changes.

Amazon Q Developer is also aware of the AWS resources in your account and can easily list specific aspects of your resources through its conversational capability. It can also help you understand the cost of AWS services used.

Q Developer not only automates many development tasks within AWS but also aids builders in troubleshooting lambda code and understanding networking-related issues.

Furthermore, you can consult Q for best practices and solutions for use cases. It can also recommend optimal EC2 instances for specific use cases. The chat capability of Q allows you to ask questions and receive responses easily, simplifying integrations with AWS support.

We will dive deeper into Q's assistance with AWS services in the *Build on AWS with support from Amazon Q Developer* section later in this chapter. A more detailed exploration of these capabilities is also provided in separate chapters in *Part 4* of the book, where we explain real-world development use cases in detail.

Amazon Q Developer provides many features, with some advanced options available in a Pro tier. In the following section, we will walk through both the free and Pro tiers and explain how you, as a user, can leverage each of them.

Amazon Q Developer tiers

Amazon Q Developer offers two tiers: free tier and Pro tier. Let's quickly look at how these two tiers work and what features are available in them.

Amazon Q Developer free tier

The free tier of Amazon Q Developer provides monthly limits for anyone logged in as an AWS IAM user or AWS Builder ID user. The specific features available to you depend on your interface and authentication method.

The best thing about the free tier is that anyone can use Amazon Q Developer in one of the supported IDEs they use for software development, even if they don't use AWS services or don't have an AWS account.

So, if you are reading this book and don't have an AWS account set up, you can quickly set up an AWS Builder ID using this link: `https://profile.aws.amazon.com`. Your AWS Builder ID represents you as an individual and is separate from any credentials and data associated with your existing AWS accounts. All you need is your personal email ID to quickly set it up.

The following screenshot highlights my AWS Builder ID page once it is set up.

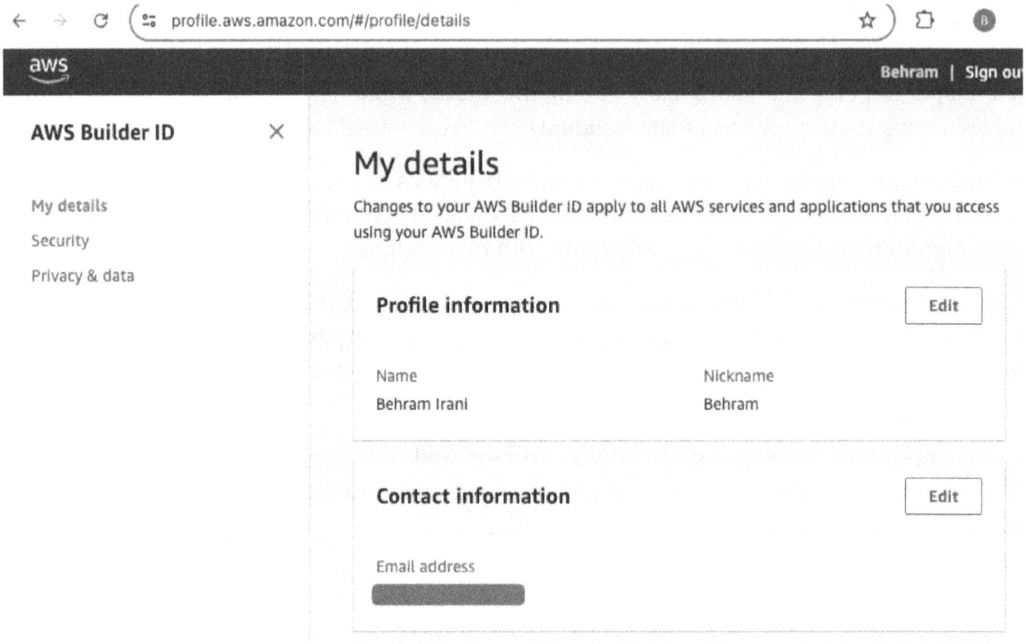

Figure 2.2 – AWS Builder ID creation

Once your AWS Builder ID is created, you can log in to one of the supported IDEs. We will cover this part when we go through the IDE setup section later in the chapter.

Amazon Q Developer Pro tier

To access Amazon Q Developer Pro, you must be an IAM Identity Center user, and your administrator must subscribe you to Amazon Q Developer Pro. As a subscriber, your usage limits are determined at an individual user level in the Amazon Q Console, Q in the IDE, and Q in Amazon CodeCatalyst.

If you are part of an organization, then access to the Pro tier will be set up by the admin team. However, if you are reading this book and want to try out some of the capabilities of Amazon Q Developer that are only available in the Pro tier, you can do so as an individual AWS user as well.

We will quickly guide you on one of the ways to access the Pro tier. Once you set up your AWS account, you will most likely assign yourself the admin role so that you can access all AWS services without provisioning additional privileges. However, if you are not the admin, make sure you have the admin role for the Amazon Q Developer service.

Once you log in to the AWS console, search for and open the Amazon Q service page. The following screenshot shows the Q Developer Pro bundle available for subscription.

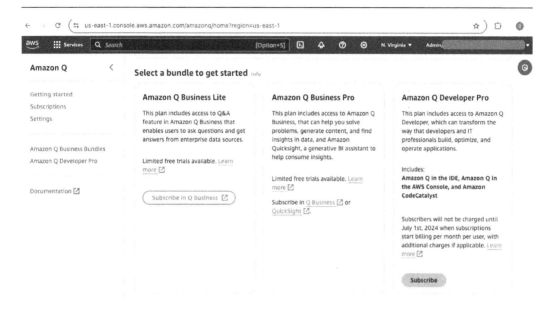

Figure 2.3 – Amazon Q Developer Pro bundle

As soon as you hit **Subscribe**, it will ask you to select a user or group from the IAM Identity Center to grant the pro subscription to. If you are a first-time user and your IAM Identity Center user is not set up, you will have to set it up before you can assign it, as seen in the following screenshot.

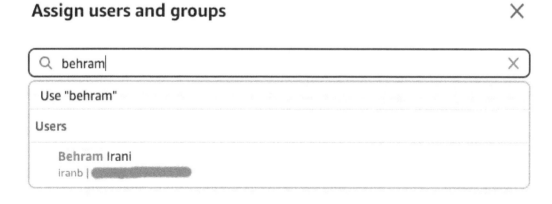

Figure 2.4 – Amazon Q Developer Pro – IAM Identity Center users and groups

Setting up users in IAM Identity Center involves several tasks, and we have included a link to the setup instructions in the *References* section of this chapter. Once the user is ready, they will appear in the dropdown menu for you to search by name and assign in the previous step. During this setup process, IAM Identity Center will also configure Amazon Q in its application settings, where the identity source and authentication settings will be found.

Once you have assigned the user in the subscription screen, the Pro tier subscription will be active, as seen in the following screenshot.

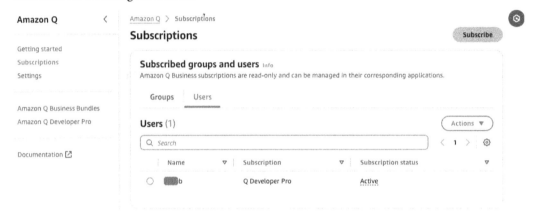

Figure 2.5 – Amazon Q Developer Pro – Active subscription for a user

Once the subscription is active, go to the Amazon Q Developer service from the AWS console, and on its settings page, you will see a full list of features available for the user to use, along with the start URL. The start URL is what we will use when authenticating in the external IDE.

The following screenshot highlights the Amazon Q Developer Pro subscription details that are ready for the user.

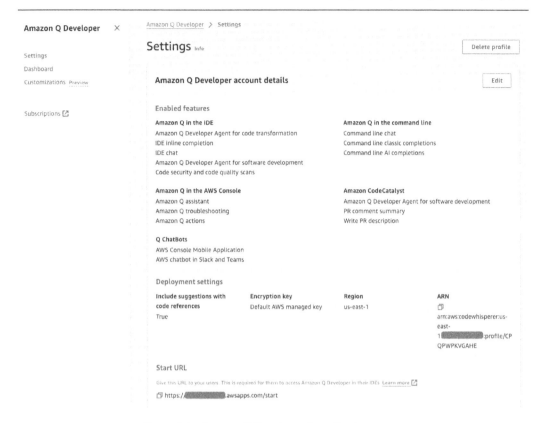

Figure 2.6 – Amazon Q Developer Pro – Settings screen

To determine the specific features included in the free versus Pro tiers of Q Developer, please refer to the pricing document, the link to which has been provided in the *References* section at the end of this chapter.

Now, let's move on to our next topic: setting up Amazon Q Developer inside your favorite IDEs.

Amazon Q Developer setup for third-party IDEs

An IDE is a software application that provides comprehensive facilities to programmers for software development. Typically, an IDE includes a source code editor, build automation tools, and a debugger. Its purpose is to streamline the coding and development process by integrating various aspects of software development into a single environment, making it more efficient and convenient for developers. Popular examples of IDEs include Visual Studio, Eclipse, and IntelliJ IDEA.

To enhance developer productivity, Amazon Q Developer seamlessly integrates with Visual Studio, Visual Studio Code, and JetBrains IDEs. Each of these IDEs has its own strengths, and developers often have a favorite go-to IDE or switch around depending on the features they require for a particular programming language. We aim to demonstrate how to enable Q in all three of them and leave the choice to our end users.

VS Code

Visual Studio Code (**VS Code**) is a standalone source code editor compatible with Windows, macOS, and Linux. It is ideal for Java and web developers and offers many extensions to support virtually any programming language.

Once you install VS Code, to start using Q, you need to install the Amazon Q extension. You can install it either by searching the Extensions section of VS Code or by installing it via the VS Code Marketplace. For further assistance in installing and setting up the Q extension for VS Code, refer to the link provided in the *References* section at the end of this chapter.

Once the extension is installed, you need to authenticate yourself. If you are using the Amazon Q Pro tier as part of your organization, your AWS account administrator will setup and enable IAM Identity Center for you to authenticate. Basically, your administrator will add you as a user and provide you with a start URL for you to log in with IAM Identity Center so that you can leverage Q as part of the organizational policies enabled.

If you are using VS Code and want to use Amazon Q as a free tier user, then you need to log in using your AWS Builder ID. Once installed and authenticated, you will see the following screenshot, where the extension for Amazon Q is installed for VS Code, and you can see it being authenticated using the Builder ID at the bottom of the screenshot.

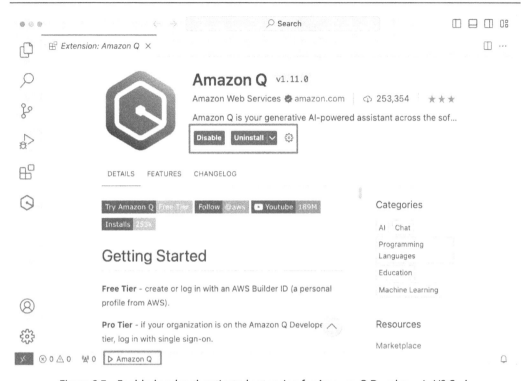

Figure 2.7 – Enabled and authenticated extension for Amazon Q Developer in VS Code

Let's quickly see how the authentication for the free and Pro tiers appears in VS Code. In the previous sections, we established an AWS Builder ID for the free tier and also set up an IAM Identity Center user who was provided with a subscription to the Pro tier.

After installing the Amazon Q extension, when you first try to authenticate to Q from the VS Code editor, you will be presented with both choices, free and Pro tier, as seen in the following screenshot.

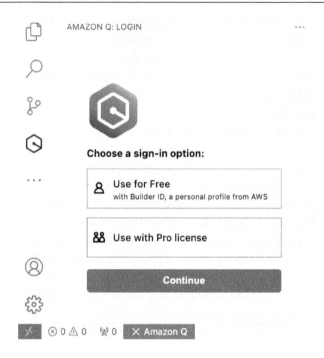

Figure 2.8 – Authentication options to Amazon Q Developer in VS Code

When you proceed with the free tier, it will open a browser screen asking you to input your AWS Builder ID credentials. It will also ask you to confirm whether you approve giving the IDE access to your data through Amazon Q, as seen in the following screenshot.

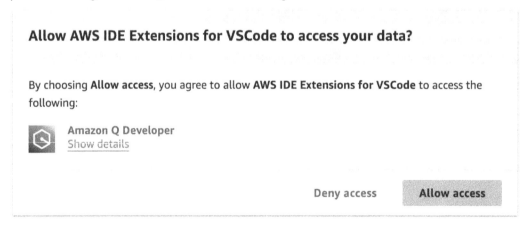

Figure 2.9 – Approve Amazon Q access to VS Code

You will see a notification in the VS Code IDE, as seen in the following screenshot, indicating that you have successfully authenticated using the AWS Builder ID and that the IDE is ready to leverage the free tier features of Amazon Q.

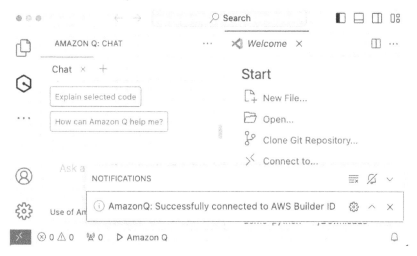

Figure 2.10 – Authenticated in VS Code using AWS Builder ID

If you want to authenticate to the Amazon Q Pro tier, you will be prompted to enter the start URL. This URL was obtained during the setup of the Pro tier in the earlier section. Simply copy the URL and paste it here, as seen in the following screenshot.

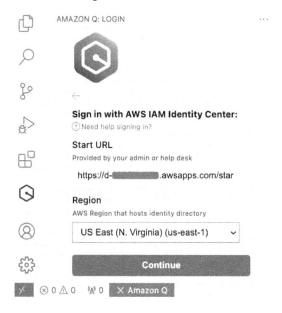

Figure 2.11 – Authenticating to Amazon Q Pro tier in VS Code IDE

After continuing, it will prompt you to authenticate again via the web browser using your IAM Identity credentials. Upon successful multi-factor authentication, VS Code will notify you, as shown in the following screenshot, that you are connected and ready to use the Amazon Q Pro tier.

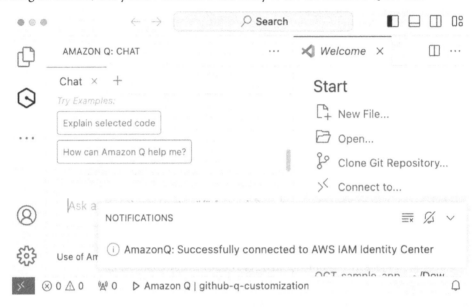

Figure 2.12 – Successfully authenticated to Amazon Q Pro tier in VS Code IDE

The setup for JetBrains IDE is identical so let's quickly take a look at that. We will not repeat all the steps again since the process is similar.

JetBrains

JetBrains provides a series of IDEs, each with its own set of support for different programming languages. For example, Java developers use IntelliJ Idea IDE whereas Python developers use PyCharm IDE. Similarly, JetBrains provides many other IDEs for other programming languages. Since we will be using Python as our primary language to describe different functionalities of Amazon Q, let's set it up for PyCharm IDE.

We need to get the Amazon Q plugin for JetBrains installed and authenticated as we did for VS Code earlier. You can either install the plugin from the Plugins section of the IDE or from the JetBrains marketplace. To further assist you in installing and setting up the plugin for JetBrains IDE, refer to the link provided in the *References* section at the end of this chapter, which also lists support for different JetBrains IDEs and their specific versions.

The following screenshot shows the plugin for Amazon Q installed inside PyCharm IDE. Keep in mind to disable other AI assistants to avoid conflicting results. If PyCharm is what you use for Python coding, you are all set to jump in to *Chapter 4* to start using Amazon Q Developer.

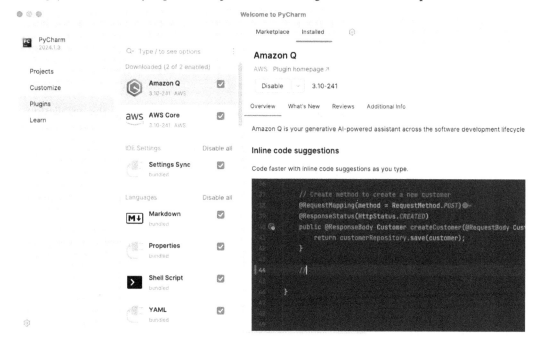

Figure 2.13 – Enabled plugin for Amazon Q Developer in PyCharm IDE

Let's look at another popular IDE that Amazon Q supports, Visual Studio.

Visual Studio

Visual Studio 2022 is a popular IDE for Windows, catering to .NET and C++ developers. It excels in constructing web, cloud, desktop, mobile apps, services, and games.

For Amazon Q to work inside the Visual Studio IDE, you will need to install the AWS Toolkit for Visual Studio. From the Visual Studio Marketplace, first install the AWS Toolkit for Visual Studio, and then there are multiple ways to authenticate into the AWS account. For detailed instructions on setting up Amazon Q with Visual Studio, refer to the link provided in the *References* section at the end of the chapter.

Amazon Q supports C, C++, and C# as programming languages in Visual Studio. And support for the command line is also available so let's look at the initial setup of Amazon Q Developer for use with the command line.

Amazon Q Developer setup for command line

In the age of sophisticated IDEs, **command-line interfaces (CLIs)** are still very popular with developers for quick tests and builds. In *Part 2* of the book, we will look at how you can use Amazon Q Developer with command lines, but first, we need to ensure that Q is installed and set up for the command line.

Things are constantly evolving, but currently, Amazon Q Developer for the command line is supported only with macOS. There are a handful of shells, terminal emulators, terminal IDEs, and over 500 CLIs supported. Always refer to the AWS documentation for newly supported environments.

Since we are using macOS with the zsh shell terminal, we will guide you through the installation steps for it:

- Download and install Amazon Q Developer for the command line. The link is in the *References* section at the bottom of this chapter.

- If you have Pro-tier access provided by your organization, you will need an IAM Identity Centre start URL provided by your organization's administrators.

- If you are a free user, it will ask you to authenticate using your Builder ID or with IAM Identity Centre. AWS Builder ID is a personal profile that grants you access to Amazon Q Developer. Builder IDs are free, and you can sign up using your email address.

Once the installation is successful, the **Automated checks** section in Amazon Q should display checkmarks, as seen in the following screenshot.

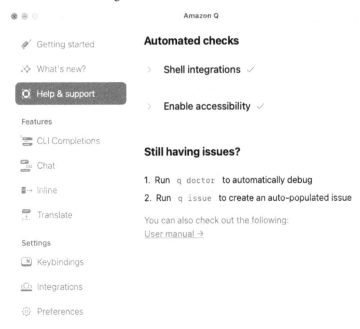

Figure 2.14 – Amazon Q installation for the command line

You can verify if all went well by using the `q doctor` command. The following screenshot confirms that Q Developer was installed correctly for the command line.

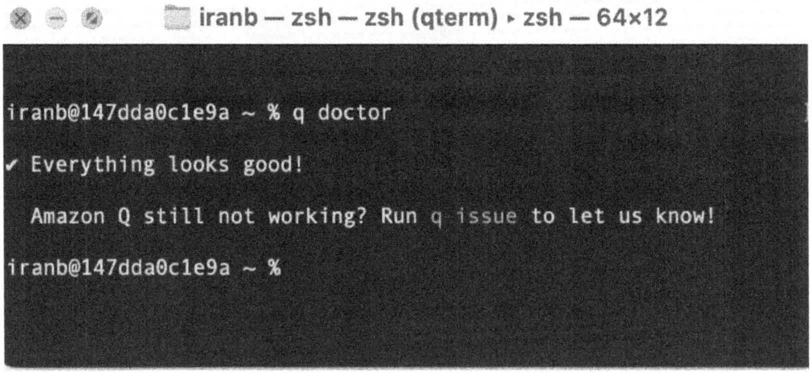

Figure 2.15 – Amazon Q Developer installation for the command line – success

Now, let's look at the initial setup of Amazon Q Developer with some of the supported AWS services and tools.

Amazon Q Developer setup for AWS coding environments

If you are an application builder, software developer, data engineer, or data scientist working with AWS services, you would frequently use builder-friendly tools such as Amazon SageMaker as a platform for building AI/ML projects, Amazon EMR as a platform for building big data processing projects, AWS Glue for building **extract, transform, and load** (**ETL**) pipelines, and AWS Lambda as a serverless compute service. All these services provide tools that help builders and developers write code.

To simplify the developer experience with these AWS services, Amazon Q provides code suggestion and code generation features inside the supported AWS tools. Let's explore all such tools and how to set them up.

Amazon SageMaker Studio

Amazon SageMaker Studio is a comprehensive platform featuring specialized tools for every stage of **machine learning** (**ML**) development. From data preparation to model building, training, deployment, and management, it provides a seamless workflow. Upload data swiftly, construct models in your preferred IDE, enhance team collaboration, optimize coding with an AI-powered companion, fine-tune and debug models, deploy and manage them in production, and automate workflows—all effortlessly unified within a single web-based interface.

Before enabling Q Developer to make Python code recommendations in SageMaker Studio, we assume that you have your SageMaker Studio environment up and running, with all the prerequisites completed and the SageMaker Domain created.

To proceed, in your SageMaker IAM execution role, just add the following IAM policy allowing Amazon Q to generate code recommendations:

```
{
   "Version": "2012-10-17",
   "Statement": [
     {
        "Sid": "CodeWhispererPermissions",
        "Effect": "Allow",
        "Action": ["codewhisperer:GenerateRecommendations"],
        "Resource": "*"
     }
   ]
}
```

Note the legacy name CodeWhisperer still being referenced in the policy statement. It may change in the future, so always refer the official documentation for updates.

If you already have a SageMaker Domain, you can find the execution role for it in the domain setting as seen in the following screenshot.

Figure 2.16 – Execution role for a SageMaker domain

> **Note**
>
> In November 2023, SageMaker Studio was updated with a brand-new experience. The previous experience, now named Amazon SageMaker Studio Classic, is still available for use. If you are already using the Classic theme, you can still enable Amazon Q Developer for in-line prompt-based code generation.
>
> However, in the new experience, along with in-line prompts, you can also enable chat-style assistance from Amazon Q. The chat feature can only be enabled with the Pro tier of Amazon Q Developer, which requires the SageMaker domain to be integrated with IAM Identity Center.

To enable chat style feature of Amazon Q inside SageMaker Studio, you can enable Q in the **App Configurations** tab of the domain details as shown in the following screenshot. The Q Profile ARN can be found in the settings page of Amazon Q Developer as shown in *Figure 2.6*.

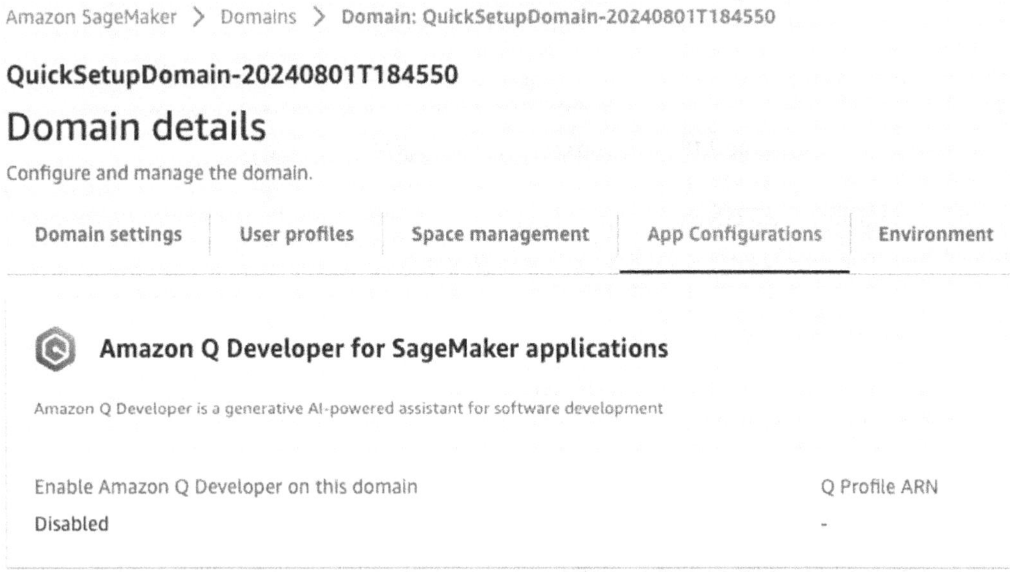

Amazon SageMaker > Domains > **Domain: QuickSetupDomain-20240801T184550**

QuickSetupDomain-20240801T184550

Domain details

Configure and manage the domain.

Domain settings	User profiles	Space management	App Configurations	Environment

◉ Amazon Q Developer for SageMaker applications

Amazon Q Developer is a generative AI-powered assistant for software development

Enable Amazon Q Developer on this domain	Q Profile ARN
Disabled	-

Figure 2.17 – Enabling Amazon Q chat in SageMaker Studio

After this, you can open a new JupyterLab notebook and see at the bottom that Amazon Q is enabled, as seen in the following screenshot. Also, SageMaker supports in-line prompts as well as chat-based code generation which is also highlighted in the screenshot.

Figure 2.18 – Amazon Q in action in SageMaker Studio

We will look into how to effectively use Amazon Q inside SageMaker Studio in *Chapter 14*. For now, we are just setting it up in all the supported tools inside AWS.

Amazon EMR Studio

EMR Studio is an IDE inside Amazon EMR service that simplifies the process for data scientists and engineers to create, visualize, and debug data engineering and data science applications using R, Python, Scala, and PySpark. Amazon Q Developer supports Python language making it easy to write Spark jobs.

To enable Amazon Q for EMR Studio, all you need to do is attach the same IAM policy that we used for SageMaker Studio to EMR. Once you launch the notebook via the workspace, Q will be enabled for use.

The following screenshot shows Q enabled inside EMR Studio Notebook and is able to generate code based on comments. Note that the old name of CodeWhisperer is still being displayed. This may eventually change to Amazon Q.

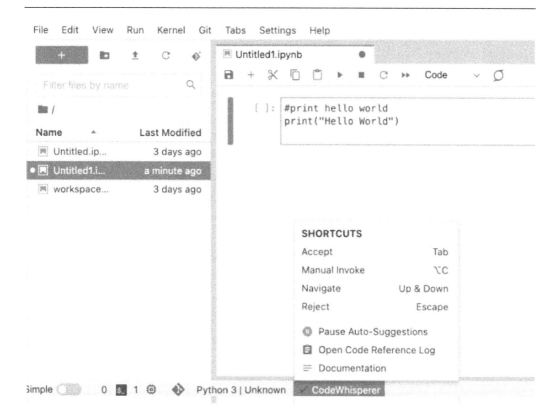

Figure 2.19 – Amazon Q enabled in Amazon EMR Studio

JupyterLab

Many data scientists and data engineers use Jupyter Notebooks for their data science projects. JupyterLab, a customizable and feature-rich application for authoring notebooks, is a key component of Project Jupyter, a non-profit, open source project that aims to deliver tools and standards for interactive computing using computational notebooks.

Amazon Q supports Python code recommendations in JupyterLab. The following commands in macOS install Q for JupyterLab 3 or 4.

```
# Use the below command if you have JupyterLab 4
pip install amazon-codewhisperer-jupyterlab-ext

# Use the below command if you have JupyterLab 3
pip install amazon-codewhisperer-jupyterlab-ext~=1.0
jupyter server extension enable amazon_codewhisperer_jupyterlab_ext
```

After installing you can authenticate using the AWS Builder ID, after which Q will start giving suggestions inside the notebook.

AWS Glue Studio

AWS Glue Studio provides a user-friendly graphical interface for effortless creation, execution, and monitoring of data integration jobs within AWS Glue. Amazon Q supports Python as well as Scala languages, which are popularly used for coding ETL pipelines using Glue Studio.

To enable Amazon Q inside Glue Studio, the same IAM policy that we used in the SageMaker Studio setup has to be attached to the Glue role. Once enabled, you can launch Glue Studio Notebook under the **ETL Jobs** and start leveraging the features of Q.

The following screenshot shows Q enabled inside Glue Studio Notebook and is able to generate code based on prompts. Note that the old name of CodeWhisperer is still being displayed. This may eventually change to Amazon Q.

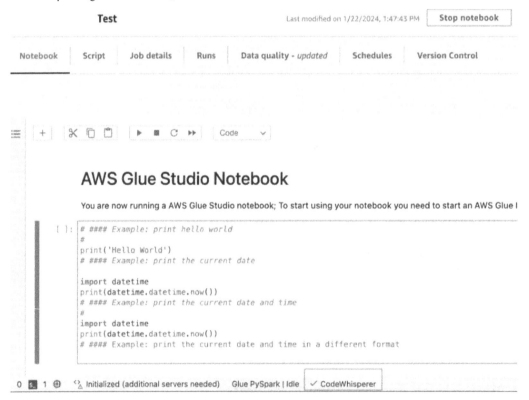

Figure 2.20 – Amazon Q enabled in AWS Glue Studio

AWS Lambda

AWS Lambda is a serverless and event-driven computing service that executes your code without the need to provision or manager servers. It provides a quick path to transform ideas into modern, production-ready, serverless applications.

Amazon Q, as of now, supports the Python and Node.js languages in AWS Lambda. After you assign the same IAM policy for Q, you can activate it by choosing the Q code suggestions option from the **Tools** menu.

The following screenshot shows the option to enable Amazon Q inside a Lambda function. Note that the old name of CodeWhisperer is still being displayed. This may eventually change to Amazon Q.

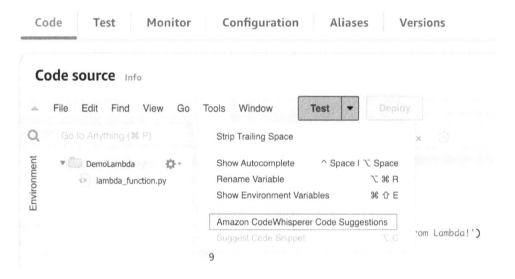

Figure 2.21 – Amazon Q enabled in AWS Lambda

The lambda editor can now accept code suggestions. Refer to the *User actions* webpage from the *References* section at the end of the chapter to test different keyboard shortcuts for different editors including lambda.

Now let's switch gears and see how Amazon Q Developer can also assist AWS builders in building solutions faster.

Build on AWS with support from Amazon Q Developer

If you're a builder in your IT department and use AWS services to solve business use cases, then Amazon Q can help boost your productivity and enhance your experiences with AWS services. Amazon Q can be accessed from the AWS Management Console, the AWS website, and even within the AWS documentation to help you reach your desired end goal faster.

Before we explore some of the capabilities of Amazon Q on AWS, let's review the permissions you may need in your AWS account before you can start using it.

Amazon Q permissions

When a user logs into the AWS console, they assume a role that has been granted certain permissions to use specific resources in the AWS services. For a user to utilize the features of Amazon Q, they must have permission to use Q features. To facilitate this, the role assumed by the user needs to have an IAM policy with Q permissions. The quickest and easiest way to get started is to attach the managed IAM policy to the role. AmazonQFullAccess is a managed IAM policy that provides full access to all features of Amazon Q.

This managed IAM policy contains wildcard (*) characters in actions and resources, allowing all features of Q to be used by all AWS resources. The following code snippet illustrates this policy:

```
{
    "Version": "2012-10-17",
    "Statement": [
      {
        "Sid": "AllowAmazonQFullAccess",
        "Effect": "Allow",
        "Action": [
          "q:*"
        ],
        "Resource": "*"
      }
    ]
}
```

In your organization, the AWS account admin most probably won't grant you full access. Typically, the wildcard character is replaced with actual actions and resources your role needs access to.

For example, to allow users to utilize the conversational feature of Q, in the `Action` section, the * will be replaced with `q:StartConversation` and `q:SendMessage`. And to use the AWS console troubleshoot feature, the `Action` will have `q:StartTroubleshootingAnalysis`, `q:GetTroubleshootingResults`, and `q:StartTroubleshootingResolutionExplanation` actions.

Now that we have the permissions sorted out, let's explore some areas where Amazon Q can assist on AWS.

Conversational Q&A capability

Amazon Q allows you to ask conversational-style questions right inside the AWS Management Console itself. AWS builders can ask a wide range of questions about architecture, services, best practices, and ask multiple follow-up questions too to get the desired guidance. Getting all the responses in the Q chat console cuts down the time needed for research and investigation, which in turn fast-tracks the application build process.

The following screenshot highlights Q's ability to provide context-based answers. The Amazon Q icon is in the top right corner of the console. We will explore many more conversational-style uses of Amazon Q for AWS in *Part 4* of the book.

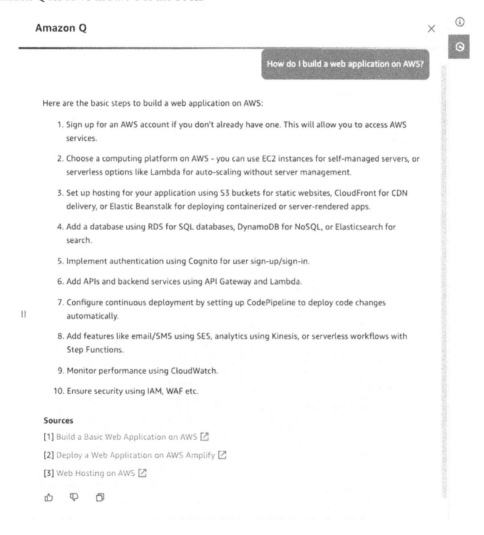

Figure 2.22 – Working with Amazon Q inside the AWS console

Amazon Q also provides targeted assistance for AWS services such as Amazon Redshift and AWS Glue by using its conversational Q&A capability. While we will briefly cover them in this introductory chapter, a more detailed discussion will be provided in *Part 4* of the book, where we delve into chapters focused on the AWS build process.

Chat to generate insights from Amazon Redshift

Amazon Redshift is a fully managed data warehousing service in the cloud, offering fast query performance using robust SQL-based analytics tools. It efficiently manages petabyte-scale data warehouses, enabling users to analyze large datasets and derive valuable insights for decision-making. The Amazon Redshift Query Editor is a browser-based tool that allows users to run SQL queries directly against their Redshift data warehouse.

The Amazon Q generative SQL feature within the Amazon Redshift Query Editor generates SQL recommendations based on natural language prompts. This enhances productivity by assisting users in extracting insights more efficiently from their data.

The provided screenshot exemplifies how Amazon Q comprehended the query posed in the chat and effectively joined the various tables necessary to complete the SQL query. You can incorporate the generated query into the notebook and validate its accuracy by testing the query for precise results.

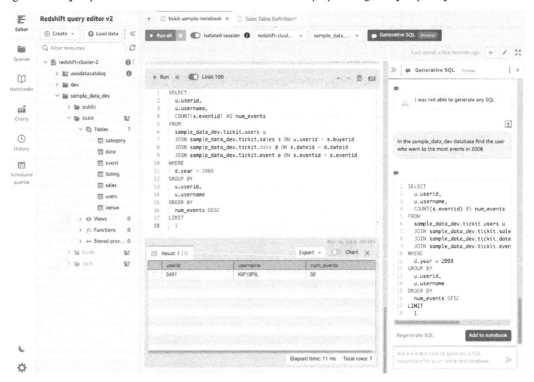

Figure 2.23 – An overview of working of Amazon Q inside Amazon Redshift Query Editor

This feature would expedite report creation and enable non-technical users to derive insights from the data without having to wait for technical expertise to generate reports. We will delve into the details of this in *Chapter 14*. Now, let's explore how Amazon Q can add value in AWS Glue notebooks, making it easier for data engineers to create ETL jobs.

Chat to generate logic for AWS Glue ETL

In the previous section, we discussed how Amazon Q can assist with auto-coding inside the Glue Studio notebook. However, sometimes you need to chat with the assistant to generate the boilerplate logic in its entirety. Using the Amazon Q chat capability simplifies job authoring, troubleshooting, and provides immediate responses to inquiries regarding AWS Glue and data integration tasks, significantly reducing time and effort.

The following screenshot demonstrates how, simply by providing Q with the use case, it was able to generate the Glue code, which you can then copy into the Glue Studio notebook for testing. This significantly saves time for data engineers who prefer to create custom scripts for their ETL jobs.

Figure 2.24 – Working with Amazon Q chat to generate Glue code

In *Chapter 14*, we will introduce a use case and provide a solution to illustrate how Q can expedite ETL job creation in AWS Glue.

Chat about AWS resources and costs

Amazon Q can now also understand Q&A within the context of the resources created in your account. You can ask questions such as "Show all my EC2 instances running in the us-west-1 region," and it will list them all for you. Amazon Q can also provide you with a cost breakdown of resources being used in the AWS account. You can ask questions such as "How much did we spend on Amazon Redshift in 2023 in the us-east-1 region?" and it will provide you with the cost structure.

For now, let's explore some other important features of Amazon Q with various AWS services.

Troubleshoot AWS console errors

AWS builders spend a significant amount of time troubleshooting issues during their development process. Amazon Q makes it easy to identify and solve errors directly from the AWS console. Instead of manually checking logs and researching solutions to errors, Q proposes possible solutions with just a click of a button.

In the following screenshot, we have a simple AWS Lambda function that prints a message. But the sample code has a bug.

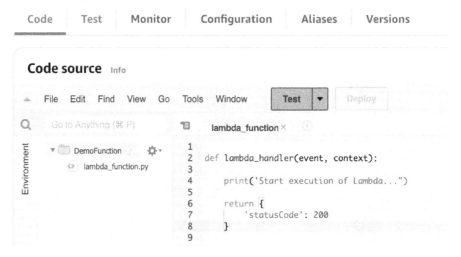

Figure 2.25 – Example of an AWS Lambda function with a bug in the code

Everything looks great, but we mistakenly ended the string inside the `print` statement with a double quote instead of a single quote. The error becomes apparent when we run the test, as seen in the following screenshot.

Code
Test
Monitor
Configuration
Aliases
Versions
Executing function: failed (logs)
Troubleshoot with Amazon Q
Details
The area below shows the last 4 KB of the execution log.
{
"errorMessage": "Syntax error in module 'lambda_function': unterminated string literal (detected at line 4) (lambda_function.py, line 4)",
"errorType": "Runtime.UserCodeSyntaxError",
"requestId": "3981c5d7-6107-4745-a95a-9f1c67cf0e46",
"stackTrace": [
" File \"/var/task/lambda_function.py\" Line 4\n print('Start execution of Lambda...\")"
]
Summary
Code SHA-256
GMjyU8LUbFvBWopcgCVcidiNCdk7VKcZToFuJxAAgkE=
Execution time
20 seconds ago (February 7, 2024 at 04:27 PM EST)
Request ID
3981c5d7-6107-4745-a95a-9f1c67cf0e46
Function version
$LATEST
Duration
1.48 ms
Billed duration
2 ms
Resources configured
128 MB
Max memory used
40 MB

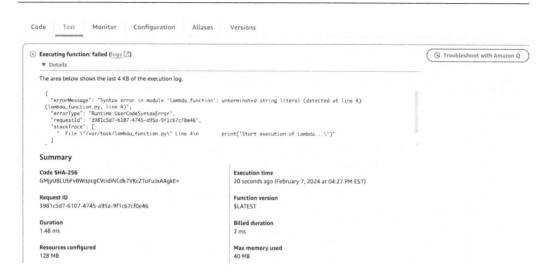

Figure 2.26 – Amazon Q troubleshoot feature with AWS Lambda – Error

Instead of manually combing through log files or researching the error on the internet, you can simply click the **Troubleshoot with Amazon Q** button on the test screen itself. Q will provide an analysis of the issue, and you can also ask it to provide a resolution. The following screenshot shows the analysis and resolution of the missing single quote.

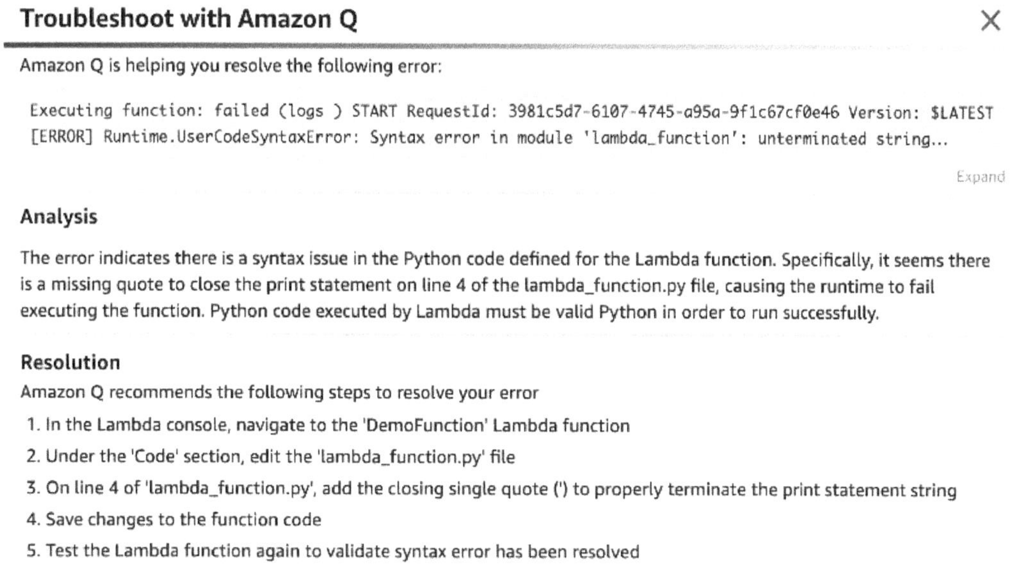

Troubleshoot with Amazon Q
Amazon Q is helping you resolve the following error:
Executing function: failed (logs) START RequestId: 3981c5d7-6107-4745-a95a-9f1c67cf0e46 Version: $LATEST
[ERROR] Runtime.UserCodeSyntaxError: Syntax error in module 'lambda_function': unterminated string...
Expand
Analysis
The error indicates there is a syntax issue in the Python code defined for the Lambda function. Specifically, it seems there is a missing quote to close the print statement on line 4 of the lambda_function.py file, causing the runtime to fail executing the function. Python code executed by Lambda must be valid Python in order to run successfully.
Resolution
Amazon Q recommends the following steps to resolve your error
1. In the Lambda console, navigate to the 'DemoFunction' Lambda function
2. Under the 'Code' section, edit the 'lambda_function.py' file
3. On line 4 of 'lambda_function.py', add the closing single quote (') to properly terminate the print statement string
4. Save changes to the function code
5. Test the Lambda function again to validate syntax error has been resolved

Figure 2.27 – Amazon Q troubleshooting feature with AWS Lambda – Error resolution

In *Part 4* of the book, we will look into details other on how to troubleshoot other complicated issues while building solutions using AWS services.

Troubleshoot networking issues

Every application builder and developer knows how nightmarish it can be to deal with networking issues. To alleviate this frustration, Amazon Q can also help troubleshoot networking issues. Amazon Q works in tandem with Amazon VPC's Reachability Analyzer to check networking connections and identify potential configuration issues.

For example, in the following screenshot, you can see that by simply asking a connectivity question to Q, it was able to suggest possible network issues.

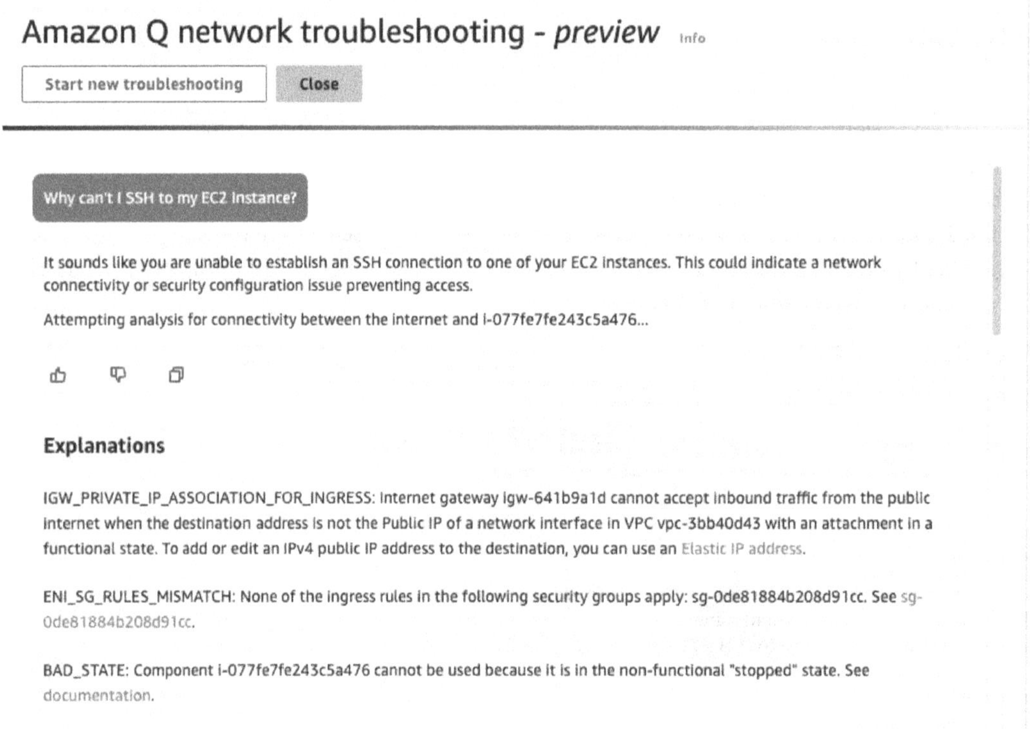

Figure 2.28 – Amazon Q networking troubleshoot

After Q identifies that the issue arises due to a network connectivity problem, it then utilizes Amazon VPC's Reachability Analyzer to analyze the full networking path, pinpointing where the issue might be occurring along the route. The following screenshot shows the path analysis from source to destination and suggests potential locations for the issue.

Figure 2.29 – Amazon Q networking troubleshoot – Path analysis

Let's move on to see some more capabilities of Amazon Q on AWS.

Optimal selection of Amazon EC2 instances

AWS builders are well familiar with Amazon **Elastic Compute Cloud** (**EC2**) instances as many of them use servers to deploy and run their applications. However, with so many different types of EC2 instances, it is difficult to know which type of instance would be best suitable for a particular workload. Of course, you can research it and choose the best one, but Amazon Q makes it easy to select the best EC2 for the job, right from the EC2 console itself.

Inside the EC2 console, where you get to choose an instance type, Amazon Q comes in handy when you click the **Get advice** link as shown in the following screenshot.

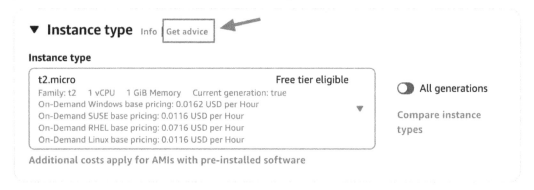

Figure 2.30 – Amazon Q – EC2 instance type advise

You can then choose your use case, workload type, priority, and CPU types, which make up the input to Amazon Q as it advises the best possible EC2 instances to use. The following screenshot shows the criteria selection section once you seek advice from Q.

Get advice on instance type selection from Amazon Q   ✕

Tell us more about your requirements to generate instance type suggestions

We will use Amazon Q, a generative AI assistant, to generate instance type suggestions

Use Case

Web Hosting   ▼

Workload type

Web/App Server   ▼

Priority

Low cost   ▼

CPU Manufacturers

No preference   ▼

Cancel   **Get instance type advice**

Figure 2.31 – Amazon Q – EC2 instance type selection criteria

Once you hit the **Get instance type advice** button, Q works its magic. The following screenshot shows the instances to use for the input criteria we provided, and it also explains what each of these EC2 instances brings to the table. It lists the sources from where the information is taken so that anyone can look at the original source of truth.

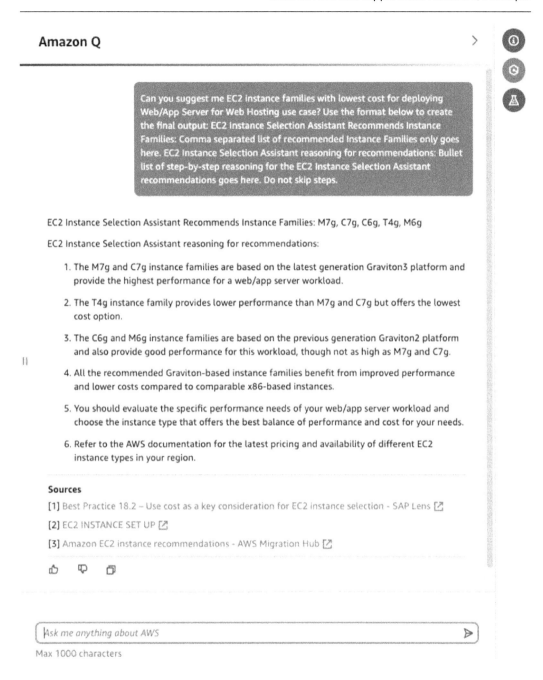

Can you suggest me EC2 instance families with lowest cost for deploying Web/App Server for Web Hosting use case? Use the format below to create the final output: EC2 Instance Selection Assistant Recommends Instance Families: Comma separated list of recommended Instance Families only goes here. EC2 Instance Selection Assistant reasoning for recommendations: Bullet list of step-by-step reasoning for the EC2 Instance Selection Assistant recommendations goes here. Do not skip steps.

EC2 Instance Selection Assistant Recommends Instance Families: M7g, C7g, C6g, T4g, M6g

EC2 Instance Selection Assistant reasoning for recommendations:

1. The M7g and C7g instance families are based on the latest generation Graviton3 platform and provide the highest performance for a web/app server workload.

2. The T4g instance family provides lower performance than M7g and C7g but offers the lowest cost option.

3. The C6g and M6g instance families are based on the previous generation Graviton2 platform and also provide good performance for this workload, though not as high as M7g and C7g.

4. All the recommended Graviton-based instance families benefit from improved performance and lower costs compared to comparable x86-based instances.

5. You should evaluate the specific performance needs of your web/app server workload and choose the instance type that offers the best balance of performance and cost for your needs.

6. Refer to the AWS documentation for the latest pricing and availability of different EC2 instance types in your region.

Sources

[1] Best Practice 18.2 – Use cost as a key consideration for EC2 instance selection - SAP Lens

[2] EC2 INSTANCE SET UP

[3] Amazon EC2 instance recommendations - AWS Migration Hub

Ask me anything about AWS

Max 1000 characters

Figure 2.32 – Amazon Q – EC2 instance selection advise

Before we wind down this introductory chapter about Amazon Q, let's quickly look at another key aspect around feature development.

Assist with AWS Support cases

You can use Amazon Q Developer to create a support case and contact AWS Support from anywhere in the AWS Management Console, including the AWS Support Center Console. Amazon Q leverages the context of your conversation to automatically draft a support case on your behalf, incorporating your recent conversation into the support case description. Once the case is created, Amazon Q can connect you to a support agent through your preferred method, including live chat within the same interface.

Assist during DevOps process in Amazon CodeCatalyst

Amazon CodeCatalyst is a service that offers development teams a unified software development service to rapidly construct, deploy, and expand applications on AWS, all while maintaining organization-specific best practices.

Amazon Q's feature development capability in Amazon CodeCatalyst acts as a generative AI assistant to which you can assign issues. Once an issue is assigned, Amazon Q analyzes its title and description, reviews the code in the specified repository, and, if possible, drafts a solution. This draft solution is then presented to users for evaluation in a pull request.

We have a whole chapter on this topic in *Part 4* of the book, so we will keep this introduction short here.

Summary

In this chapter, we covered what Amazon Q Developer is and how it can assist developers and application builders in their day-to-day tasks. We also briefly explored some of its features, along with considerations for setting it up.

Next, we walked through the setup of Amazon Q for command-line interfaces, external IDEs such as VS Code and JetBrains IDEs, as well as AWS services, IDEs, and notebooks such as Amazon SageMaker Studio, Amazon EMR Studio, AWS Glue Studio, and AWS Lambda.

We examined its benefits for AWS builders, highlighting how Amazon Q can be leveraged from the AWS console itself to assist with various activities. On a high level, we covered how Amazon Q can help with conversational Q&A style chat, console issues, network troubleshooting, EC2 instance selection, and also during the DevOps process in Amazon CodeCatalyst.

In the upcoming chapters in *Part 2* of the book, we will delve into detail about auto code generation techniques and how Amazon Q Developer can assist developers in this process.

References

- Amazon Q home page: `https://aws.amazon.com/q/`

- Amazon Q Developer pricing: `https://aws.amazon.com/q/developer/pricing/`

- Setting up Amazon Q Developer: `https://docs.aws.amazon.com/amazonq/latest/qdeveloper-ug/getting-started-q-dev.html`

- Amazon Q Developer install for macOS command line: `https://desktop-release.codewhisperer.us-east-1.amazonaws.com/latest/Amazon%20Q.dmg`

- Install Amazon Q Developer extension/plugin for your IDEs: `https://docs.aws.amazon.com/amazonq/latest/qdeveloper-ug/q-in-IDE-setup.html`

- User actions for Amazon Q Developer in different IDEs: `https://docs.aws.amazon.com/amazonq/latest/qdeveloper-ug/actions-and-shortcuts.html`

- AWS IAM Identity Center: `https://docs.aws.amazon.com/singlesignon/latest/userguide/what-is.html`

- Amazon Q Developer tiers: `https://docs.aws.amazon.com/amazonq/latest/qdeveloper-ug/getting-started-q-dev.html`

Part 2: Generate Code Recommendations

In this part, we will look at many of the key features of Amazon Q Developer that can assist developers during the software development life cycle. These features can be used within many of the supported IDEs and assist with various programming languages.

This part contains the following chapters:

- *Chapter 3, Understanding Auto-Code Generation Techniques*
- *Chapter 4, Boost Coding Efficiency for Python and Java with Auto-Code Generation*
- *Chapter 5, Boost Coding Efficiency for C and C++ with Auto-Code Generation*
- *Chapter 6, Boost Coding Efficiency for JavaScript and PHP with Auto-Code Generation*
- *Chapter 7, Boost Coding Efficiency for SQL with Auto-Code Generation*
- *Chapter 8, Boost Coding Efficiency for Command-Line and Shell Script with Auto-Code Generation*
- *Chapter 9, Boost Coding Efficiency for JSON, YAML, and HCL with Auto-Code Generation*

3

Understanding Auto-Code Generation Techniques

In this chapter, we will look at the following key topics:

- What is a prompt?
- Single-line prompts for auto-code generation
- Multi-line prompts for auto-code generation
- Chain-of-thought prompts for auto-code generation
- Chat with code assistant for auto-code generation
- Common building methods of auto-code generation

With the growth in **large language model** (LLM) applications, one of the interesting use cases, auto-code generation based on user comments, has become popular. The last few years have given rise to multiple code assistants for developers, such as GitHub Copilot, Codex, Pythia, and Amazon Q Developer, among many others. These code assistants can be used to get code recommendations and, in many cases, generate error-free code from scratch, just by passing a few plain text comments that describe what the user requires from the code.

Many of these code assistants are now backed by LLMs. LLMs are pretrained on large publicly available datasets, including public code bases. This training on large corpora of data helps code assistants generate more accurate, relevant code recommendations. To improve the developer's code-writing experience, these code assistants can not only easily integrate with different **integrated development environments** (IDE) and code editors but are also readily available with different services offered by most of the cloud providers, with minimal configuration.

Overall, auto-code generation is a process in which the developer has the ability to interact with different code assistants from any supported code editor, using simple plain text comments, to get code recommendations in real time for different supported coding languages.

Before we go deeper into auto-code generation with the help of code assistants later in this chapter, let's look at the key concept of prompts related to generative AI.

What is a prompt?

As specified earlier, LLMs are pretrained on publicly available large datasets, which makes them very powerful and versatile. These LLMs typically have billions of parameters that can be used to solve multiple tasks out of the box, without the need for additional training.

Users just need to ask the right question with relevant context to get the best output from an LLM. The plain text comments/questions that act as an instruction to an LLM are called **prompts** and the technique of asking the right questions with corresponding context is called **prompt engineering**. While interacting with LLMs, providing prompts with precise information and, if required, supplementing with additional relevant context, is very important in order to get the most accurate results. The same is true while interacting with code assistants as most of the code assistants are integrated with an LLM. As a user, while interacting with a code assistant, you should provide prompts with simple, specific, and relevant context, which helps to generate quality code with high accuracy.

The following diagram shows the integration of a code assistant with an LLM.

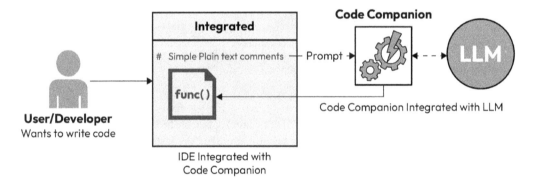

Figure 3.1 – Overview of code assistant integration with an LLM

There are multiple ways to interact with code assistants to get the desired outcomes. In the following sections, we will explore a few of those techniques. To illustrate these techniques, we will leverage the Python programming language inside JetBrains' PyCharm IDE that has been set up to work with Amazon Q Developer (refer *Chapter 2* for the setup).

> **Note**
>
> Amazon Q Developer uses LLMs in the background to generate the code. LLMs, by nature, are non-deterministic, so you may not get exactly the same code blocks as shown in the code snapshots. However, logically, the generated code should meet the requirements.

Single-line prompts for auto-code generation

A single-line prompt for auto-code generation refers to the technique in which the user, using a code assistant, specifies requirements in a single line of plain text, with the expectation of getting relevant lines of code generated in an automated manner.

Here are some key points about single-line prompts for auto-code generation:

- In the single-line prompt technique, instead of specifying complex technical details, the user needs to effectively summarize the requirements at a high level in plain text.

- Code assistants integrated with LLMs are trained to understand these single-line prompts in order to translate them into executable and mostly error-free code.

- Based on the single-line prompt with instructions, context, and specific requirements, code assistants will generate code that can range from a single line to multiple lines to more complex functions and classes to meet the intended functional requirements.

- A single-line prompt for auto-code generation is useful when the code requirement is relatively simple and can easily be described in a sentence.

- A single-line prompt for auto-code generation can significantly reduce time, compared to manually writing the same code.

- Since a single-line prompt for auto-code generation does not need much prompt-engineering experience, it is most commonly used by users who have little to no coding experience. In plain text, they can provide *what* the code should do instead of *how* to make it happen using actual code.

- In most cases, as simple requirements are used in a single-line prompt, the generated code may not require extensive reviews.

Summary – single-line prompts for auto-code generation

In summary, a single-line prompt for auto-code generation is a technique where the user describes relatively simple code requirements using natural language in plain text format; the code assistant then uses LLMs to auto-generate single or multiple lines of code. It makes coding much simpler and accessible to more people, especially users who have relatively little or no coding experience or may be new to a specific coding language.

The following is an example of a single-line prompt for auto-code generation in a PyCharm IDE with Amazon Q Developer enabled as the code assistant, followed by the response:

```
Prompt:
# generate code to display hello world in python
```

Figure 3.2 – Auto-code generation with a single-line prompt in PyCharm IDE with Amazon Q Developer

Observe that Amazon Q Developer generated a single line of code as our requirement was simple and could be easily achieved within a single-line prompt. We will look at more complex examples in our next chapters. In this chapter, we are just explaining the different techniques of auto-code generation.

> **Important note**
>
> You may find typos in a few prompts and screenshots in *Part 2* of the book. We have purposely not corrected them to highlight that Amazon Q Developer understands the underlying meaning of what's being asked, even with incorrect grammar in the prompts.

Multi-line prompts for auto-code generation

Multi-line prompts for auto-code generation refer to the technique where a user can define requirements using natural language text in a single prompt, which can span multiple sentences. Based on the information provided in each of the sentences, the code assistant then tries to understand the correlation among them in order to grasp the requirements and generate the desired code.

The code assistant will mainly rely on key phrases from each of the sentences; it will then use each of those key phrases to form a relationship among all the sentences. This guides code assistants to use LLMs to form a sequence of code lines to meet the requirement. In most cases, code assistants will generate multiple lines of code.

Here are some key points about multi-line prompts for auto-code generation:

- Multi-line prompts for auto-code generation are useful when the code requirement is relatively complex, has multiple steps, and cannot be easily described in one simple sentence, as it may require more context.

- Code assistants use each sentence from the multi-line prompt to extract key phrases, then use an LLM to understand the relationship across sentences and key phrases. This generates the interpretation of the code fragments and corresponding correlations to define end-to-end code requirements.

- Due to the complexity of the requirements, better quality and more targeted code is usually achieved by providing context in multiple simple, concise sentences. These sentences can provide details such as functionality requirements, architectural constraints, and platform specifications.

- Code generated by multi-line prompts is more customized, elaborative, and relatively complex versus single-line prompts.

- Generated code may require code review and thorough testing, and, depending on the complexity of the requirements, some code refinement may be needed before being promoted to the next project life cycle.

- Multi-line prompts for auto-code generation require a good grasp of prompt engineering. It's important to include key details without being overly verbose or ambiguous and to understand the generated code. As a result, this technique is typically used by users with coding experience.

- The accuracy of the code is highly dependent on the maturity/training of the code assistants and LLM used to generate the code.

> **Summary – multi-line prompts for auto-code generation**
>
> In summary, multi-line prompts for auto-code generation is a technique where users describe relatively complex code goals in plain natural language text with multiple sentences. Then, the code assistant can use the key phrases from those sentences to form relationships around them, which helps understand the user requirements and turn them into multiple lines of code. It requires some degree of experience in prompt engineering, code reviews, and in some cases, rewriting the generated code.

The following is an example of a multi-line prompt for auto-code generation.

We will use an auto-generated code block for the following requirements:

1. The code needs to be generated with a specific language version – in this case, Python 3.6.0.

2. The code needs to read the `/user/data/input/sample.csv` file.

3. The code needs to load data in a pandas DataFrame with a specific name, `csv_data`.

4. The code needs to display a sample of 50 rows with a column header.

We write the prompt as follows:

```
"""
Generate code in python 3.6.0 version.
Read CSV file from /user/data/input/sample.csv in a pandas dataframe
named csv_data.
Use csv_data dataframe to display sample 50 records with corresponding
column details.
"""
```

We get the following output:

Figure 3.3 – Auto-code generation with a multi-line prompt in the PyCharm IDE with Amazon Q Developer

Observe that Amazon Q Developer generated multiple lines of code with two functions – `read_csv_file()` to read the CSV file and `display_sample_records()` to display 50 records. Then, it created `__main__` to call both functions to generate an end-to-end script.

Overall, the preceding code does meet the requirements but, based on user/corporate preferences, may need some modification and/or tuning. In the next section, we will learn about another technique suitable for experienced developers. This technique is particularly beneficial when developers are familiar with the code flow and require assistance in generating the code.

Chain-of-thought prompts for auto-code generation

Chain-of-thought prompts for auto-code generation refer to a technique where the user employs a combination of single-line or multi-line prompts to provide step-by-step instructions. Code assistants then use an LLM to automatically generate code for each step individually. Users can use multiple natural language prompts that can link together to solve complex requirements. These prompts can be chained together to guide the model to produce relevant targeted code. It's an effective technique to divide complex coding tasks into smaller code fragments by providing simple prompts to the code assistant one at a time. The code assistant can use each prompt to generate more tailored code fragments. Ultimately, all the code fragments can be used as building blocks to solve the complex requirement.

Here are some key points about chain-of-thought prompts for auto-code generation:

- Chain-of-thought prompts are a technique in which users divide complex requirements into smaller, more manageable single-line or multi-line prompts.

- Chain-of-thought prompts can also be useful for code customization. Users can effectively use this technique to obtain customized code by providing specific information in a prompt, such as generating code with relevant variable names, specific function names, step-by-step logic flow, and so on.

- Code assistants leverage advances in LLMs to generate the code fragments to meet each individual prompt, which acts as a part of the final end-to-end code.

- Chain-of-thought prompts can be used to generate code for a wide range of tasks, such as creating a new project, implementing a specific feature, customizing the code to meet custom standards, improving flexibility, accuracy of the code, and code organization.

- Generated code may require code review and integration testing to verify whether all the individual code fragments combined meet the end-to-end requirements. Depending on the outcome of test cases, users may need to readjust the prompts or rewrite some of the code blocks before being promoted to the next project life cycle.

- This technique can be used by more experienced users as they are mainly responsible for tracking the code generation flow to meet the final goal.

- The accuracy of the overall code is highly dependent on the prompts provided by users, due to which users need to have some background in prompt engineering to generate accurate end-to-end code.

Summary – chain-of-thought prompts for auto-code generation

In summary, chain-of-thought prompts for auto-code generation are a technique in which users describe complex code requirements in smaller, easier, and more manageable single-line or multi-line prompts of plain, natural language text. The code assistant uses each of these prompts to generate specific code based on the information provided in a prompt. In the end, the output from all these individual prompts combined generates the final code. This technique is effective in creating highly tailored code. User needs to carry out integration tests to verify that the code meets the end-to-end functionality; based on the test case results, users may need to adjust the prompts and/or rewrite some of the code fragments of the final code.

The following is an example of chain-of-thought prompts for auto-code generation.

We will use an auto-generated code block with multiple prompts for the following requirements:

1. Use the Python coding language.
2. The code must check whether the `/user/data/input/sample.csv` file is present.
3. Create a function named `read_csv_file()` to read the CSV files. Also, try to use a specific `read_csv` from pandas to read records.
4. Use the `read_csv_file()` function to read the `/user/data/input/sample.csv` file.
5. Display a sample of 50 records with corresponding column details from the file.

Let's break down the preceding requirements into three separate prompts.

Note that, for simplicity, in this example, we will use single-line prompts for auto-code generation but based on the complexity of your requirements, you can have a combination of single-line and multi-line prompts to achieve chain-of-thought prompts. The code assistant generates multiple lines of code based on the prompts, accompanied by corresponding inline comments. This simplifies the understanding of the code for users.

Here is the first prompt followed by the output:

```
Prompt 1 :
# Generate code to Check if /user/data/input/sample.csv exists
```

```python
test-prompts-3.py ×
1    """
2    Generate code to Check if /user/data/input/sample.csv exists
3    """
4    import os
5    def check_file_exists(file_path):
6        """
7        Check if a file exists at the given file path.
8
9        Args:
10           file_path (str): The path to the file to check.
11
12       Returns:
13           bool: True if the file exists, False otherwise.
14       """
15       if os.path.exists(file_path):
16           return True
17       else:
18           return False
19       # return os.path.exists(file_path)
20
21   file_path = "/user/data/input/sample.csv"
22
```

Figure 3.4 – Prompt 1: auto-code generation with a chain-of-thought
prompt in the PyCharm IDE with Amazon Q Developer

Observe that Amazon Q Developer created multiple lines of code that included a function called
check_file_exists() with one parameter to get file_path, which checks whether the file
exists or not and returns True/False. It also added the next file_path variable with an assigned
value as the /user/data/input/sample.csv path.

Here is the second prompt followed by the output:

```
Prompt 2 :
# generate function read_csv_file() and read /user/data/input/sample.
csv using pandas read_csv() method.
```

test-prompts-3.py ×

```
23
24    """
25    Generate function read_csv_file()
26    and read /user/data/input/sample.csv using pandas read_csv() method.
27    """
28    import pandas as pd
29
30
31    def read_csv_file(file_path):
32        """
33        Read a CSV file using pandas.
34
35        Args:
36            file_path (str): The path to the CSV file.
37
38        Returns:
39            pandas.DataFrame: The DataFrame containing the data from the CSV file.
40        """
41        return pd.read_csv(file_path)
```

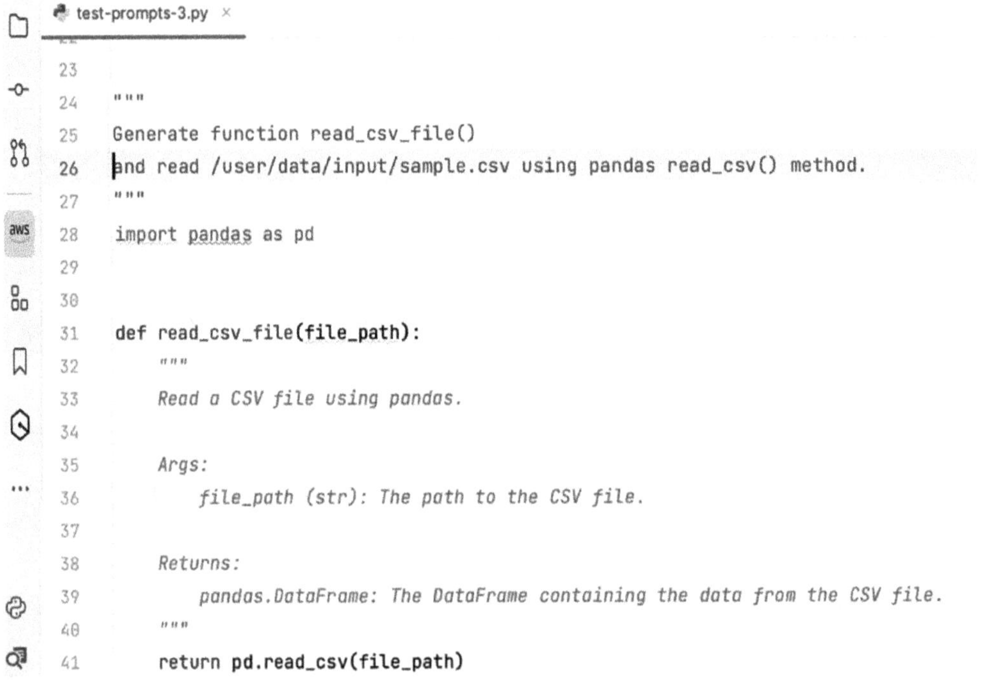

Figure 3.5 – Prompt 2: auto-code generation with a chain-of-thought
prompt in the PyCharm IDE with Amazon Q Developer

Here, also observe that Amazon Q Developer created multiple lines of code. As instructed, it created a function called `read_csv_file_file_exists()` with one parameter to get `file_path` and used the `read_csv` method to read the file.

Here is the third prompt followed by the output:

```
Prompt 3 :
# Display sample 50 records with corresponding column details.
```

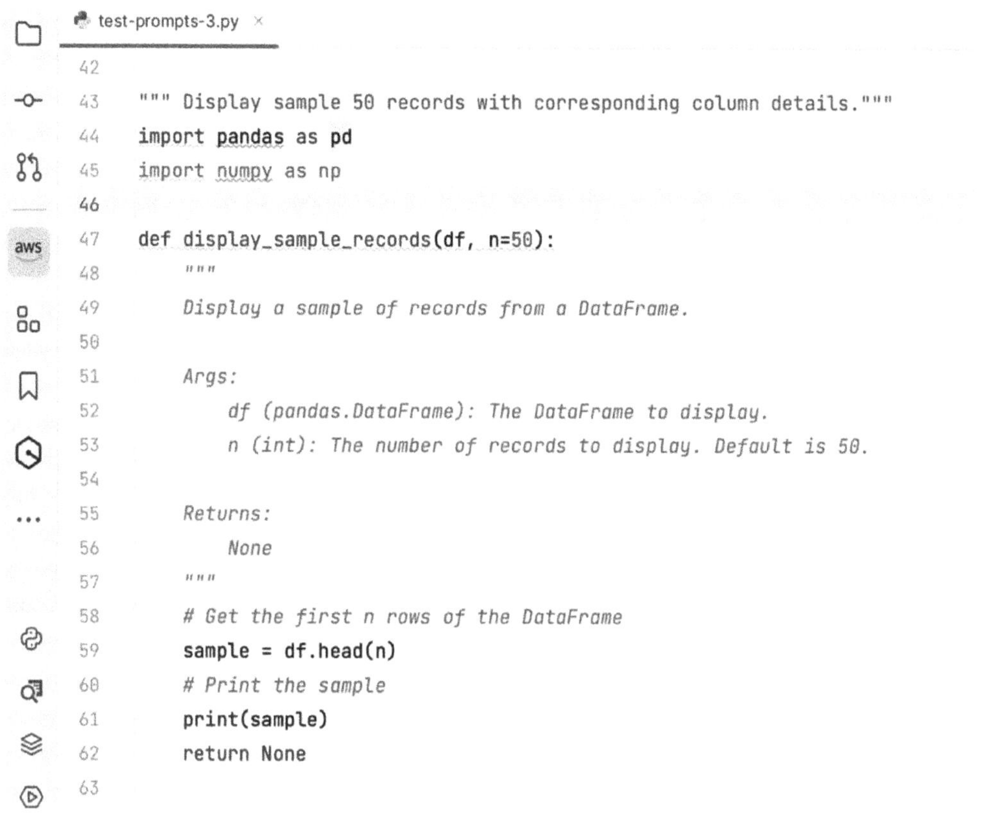

```
42
43    """ Display sample 50 records with corresponding column details."""
44    import pandas as pd
45    import numpy as np
46
47    def display_sample_records(df, n=50):
48        """
49        Display a sample of records from a DataFrame.
50
51        Args:
52            df (pandas.DataFrame): The DataFrame to display.
53            n (int): The number of records to display. Default is 50.
54
55        Returns:
56            None
57        """
58        # Get the first n rows of the DataFrame
59        sample = df.head(n)
60        # Print the sample
61        print(sample)
62        return None
63
```

Figure 3.6 – Prompt 3: auto-code generation with a chain-of-thought
prompt in the PyCharm IDE with Amazon Q Developer

Finally, here, observe how Amazon Q Developer created multiple lines of code with a `display_sample_records()` function to display the 50 sample records.

Let's move on to the next technique for auto-code generation.

Chat with code assistant for auto-code generation

Many of the code assistants allow users to use a chat-style interaction technique to get code recommendations and auto-generate error-free code. Some of the code assistant examples include but are not limited to Amazon Q Developer, ChatGPT, and Copilot. Just like asking questions to your teammates to get recommendations, users can interact with code assistants to ask questions and get recommendations/suggestions related to your questions. In addition, many of the code assistants will also provide the details in a step-by-step walk-through for the information associated with the answer, and will also provide some reference links to get more context. Users can review the details and can choose to integrate the code into the main program and/or update existing code.

Here are some key points about chatting with a code assistant for auto-code generation:

- Chatting with a code assistant helps users to engage in a question-and-answer style interaction in real time.

- Chatting with a code assistant allows users to receive suggestions and recommendations by directly asking questions and stating their requirements in natural language.

- This technique can be used by users of any experience level to obtain information and suggestions, but it requires some experience to understand and review the recommended code. It can be utilized to obtain responses for a variety of complex code issues, learn new technology, generate detailed designs, retrieve overall architecture details, discover coding best practices and code documentation, perform debugging tasks, and future support for the generated code.

- During coding, similar to other prompting techniques, it allows users to describe complex requirements in smaller, more manageable step-by-step questions to generate code. Additionally, the code assistant tracks the questions and corresponding answers to improve responses for future questions by recommending the most relevant options.

- In most cases, users are responsible for reviewing and testing to verify that the recommended code truly meets the overall needs and integrates within the overall code base.

- To help users understand the code, the code assistant also provides a detailed flow associated with the answer and reference links to obtain additional context.

- Many of the advanced code assistants, such as Amazon Q Developer, also gather context from current files open in the IDE. They automatically retrieve information related to the coding language used and the location of the file to provide more relevant answers. This assists the code assistant in handling questions related to updating existing code, software support, software development, best practices, and so on.

- The accuracy of the overall code is highly dependent on the user's ability to ask relevant and accurate questions with specific details.

> **Summary – chatting with a code assistant for auto-code generation**
>
> In summary, users at any level of experience can utilize chatting with the code assistant for auto-code generation. This technique involves interacting with the code assistant in a question-and-answer style in real time to receive suggestions and/or recommendations for a wide variety of use cases, including complex code, detailed designs, overall architecture, coding best practices, code documentation, future support, code updates, and understanding the code, among others. In most cases, users are responsible for reviewing and integrating the suggested code into the main code base during the coding phase. To assist users, code assistants can provide details related to the flow associated with the answer and reference links to obtain more context.

The following is a generic example of chatting with a code assistant for auto-code generation.

The requirement is to get generic help from Amazon Q Developer to understand the approaches to debug issues with a Python function and, more specifically, for `read_csv_file()`:

```
Q : How do I debug issues with my read_csv_file()python function?
```

We get the following result:

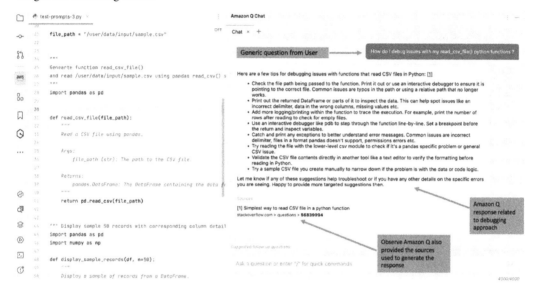

Figure 3.7 – Chatting with a code assistant technique for auto-code generation using Amazon Q Developer

Here is the response with step-by-step details related to our question:

Here are a few tips for debugging issues with functions that read CSV files in Python: [1]

- Check the file path being passed to the function. Print it out or use an interactive debugger to ensure it is pointing to the correct file. Common issues are typos in the path or using a relative path that no longer works.
- Print out the returned DataFrame or parts of it to inspect the data. This can help spot issues like an incorrect delimiter, data in the wrong columns, missing values etc.
- Add more logging/printing within the function to trace the execution. For example, print the number of rows after reading to check for empty files.
- Use an interactive debugger like pdb to step through the function line-by-line. Set a breakpoint before the return and inspect variables.
- Catch and print any exceptions to better understand error messages. Common issues are incorrect delimiter, files in a format pandas doesn't support, permissions errors etc.
- Try reading the file with the lower-level csv module to check if it's a pandas specific problem or general CSV issue.
- Validate the CSV file contents directly in another tool like a text editor to verify the formatting before reading in Python.
- Try a sample CSV file you create manually to narrow down if the problem is with the data or code logic.

Let me know if any of these suggestions help troubleshoot or if you have any other details on the specific errors you are seeing. Happy to provide more targeted suggestions then.

Figure 3.8 – Chatting with a code assistant for auto-code generation – Amazon Q Developer response

At the bottom of the response, observe that Amazon Q Developer also provided the sources used to provide the details:

Sources

[1] Simplest way to read CSV file in a python function
stackoverflow.com > questions > **56839994**

Figure 3.9 – Chatting with a code assistant for auto-code generation – Amazon Q Developer sources

To make it easy for the user, if you just hover over the sources, Amazon Q Developer will show exact details from the source in the same window:

Amazon Q Chat

```
ValueError: I/O operation on closed file.
>>>
```

My Script Code

```
import csv
 #Function for reading .csv file
    def csvReader(filename):
        with open(filename) as file:
            reader = csv.reader(file, delimiter =' ')
        for row in reader:
            print(row[0])

    #Function that reads file
    csvReader('test.csv')
```

You are trying to read from the reader object after the with block, at which time the file will be closed. A file that is opened using the with block will get closed on exiting the with block. Put your reader code inside the with block like below.

```
import csv
 #Function for reading .csv file
def csvReader(filename):
    with open(filename) as file:
        reader = csv.reader(file, delimiter =' ')
        for row in reader:
            print(row[0])

#Function that reads file
csvReader('test.csv')
```

[1] Simplest way to read CSV file in a python function
stackoverflow.com > questions > **56839994**

Figure 3.10 – Chatting with a code assistant for auto-code
generation – Amazon Q Developer source details

Now, let's get into the topic of different building methods of auto-code generation.

Common building methods of auto-code generation

As described in the previous sections, different levels of experienced users can make use of code assistants to generate the desired code to meet functional requirements. In this section, we will walk through some common building methods useful for auto-code generation, depending on the complexity of the requirements. We will use Amazon Q Developer integration with JetBrains' PyCharm IDE to illustrate how a code assistant can help users automate the generation of code fragments and/or get recommendations from the system.

Now, let's begin with each of the code-building methods used in the auto-code generation technique.

Single-line code completion

Code assistants that leverage LLMs can keep track of all the prompts provided as input by a user. At runtime, code assistants use all the input information to suggest relevant code. Here is a simple demonstration of how Amazon Q Developer helps users with single-line code completion.

When users start typing code in an Amazon Q Developer-enabled environment, it can understand the context of the current and previous inputs. It will start suggesting the next code block to complete the existing line or recommend the next line that may follow after the current one. The following screenshot highlights this method:

Figure 3.11 – Single-line code completion in the PyCharm IDE using Amazon Q Developer

Observe that when the user starts typing the DataFrame name, `csv_data`, Amazon Q Developer suggests using the `read_csv_file()` function, which is defined in the script.

Full function generation

One of the basic building blocks of programming is functions. Typically, a function is a reusable multi-line block of code, defined anywhere in the program, to perform a specific task. Functions do not run unless called. A function can accept parameters or arguments that can be passed when being called in your script. Optionally, a function can return data to the calling statement. Code assistants can help users write an entire function body. Users just need to provide the information about the functionality they need from a function, and optionally, the programming language, using any of the previous auto-code generation techniques.

Now, let's look at an example of how users can auto-generate a function.

The following is a simple demonstration of how Amazon Q Developer helps users with the generation of a full function:

1. The user needs to write a simple Python function named `read_csv()` that takes `file_path` as a parameter and returns a sample of 50 records

2. Call the `read_csv()` function to read a CSV file from the `/user/data/input/sample.csv` path

We will use the chain-of-thought prompt technique to generate the preceding code. Here is the first prompt followed by the output:

```
Prompt 1 :
" " "
write a python function named read_csv() which takes file_path as a
parameters and returns sample 50 records.
" " "
```

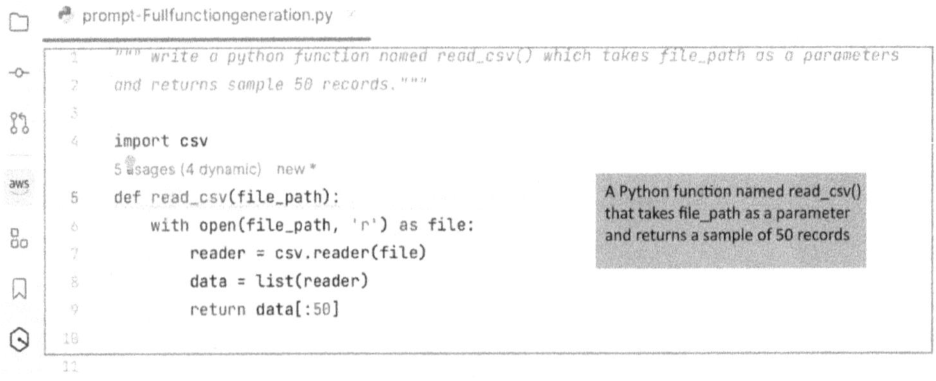

Figure 3.12 – Function generation in the PyCharm IDE using
Amazon Q Developer – generating the function

Observe, as instructed in the single-line prompt, that Amazon Q Developer has created a Python function named `read_csv()` that takes `file_path` as a parameter and returns a sample of 50 records.

Now, let's look at how you can get the function statement logic auto-generated, followed by the output:

```
Prompt 2 :
"""
call read_csv by passing file path as /user/data/input/sample.csv
"""
```

```
11
12    """ call read_csv by passing file path as /user/data/input/sample.csv """
13    file_path = "/user/data/input/sample.csv"
14    data = read_csv(file_path)
15    print(data)
16                                          Read /user/data/input/sample.csv using the
                                            preceding function
```

Figure 3.13 – Full function generation in PyCharm IDE using Amazon Q Developer – calling the function

Observe, as instructed in Amazon Q Developer, that the call logic is generated for the `read_csv()` function to read data from the `/user/data/input/sample.csv` file.

Block completion

During the logical flow of a program, users need to run certain code blocks based on conditions and/or need to run certain lines in loops. The most common code blocks to achieve these functionalities are `if` conditions, `for` loops, `while` conditions, and `try` blocks. Code assistants are trained to complete and suggest possible code to write these conditional and loop statements.

Now, let's see how Amazon Q Developer can help users suggest possible `if` conditions in the previously created function in the *Full function generation* example (reference *Figure 3.13*). As Amazon Q Developer understands the context of the code, it's able to understand the functionality associated with the `read_csv()` function. So, start typing `if` inside the `read_csv()` function to show our intent to write a conditional block:

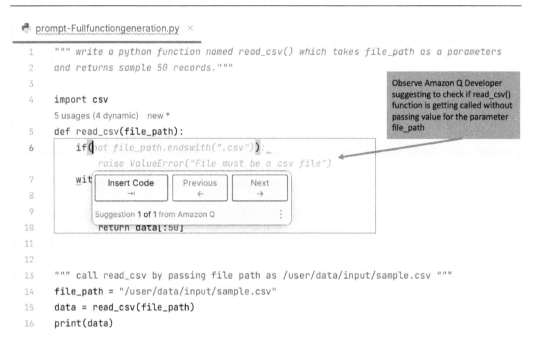

Figure 3.14 – Block completion in the IDE using Amazon Q Developer

Observe that, once the user starts typing `if`, Amazon Q Developer understands that `file_path` is a mandatory parameter for the `read_csv()` function and expects to have a `.csv` file; based on this understanding, it suggests adding an error handling condition to check whether the passed parameter has a `.csv` file.

Line-by-line recommendations

Generally, user requirements can be complex and users may not be able to define all of them using a combination of multiple prompts. In a few cases, code assistants may not be able to generate the script as expected by the user in a single code generation. In addition, if the user decides to update the existing code, then Amazon Q Developer provides line-by-line recommendations. Amazon Q Developer tries to understand the context of the script and predict relevant next lines that might be logically useful.

Now, let's use the previous script from the *Block completion* section (reference *Figure 3.14*) to check whether Amazon Q Developer can recommend the next line of code. To illustrate the functionality, let's delete the last line, `print(data)`. Then, go to the last line of the script and press *Enter*; now, Amazon Q Developer will try to predict the next logical functionality for the script:

```
 prompt-Fullfunctiongeneration.py  ×

1     """ write a python function named read_csv() which takes file_path as a parameters
2     and returns sample 50 records."""
3
4     import csv
      5 usages (4 dynamic)  new *
5     def read_csv(file_path):
6         if(not file_path.endswith(".csv")):
7             raise ValueError("File must be a csv file")
8         with open(file_path, 'r') as file:
9             reader = csv.reader(file)
10            data = list(reader)
11            return data[:50]        Observe Amazon Q Developer
12                                    suggesting next line to print dataframe
13
14    """ call read_csv by passing file path as /user/data/input/sample.csv """
15    file_path = "/user/data/input/sample.csv"
16    data = read_csv(file_path)
17    print(data)
18
      ┌─────────────────────────────────────────┐
      │  Insert Code      Previous      Next      │
      │     →|               ←            →        │
      │                                           │
      │  Suggestion 1 of 1 from Amazon Q        ⋮ │
      └─────────────────────────────────────────┘
```

Figure 3.15 – Line-by-line recommendations in the IDE using Amazon Q Developer

Observe that, in this script, Amazon Q Developer is suggesting the print DataFrame statement, `print(data)`, which makes logical sense as we read the data from the `sample.csv` file in a DataFrame named `data`.

Cross-reference existing code

One of the common best coding practices is to use existing functions from different files in the same repository of code. Many developers like to create reusable functions and save them in a common file that can be referenced in other scripts. Code assistants understand existing scripts and functions present across the repository. This allows them to help users recommend new code referencing functions from existing files. Users can write simple single-line prompts to suggest Code assistants to generate the code referencing existing functions from a particular program file.

For example, Amazon Q Developer has functionalities available that can cross-reference the existing code during new code generation. Let's assume the user already has the function saved in the `prompt-Fullfunctiongeneration.py` file (reference *Figure 3.14*) and wants to use the existing function, `read_csv`, in a new script. Here is the prompt followed by the output:

```
Prompt:
# read test.csv file using read_csv function from prompt-
Fullfunctiongeneration.py file
```

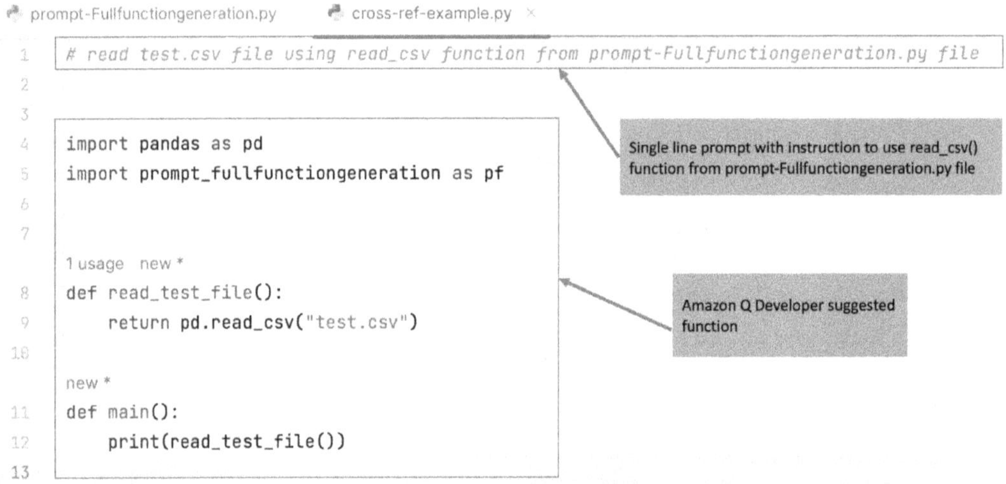

Figure 3.16 – Cross-referencing the existing code in the IDE using Amazon Q Developer

Observe that Amazon Q Developer imported all the functions from `prompt_fullfunctiongeneration` and then used the `read_csv()` function and generated code to read the `test.csv` file.

Generating sample data

During the development process, developers commonly need sample data that may or may not be readily available. Code assistants can help users generate sample data in different ways. We will explore a couple of methods by which Amazon Q Developer can help users generate the sample data.

Let's assume that a user already has some sample data and wants to create additional records to bump up the volume of data. Just like many other code assistants, Amazon Q Developer understands the structure and format of the existing sample data to generate the next sample record. In the following sample example, the user has one record in JSON format with sr_nbr, name, age, and city as attributes. Once the user hits *Enter* at the end of the first record, Amazon Q Developer will start generating sample records. The following screenshot highlights this feature:

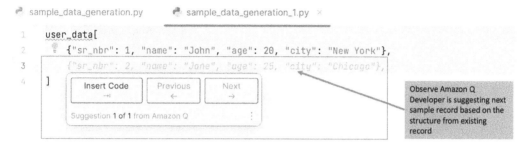

Figure 3.17 – Generating sample data in PyCharm IDE using Amazon Q Developer

Observe that Amazon Q Developer is suggesting the next sample record based on the structure from an existing record, {"sr_nbr": 2, "name": "Jane", "age": 25, "city": "Chicago"}.

Let's consider another example and assume that the user does not have a sample dataset but is aware of the attributes needed in the sample data. The user needs to create a fixed number of sample records. In this case, the user can use prompts to generate the sample data. In the following sample example, the user has entered a prompt to generate 50 records in JSON format with sr_nbr, name, age, and city as the desired attributes. With this prompt, Amazon Q Developer suggested a random data generation function.

Here is the prompt followed by the output:

```
Prompt:
"""
Generate 50 records for user_data json with sr_nbr, name, age, and
city
"""
```

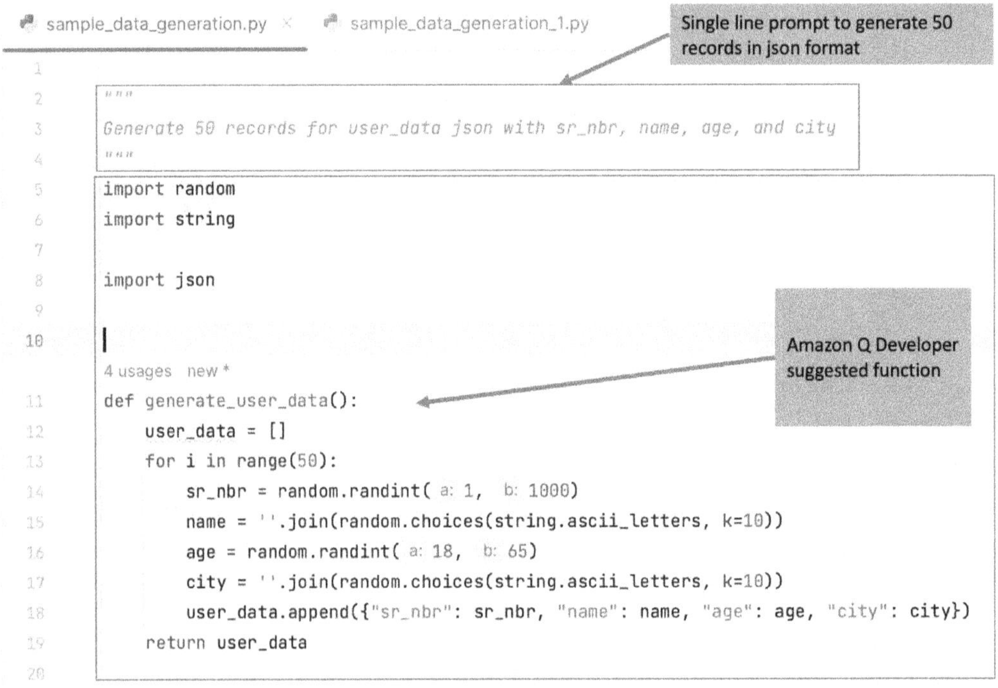

Figure 3.18 – Generating sample data in the PyCharm IDE using Amazon Q Developer – function logic

Observe that based on the requirement specified in the prompt, Amazon Q Developer generated the generate_user_data() function with a range of 50 to generate sample data with the sr_nbr, name, age, and city attributes in a JSON file format.

In addition, if you press *Enter* at the end of the function, you will observe that Amazon Q Developer will use a line-by-line recommendation to suggest end-to-end code and also save data in a user_data. json file. The following screenshot highlights this feature:

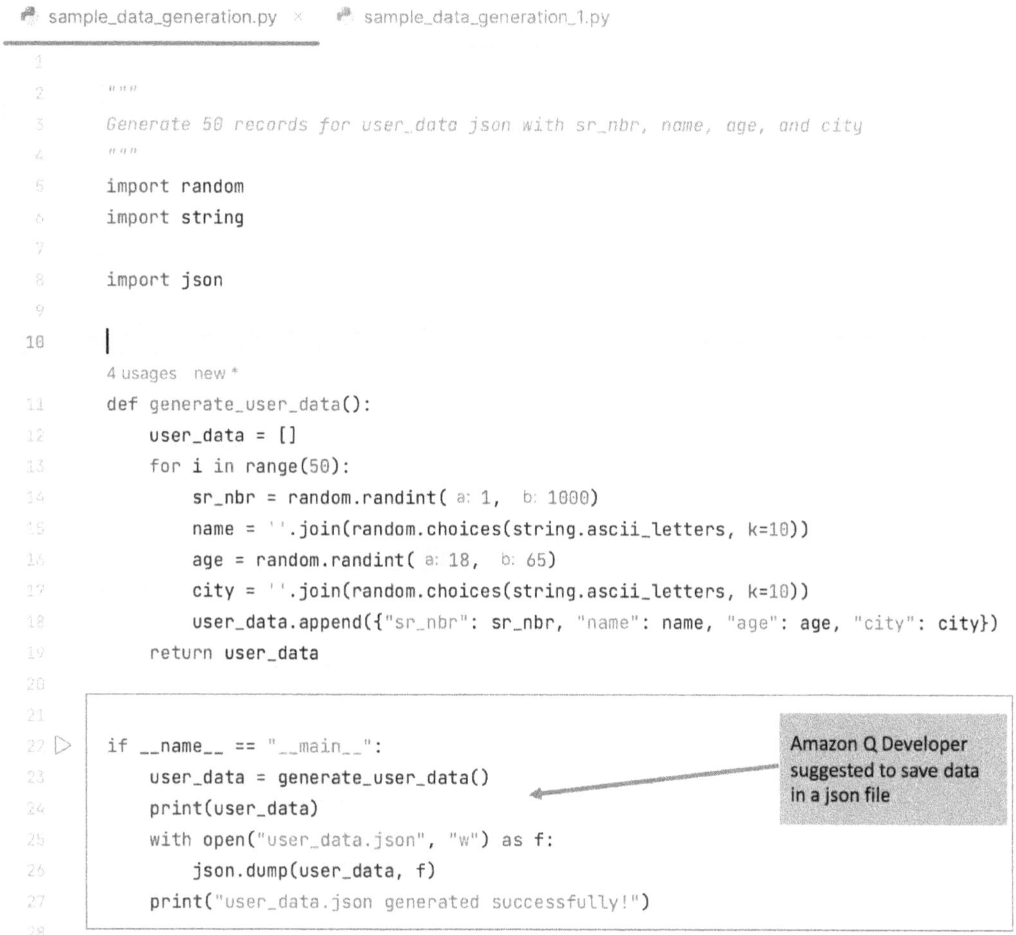

sample_data_generation.py ×   sample_data_generation_1.py

```
1
2      """
3      Generate 50 records for user_data json with sr_nbr, name, age, and city
4      """
5      import random
6      import string
7
8      import json
9
10     |
       4 usages  new *
11     def generate_user_data():
12         user_data = []
13         for i in range(50):
14             sr_nbr = random.randint( a: 1,   b: 1000)
15             name = ''.join(random.choices(string.ascii_letters, k=10))
16             age = random.randint( a: 18,   b: 65)
17             city = ''.join(random.choices(string.ascii_letters, k=10))
18             user_data.append({"sr_nbr": sr_nbr, "name": name, "age": age, "city": city})
19         return user_data
20
21
22 ▷  if __name__ == "__main__":
23         user_data = generate_user_data()
24         print(user_data)
25         with open("user_data.json", "w") as f:
26             json.dump(user_data, f)
27         print("user_data.json generated successfully!")
28
```

Amazon Q Developer suggested to save data in a json file

Figure 3.19 – Generating sample data in the IDE using Amazon Q Developer – script

Writing unit tests

Testing the scripts is an imperative part of coding. Testing code needs to be done at different phases, such as unit testing during or toward the end of coding, integration testing across multiple scripts, and so on. Let's explore the options available with code assistants to support unit test creation. As discussed earlier, code assistants can understand the context of scripts. This helps developers reference existing program files during unit test creation.

Let's assume a user wants to create unit tests for the previously generated code to generate sample data (reference *Figure 3.18*). The user can use a simple single-line prompt to create the unit tests. Amazon Q Developer can analyze the code present in the function file and generate the basic unit test cases. The following prompt and output showcases this capability:

```
Prompt:
"""
Create unit tests for the generate_user_data() function from sample_
data_generation.py file
"""
```

Figure 3.20 – Writing unit tests in the PyCharm IDE using Amazon Q Developer

Observe that Amazon Q Developer used the input prompt to reference the `generate_user_data()` function from the `sample_data_generation.py` file and, based on the functionality, generated a basic unit test case. In addition, it also created a full end-to-end script that can be used to run the unit test case.

Explaining and documenting code

During the life cycle of the project, many developers work on the same script to add code or update existing code. A lack of documentation can make it really difficult for everyone to understand the end-to-end logic in the script. To help developers understand existing code, many of the code assistants have different mechanisms to generate explanations of the code in natural language, and plain text format. Users can use the outcome of these options to create documentation or embed details in the comments section of the script.

Amazon Q Developer can help users generate the documentation with respect to the existing script. Once the script is open in an IDE, just type the `Explain` command in Q's chat session. Amazon Q Developer will analyze the whole script and will return the natural language text, explaining the functionality of the script. For example, let's use the previously created script, `prompt-Fullfunctiongeneration.py` (reference *Figure 3.14*), and ask Amazon Q Developer to explain the code.

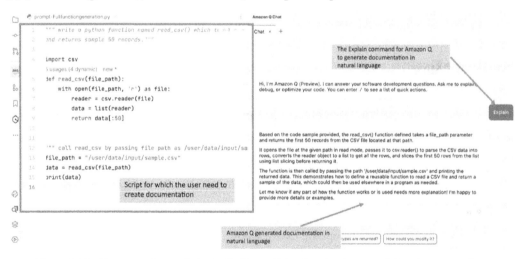

Figure 3.21 – Documenting code in the PyCharm IDE using the Amazon Q Developer interface

Observe that the user does not need to specify the script name, as Amazon Q Developer can automatically use the script open in the editor when you type the `Explain` command. The following screenshot highlights the code explanation provided by Amazon Q Developer:

Based on the code sample provided, the read_csv() function defined takes a file_path parameter and returns the first 50 records from the CSV file located at that path.

It opens the file at the given path in read mode, passes it to csv.reader() to parse the CSV data into rows, converts the reader object to a list to get all the rows, and slices the first 50 rows from the list using list slicing before returning it.

The function is then called by passing the path '/user/data/input/sample.csv' and printing the returned data. This demonstrates how to define a reusable function to read a CSV file and return a sample of the data, which could then be used elsewhere in a program as needed.

Let me know if any part of how the function works or is used needs more explanation! I'm happy to provide more details or examples.

Figure 3.22 – Documenting code in the IDE using Amazon Q Developer – documentation

Observe that Amazon Q Developer generated documentation in natural language. It was able to provide the exact functionality of the `read_csv()` function; then, it explained that the script used the function to read the `/user/data/input/sample.csv` file. In this way, any code can easily be documented without having to manually understand the code and type the whole explanation, thus saving time and improving productivity.

Updating existing code

Similar to the previous examples, developers often inherit previously developed code. Users need to update the code based on newly available libraries and known issues and to improve coding standards.

Generally, updating existing code falls into one of these three categories:

- **Refactoring**: Users need to update existing code to simplify it, making it easy to understand, and/or add additional exceptions to handle errors and so on

- **Fixing**: Users need to update existing code to fix bugs, which may be known or unknown

- **Optimizing**: Users need to update existing code to improve execution efficiency and performance tune

To help developers complete the tasks just mentioned, many code assistants offer options. Users can use the outcomes of these options to update existing code and improve coding standards.

Let's look at how Amazon Q Developer can help users update existing code. Similar to the `Explain` command discussed in the previous section, Amazon Q Developer has the `Refactor`, `Fix`, and `Optimize` commands. Once the script is open in an IDE, simply type any of these commands to get recommendations from Amazon Q Developer. Based on the code quality, Amazon Q Developer can provide multiple different recommendations and a direct option to insert the fragment into the existing script.

As seen in the following screenshot, let's use the previously created script, `prompt-Fullfunctiongeneration.py` (reference *Figure 3.15*), and ask Amazon Q Developer to refactor the code:

Figure 3.23 – Updating existing code in the PyCharm IDE using
the Amazon Q Developer interface: Refactor

Observe that Amazon Q Developer can automatically use the script open in the editor when you type the `Refactor` instruction and suggest multiple recommendations to consider while refactoring. Let's look at a couple of them in the following section.

In the following screenshot, Amazon Q Developer suggests adding hints to the `file_path` function parameter and returning parameters as a list to improve overall readability:

1. Add type hints to the function parameters and return value:

```python
from typing import List

def read_csv(file_path: str) -> List[List[str]]:
    ...
```

python   ⊡ Insert at cursor   ⧉ Copy

Figure 3.24 – Updating existing code in the PyCharm IDE using
the Amazon Q Developer interface: Refactor 2

As seen in the following screenshot, Amazon Q Developer suggests including additional exception handling for the `file_path` parameter to check whether the file is of the CSV type and whether the file exists:

2. Check for valid CSV before opening the file:

```python
import os

def read_csv(file_path: str) -> List[List[str]]:

    if not file_path.endswith('.csv'):
        raise ValueError('File must be CSV')

    if not os.path.isfile(file_path):
        raise FileNotFoundError('File not found')

    ...
```

python   ⌐↳ Insert at cursor   🗗 Copy

Figure 3.25 – Updating existing code in the PyCharm IDE using
the Amazon Q Developer interface: Refactor 3

We will explore other options and dive deep into the examples in *Chapter 12*.

Feature development

Code assistants can help developers develop features simply by describing them in natural language and specifying key phrases. As a user, you only need to provide keywords/phrases related to the functionality of the feature, and code assistants can generate the end-to-end code.

Let's look at how Amazon Q Developer can help users develop features. Amazon Q Developer leverages the context of the current project to generate a comprehensive implementation plan and specifies the necessary code changes. To initiate feature development, the user just needs to open a file within their project and type /dev in the **Amazon Q Chat** panel. The user can also specify the desired feature after /dev.

For example, let's ask Amazon Q Developer to implement a binary search feature. This is highlighted in the following screenshot:

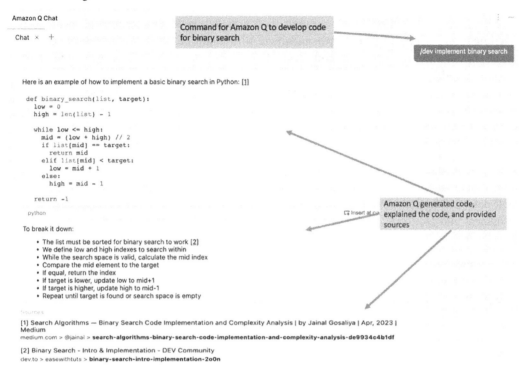

Figure 3.26 – Feature development in the PyCharm IDE using the Amazon Q Developer interface

Observe that, based on the command, Amazon Q Developer generated code for the binary search feature, provided details of the code flow/algorithm, and referenced sources to generate the features.

We will dive deep into detailed examples in *Chapter 12*.

Code transformation

When upgrading versions, depending on the programming language used, users may have to make various adjustments in their code to ensure compatibility with the latest version. Code assistants offer assistance to developers in upgrading their code, helping them transition from older versions to the most recent ones.

For example, Amazon Q Developer has the capability to upgrade the code language version. Users can just open the existing old version code and use the /transform command to upgrade the code version directly within their files.

We will dive deep into further details with examples in *Chapter 12*.

Summary

In this chapter, we covered the integration of code assistants with LLMs to assist users with auto-code generation. Then, we explored three commonly used auto-code generation prompting techniques: single-line prompts, multi-line prompts, and chain-of-thought prompts. We introduced each of these prompting techniques for auto-code generation, along with potential use cases, limitations, and required coding experience. Sample code examples were used in JetBrains' PyCharm IDE with Amazon Q Developer enabled. Additionally, we introduced the *chat with code assistant* technique in the auto-code generation process, where users interact with the code assistant in a simple question-and-answer style session. Amazon Q Developer was utilized to obtain general recommendations for coding/debugging.

We then discussed some of the common building methods of code generation, such as single-line code completion, full function generation, block completion, line-by-line recommendations, generating sample data, writing unit tests, explaining and documenting code, updating existing code, feature development, and code transformation. Furthermore, we explored these functionalities with Amazon Q Developer in JetBrains' PyCharm IDE to support the most common code-building methods.

Next, we'll start *Part 2* of the book. In this part, *Chapters 4* to *9* will walk you through how Amazon Q Developer can boost developer productivity by auto-generating code for many supported programming languages. Based on your expertise or preference, feel free to navigate directly to the chapters that interest you the most.

In the next chapter, we will delve deeper into how to utilize these techniques and building methods within the IDE environment for the Python and Java languages. Additionally, we will create a sample Python application with the assistance of Amazon Q Developer.

References

- Amazon Q Developer single-line code completion: https://docs.aws.amazon.com/amazonq/latest/qdeveloper-ug/single-line-completion.html

- Amazon Q Developer full function generation: https://docs.aws.amazon.com/amazonq/latest/qdeveloper-ug/full-function-generation.html

- Amazon Q Developer block completion: https://docs.aws.amazon.com/amazonq/latest/qdeveloper-ug/code-block.html

- Amazon Q Developer line-by-line recommendations: https://docs.aws.amazon.com/amazonq/latest/qdeveloper-ug/line-by-line-1.html

- Amazon Q Developer chat with Amazon Q Developer in IDEs: https://docs.aws.amazon.com/amazonq/latest/aws-builder-use-ug/q-in-IDE-chat.html

- Explain and update code with Amazon Q Developer: https://docs.aws.amazon.com/amazonq/latest/aws-builder-use-ug/explain-update-code.html

- Feature development with Amazon Q Developer: `https://docs.aws.amazon.com/amazonq/latest/aws-builder-use-ug/feature-dev.html`

- Code transformation with Amazon Q Developer: `https://docs.aws.amazon.com/amazonq/latest/aws-builder-use-ug/code-transformation.html`

- Prompt engineering guide – generating code: `https://www.promptingguide.ai/applications/coding`

4

Boost Coding Efficiency for Python and Java with Auto-Code Generation

In this chapter we will look at the following key topics:

- Overview of the use case for weather data analysis
- Python auto-code generation using Amazon Q Developer for weather data analysis
- Java auto-code generation using Amazon Q Developer for weather data analysis

In the previous chapter, we laid the foundation for different auto-code generation techniques to interact with AI-powered code assistants. We discussed common prompting techniques such as single-line, multi-line, chain-of-thought, and chat with code assistant, as well as common code-building methods.

In this chapter, we will look at how you can use Amazon Q Developer to suggest code in a variety of IDEs. We will start with the two most prominent programming languages used by developers, Python and Java, to demonstrate how auto-code can be generated using techniques from *Chapter 3*. We will also see how Amazon Q Developer adds value during the code development process by enabling a chat with the code assistant technique.

We believe that demonstrating the effectiveness and ease of use of AI-powered code assistants using a sample application will be impactful.

In the next section, let's first define the sample use case that we will use for both Python and Java scripts.

An overview of the use case for weather data analysis

Many customers are keen to understand the weather patterns for a specific city, which holds significant importance for numerous applications and can be regarded as a crucial data source across a wide variety of applications. Weather data applications in enterprises can serve various purposes across different industries.

Here's a general overview of how weather data applications can be utilized in some of the enterprises:

- **Risk management and insurance**: Insurance companies can use weather data to assess and mitigate risks connected with weather-related events such as hurricanes, floods, or wildfires. By analyzing historical weather patterns and forecasts, insurers can better understand potential risks and adjust their policies accordingly.

- **Supply chain optimization**: Weather data can help optimize supply chain operations by providing insights into weather conditions that may impact transportation, logistics, and distribution networks. Enterprises can use weather forecasts to anticipate disruptions and optimize routes and inventory management accordingly.

- **Energy management**: Energy companies can leverage weather data to optimize energy production and distribution. For example, renewable energy companies can use weather forecasts to predict solar or wind energy generation, helping them better plan and manage their energy resources.

- **Finance industry**: In the finance industry, weather data can be leveraged in various ways to enhance decision-making processes and improve risk management strategies. In retail banking, weather patterns directly influence consumer behavior and spending habits, thereby affecting banking operations. Additionally, in the real estate sector, weather data holds substantial value, particularly concerning property insurance and mortgage lending.

- **Agriculture and farming**: Agriculture enterprises can use weather data to optimize crop planning, irrigation schedules, and pest management. By analyzing weather patterns and forecasts, farmers can make data-driven decisions to improve crop yields and minimize risks connected with weather-related events.

- **Retail and marketing**: Retailers can use weather data to optimize marketing campaigns and inventory management. For example, retailers can adjust their promotions and inventory levels based on weather forecasts to capitalize on consumer behavior changes that are influenced by weather conditions.

- **Construction and infrastructure**: Construction companies can use weather data to plan construction projects more effectively and minimize weather-related delays. By integrating weather forecasts into project planning and scheduling, construction enterprises can optimize resource allocation and reduce project risks.

- **Tourism and hospitality**: Enterprises in the tourism and hospitality industry can use weather data to optimize operations and enhance customer experiences. For example, hotels and resorts can adjust pricing and marketing strategies based on weather forecasts to attract more visitors during favorable weather conditions.

Overall, weather data applications in enterprises can provide valuable insights and help businesses make informed decisions across various industries, ultimately driving efficiency, reducing risks, and improving competitiveness. There are many vendors, such as OpenWeatherMap, The Weather Company, AerisWeather, WeatherTAP, AccuWeather, and Yahoo Weather, that provide weather data to enterprises.

Application requirements – weather data analysis

For our application, we will use weather data provided by OpenWeatherMap, which offers a rich set of APIs. With **api.openweathermap.org**, you gain access to a comprehensive source of weather information, enabling you to seamlessly integrate real-time and forecasted weather data into your applications. OpenWeatherMap's API offers a wide range of weather data, including current weather conditions, forecasts, historical data, and more. Whether you're building a weather app, optimizing logistics, or planning outdoor events, our API provides the data you need to make informed decisions.

Let's define the requirements for this simple and versatile application. We will use the same application requirements for both Python and Java scripts.

Application requirements – weather data analysis

Business requirement: Analysts are interested in getting weather forecasting for a country and city. They would like to see the chart visualizing temperature changes.

User inputs: The application should accept a country and city name as parameters from users.

Data fetch: Based on the inputs provided, the application requests the weather data from `http://api.openweathermap.org/data/2.5/forecast` using an API key.

Data table: Shows the table with datetime and corresponding temperature in Fahrenheit.

Data visualization: Creates simple graphs to plot the temperature in Fahrenheit and date based on dataset sourced from OpenWeatherMap.

Prerequisites to access the OpenWeatherMap API

OpenWeatherMap provides options to request the data using API calls. As a prerequisite, you need to create an account. Here is the summary of the steps; for additional information, reference `https://openweathermap.org/`.

To get weather data from OpenWeatherMap, you need to follow these steps:

1. **Sign up**: Go to the OpenWeatherMap website (`https://openweathermap.org/`) and sign up for an account. Once you have signed up and logged in, move on to the next step.

2. **Understand the API documentation**: OpenWeatherMap provides various APIs for accessing weather data, including current weather data, forecast weather data, historical weather data, and more. Review the API documentation to understand how to use the different endpoints and parameters to fetch the weather data you need. We will use the `api.openweathermap.org` API to collect the dataset. Based on the number of times we call the API in this chapter, it will stay within the free tier.

3. **Get an API key**: You can generate an API key from your account dashboard or use the default key.

Figure 4.1 – An OpenWeatherMap API key

Note down the API key as we will need it in the next sections to call APIs.

Python auto-code generation using Amazon Q Developer for weather data analysis

As we have defined the use case (problem statement) and completed the prerequisites, let's utilize various auto-code generation techniques to get the solution. To illustrate these techniques, we will leverage the Python programming language inside JetBrains' PyCharm IDE, which has been set up to work with Amazon Q Developer. Please refer to *Chapter 2* for detailed steps on setting up Amazon Q Developer with JetBrains' PyCharm IDE.

Solution blueprint for weather data analysis

As an experienced code developer or data engineer, you will need to convert the preceding business objectives into technical requirements by defining reusable functions:

1. Write a Python script for weather data analysis.

2. Write a function to get weather data using API key for user entered country and city from `http://api.openweathermap.org/data/2.5/forecast`.

3. Convert the return dates from an API call to the UTC format. (Note that the OpenWeatherMap API will return dates associated with the next 40 hours from the time of the request.)

4. Convert the returned temperatures from the API call from Celsius to Fahrenheit.

5. Write a function to show the date in UTC and the temperature in Fahrenheit as a table.

6. Write a function to plot temperature on the *y* axis and date on the *x* axis based on `temperature_data`.

7. Accept user inputs for country and city names.

8. Use the user-provided country and city names to call the `get_weather_data()` function.

9. Create a table with the datetime and temperature (in Fahrenheit) based on weather data.

10. Plot a graph for the specified city with `Temperature (°F)` on the *y* axis and `Date` on the *x* axis based on weather data.

11. Generate the documentation for the script.

To achieve the overall solution, we will mainly use **chain-of-thought prompts** to obtain the end-to-end script and a combination of single-line and multi-line prompts for individual code fragments. We will also chat with the code assistant for documentation.

> **Note**
>
> The output of the AI-powered code assistant is non-deterministic, so you may not get the exact same code that follows. You may also need to modify some parts of the code to meet the requirements. Additionally, auto-generated code may reference packages that you will need to manually install. To install missing packages in JetBrains' PyCharm IDE, please refer to the instructions at `https://www.jetbrains.com/help/pycharm/installing-uninstalling-and-upgrading-packages.html`.

Let's proceed with step-by-step solutions.

Requirement 1

Write a Python script with the latest Python version.

Use JetBrains' PyCharm IDE to create a `book_weather_data.py` file and make sure that Amazon Q Developer is enabled.

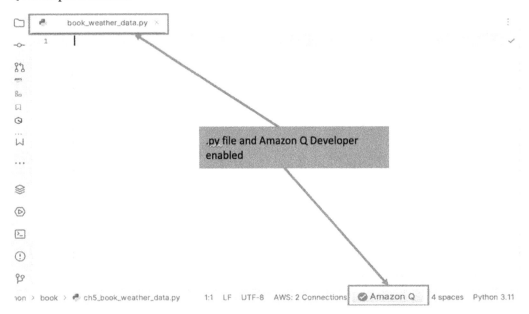

Figure 4.2 – JetBrains' PyCharm IDE with the .py file and Amazon Q Developer enabled

The previous step will generate a file with the `.py` extension, which Amazon Q Developer will recognize when generating code. Therefore, there is no need to include the language name Python in your prompts.

Requirements 2, 3, and 4

Let's combine requirements 2, 3, and 4 to create a multi-line prompt:

- Write a function to get weather data based on user's selection of a country and a city from `http://api.openweathermap.org/data/2.5/forecast`
- Convert the return dates from an API call to the UTC format.
- Convert the returned temperatures from the API call from Celsius to Fahrenheit.

We are using the multi-line prompt technique. As a reminder, in this technique, we can instruct our code assistant to generate the code based on our specific requirements.

We write the prompt as follows:

```
'''
Write function get_weather_data() to get weather data from http://api.
openweathermap.org/data/2.5/forecast based on country and city.
Convert date to UTC format and Convert temparture from Celsius to
Fahrenheit.
Then return date and temperature as temperature_data
'''
```

Note that as part of the multi-line prompt, we have provided specific instructions for the code assistant to follow:

- The function name to use is get_weather_data()

- The function has two input parameters: country and city

- Get data from http://api.openweathermap.org/data/2.5/forecast

- Return date and temperature as temperature_data

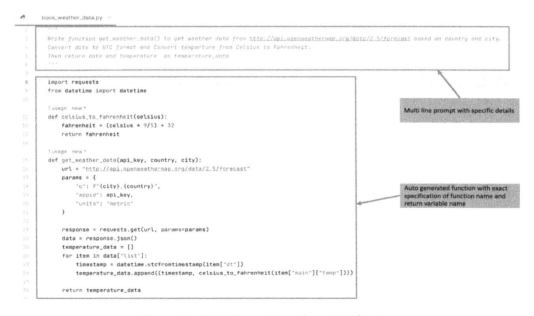

Figure 4.3 – The Python get_weather_data() function

Observe that the code assistant followed the specific instructions provided in the prompt. In addition, it also determined that api_key is required to get the data from http://api.openweathermap. org/data/2.5/forecast, so it added that as one of the parameters.

Requirement 5

Write a function to show the date in UTC and temperature in Fahrenheit as a table.

We are using the **single-line prompt** technique, as the requirement is easy and can easily be described in a single line. Note that as part of the single-line prompt, we have provided specific instructions for the code assistant to follow:

- The function name to use is `display_weather_table()`.

- Use `temperature_data` to show the table. This is the chain-of-thought prompt, that we are connecting previously the defined return results from the `get_weather_data()` function as the input for this function.

We write the prompt as follows:

```
'''
write function display_weather_table() to show temperature_data
'''
```

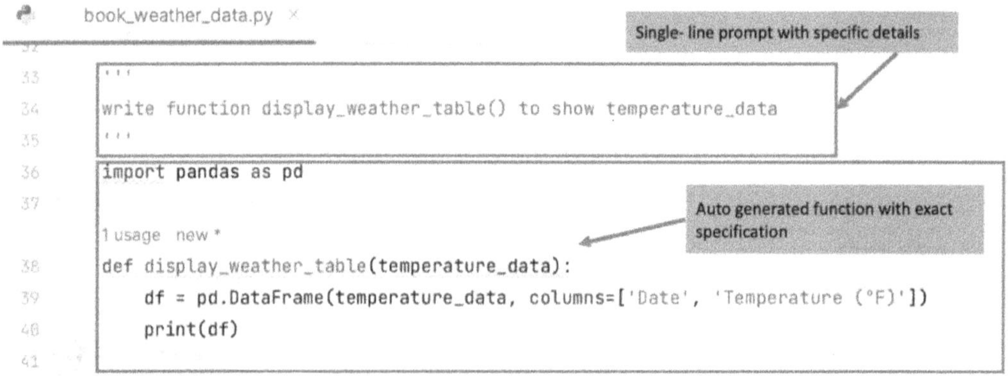

Figure 4.4 – The Python display_weather_table () function

Requirement 6

Write a function to plot `temperature` on the *y* axis and `date` on the *x* axis based on `temperature_data`.

We are using the single-line prompt technique, as the requirement is easy and can easily be described in a single line. Note that as part of the single-line prompt, we have provided specific instructions for the code assistant to follow:

- Function name to use is `plot_temperature_graph()`

- Use `temperature_data` to show the table. This is the chain-of-thought prompts as we are connecting previously defined return result from `get_weather_data()` function as input for this function.

We write the prompt as follows:

```
'''
Write function plot_temperature_graph() to plot temperature on Y axis
and date on X axis based on temperature_data
'''
```

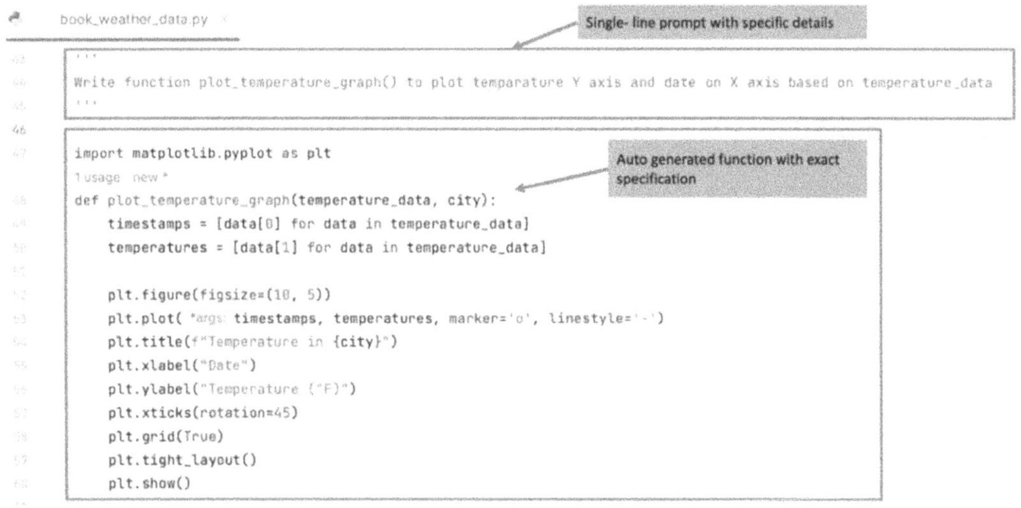

Figure 4.5 – The Python plot_temperature_graph () function

Requirements 7, 8, 9, and 10

Let's combine requirements 7, 8 ,9 , and 10 to create a multi-line prompt technique:

- Accept user inputs for country and city name.

- Use a user-provided country and city name to call the `get_weather_data()` function.

- Create a table with datetime and temperature (in Fahrenheit) based on the weather data.

- Plot a graph for the specified city with `Temperature (°F)` on the Y-axis and `Date` on the X-axis based on weather data

We are using the chain-of-thought prompt technique to link all the previously defined functions together.

We write the prompt as follows:

```
'''
Accept country and city from User and call get_weather_data() and
display_weather_table() and plot_temperature_graph() functions.
'''
```

We will get the following output:

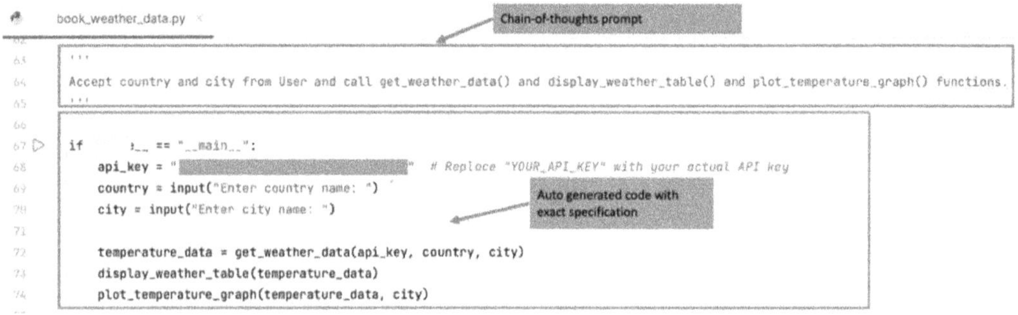

Figure 4.6 – The Python code to get the user inputs and display weather data

Note that in the preceding prompt, I have not included error handling for the input data. However, I encourage you to experiment with the prompt by adding more context to instruct Amazon Q Developer to suggest code with additional error handling.

Now let's ensure that the script is running as expected. Run the code and enter a country and city to get the weather data.

For testing, I will use the following values, but feel free to choose your own:

- Country name: `US`
- City name: `Atlanta`

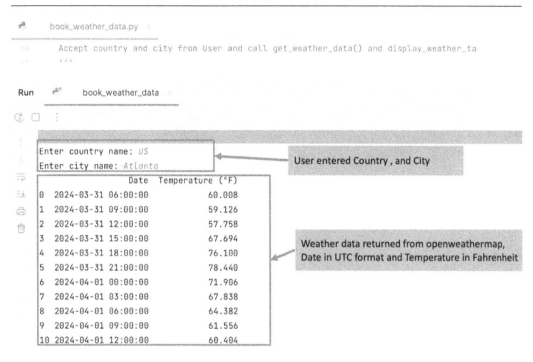

Figure 4.7 – The Python output table with user input country, city, and weather information

Now let's review the second part of the output related to the chart plotted for the date and corresponding temperature.

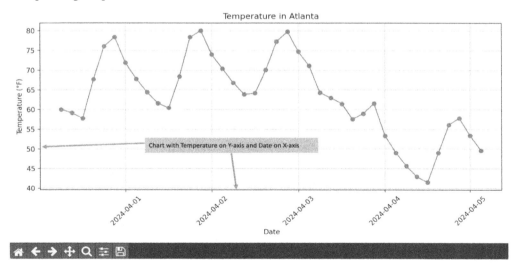

Figure 4.8 – The Python output chart with the date and temperature

Requirement 11

Generate the documentation for the script.

Let's use Amazon Q Developer chat with the code assistant technique to generate documentation. Remember that Amazon Q Developer supports the Explain prompt (for more details, see *Chapter 3*)

Alternatively, you can highlight the entire code, then right-click and select the **Amazon Q** option from the pop-up menu followed by **Explain Code**.

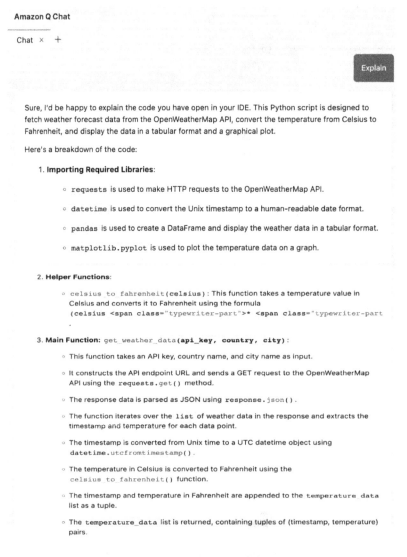

Figure 4.9 – The Python code documentation from Amazon Q Developer

The preceding screenshot shows the first part of the code explanation provided by Amazon Q Developer.

Now let's use a few suggested prompts for additional documentation.

As you can see, Amazon Q Developer analyzed the script that was open in the editor window. It then attempted to comprehend the code to derive its logic. Finally, it consolidated all its findings to generate documentation aimed at helping users understand the code better.

Figure 4.10 – The Python Amazon Q Developer suggested prompts

Note that Amazon Q Developer prompts for additional suggestions to obtain further in-depth documentation. Please feel free to explore further. Application developers can use Amazon Q Developer for code improvements just by using the chat with code assistant technique.

Let's ask Amazon Q Developer to provide updated or improved code for our previously generated `display_weather_table()` function:

```
code to improve
def display_weather_table(temperature_data):
    df = pd.DataFrame(temperature_data, columns=['Date',
        'Temperature (°F)'])
    print(df)
```

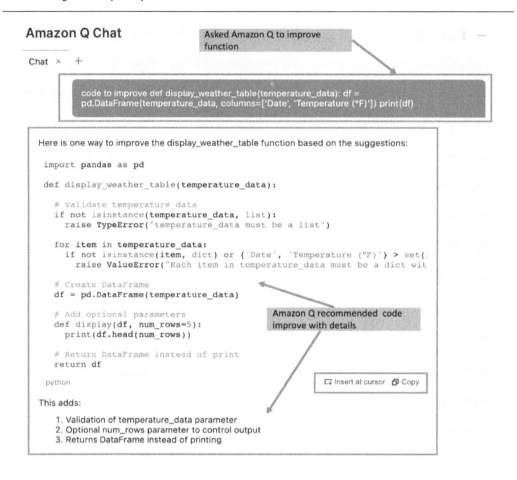

Figure 4.11 – The Amazon Q Developer suggested improvements in Python

As you can see in the preceding snapshot, Amazon Q Developer provided new code with additional validation, the use of a dataframe, and recommended changes. You can update the code fragment simply by using either `Insert at cursor` or `Copy`.

I have illustrated a simple use case with one of the code fragments, but by following the preceding steps, you can get recommended updated code from Amazon Q Developer regardless of your use case. Additionally, you can adjust the prompts to instruct Amazon Q Developer to generate code with error and exception handling.

Use-case summary

As illustrated, we used a combination of chain-of-thought, single-line, and multi-line prompts to create an end-to-end Python script for a weather data application. We utilized JetBrains' PyCharm IDE, which was set up to work with Amazon Q Developer. For the script, we used Amazon Q Developer with specific prompts to auto-generate functions for retrieving weather data from OpenWeatherMap, converting temperature from Celsius to Fahrenheit, converting dates to UTC format, and plotting charts.

Additionally, we leveraged Amazon Q Developer's chat feature to generate detailed documentation and receive code improvement recommendations for our weather data application. Through this approach, we showcased how effectively integrating various prompt types and leveraging the capabilities of Amazon Q Developer can streamline the development process.

By adjusting prompts or simply instructing Amazon Q Developer in a chat-style interface, readers can further direct Amazon Q Developer to generate code with enhanced error and exception handling, making the solution robust and versatile for different use cases. This example demonstrates the power of combining advanced IDEs and intelligent code assistants to build sophisticated applications efficiently.

In the next section, let's implement the same application using the Java language.

Java auto code generation using Amazon Q Developer for weather data analysis

As we have defined the use case and completed the prerequisites, let's use different auto-code generation techniques to achieve the use case. To illustrate this, we will leverage Java programming language inside **Visual Studio Code** (**VS Code**) IDE that has been set up to work with Amazon Q Developer. Please refer to *Chapter 2* for detailed steps to help you set up Amazon Q Developer with VS Code IDE.

Solution blueprint for weather data analysis

As an experienced code developer or data engineer, you will need to convert the preceding business objectives into technical requirements by defining reusable functions:

1. Write a Java script for weather data analysis.

2. Accept user inputs for country and city name.

3. Get weather data using an API key for a user-entered country and city from `http://api. openweathermap.org/data/2.5/forecast`.

4. Convert the return dates from an API call to the UTC format. (Note that, the OpenWeatherMap API will return dates associated with the next 40 hours from the time of the request.)

5. Convert the returned temperatures from the API call from Celsius to Fahrenheit.

6. Show the date in UTC and temperature in Fahrenheit as a table.

7. Plot a graph for the specified city with `Temperature (°F)` on the *y* axis and `Date` on the *x* axis based on weather data.

8. Generate the documentation for the script.

To achieve the overall solution, we will mainly use a combination of single-line and multi-line prompts for individual code fragments, as well as chat with the code assistant for documentation.

> **Note**
>
> The output of the AI-powered code assistant is non-deterministic, so you may not get the exact same code that follows. You may also need to modify some parts of the code to meet the requirements. Additionally, auto-generated code may reference packages or methods that you will need to manually install. To install missing packages in VS Code IDE, please reference `https://code.visualstudio.com/docs/java/java-project`.

Let's proceed with step-by-step solutions.

Requirement 1

Write a Java script for weather data analysis.

Use VS Code IDE to create a new Java project. Create the `book_weather_data.java` file and make sure that Amazon Q Developer is enabled.

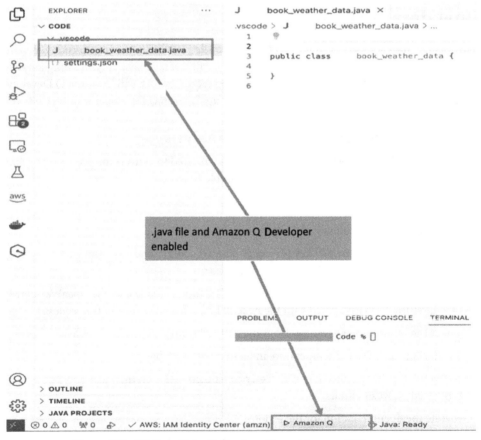

Figure 4.12 – VS Code IDE with the .java file and Amazon Q Developer

The previous step will create a file with a `.java` extension, which Amazon Q Developer will reference, so there's no need to include the language name Java in your prompts.

Requirement 2

Accept user inputs for country and city name.

We are using the single-line prompt technique, as it's a very straightforward requirement.

We write the prompt as follows:

```
// Accept user input for country name and city name
```

We will then get the following output:

```
1
2     // Accept user input for country name and city name          Single line prompt
3
4     import java.io.BufferedReader;
5     import java.io.InputStreamReader;
6     import org.jfree.ui.ApplicationFrame;
7
8     public class     book_weather_data extends ApplicationFrame {
9         private static final long serialVersionUID = 1L;
10
11        public     book_weather_data (String title) {
12            super(title);                                 Auto generated multiple line
13        }                                                 of code
14
      Run | Debug
15        public static void main(String[] args) {
16            try {
17                // Accept user input for country name and city name
18                BufferedReader reader = new BufferedReader(new InputStreamReader(System.in));
19                System.out.print(s:"Enter country name: ");
20                String country = reader.readLine();
21                System.out.print(s:"Enter city name: ");
22                String city = reader.readLine();
23
24
25
26            } catch (Exception e) {
27                e.printStackTrace();
28            }
29        }
30    }
```

Figure 4.13 – Java accepts the user input

Requirements 3, 4, and 5

Let's combine requirements 3, 4, and 5, as they make up one logical unit.

- Get weather data based on countries and cities from `http://api.openweathermap.org/data/2.5/forecast`.

- Convert the return dates from the API call to the UTC format.

- Convert the returned temperatures from the API call from Celsius to Fahrenheit.

We are using the multi-line prompt technique to describe the logic. We write the prompt as follows:

```
//Get weather data from http://api.openweathermap.org/data/2.5/
forecast based on country and city.
//Convert date to UTC format and Convert temperature from Celsius to
Fahrenheit.
```

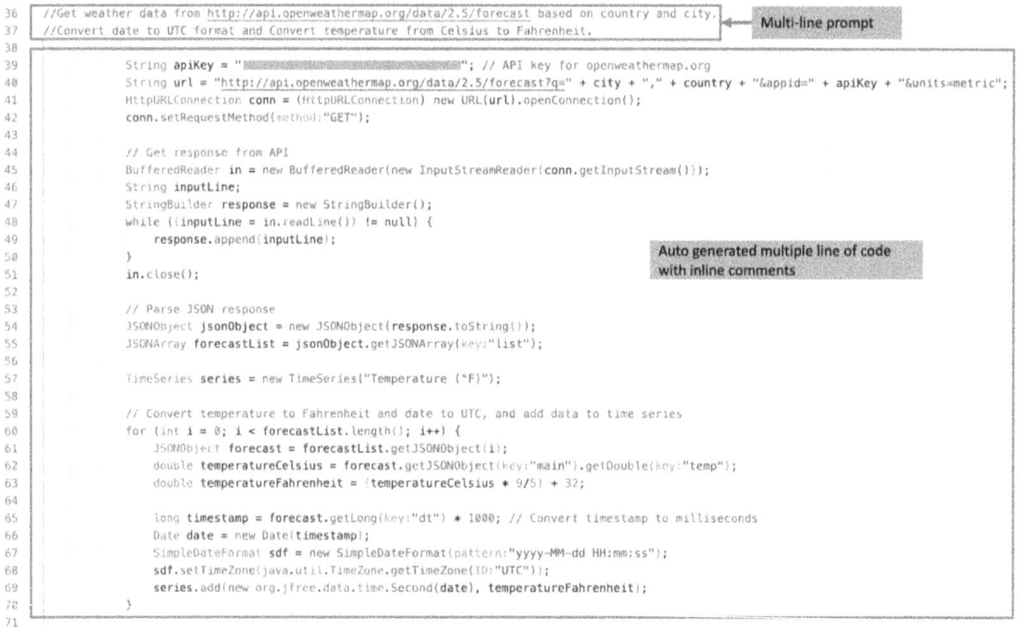

Figure 4.14 – The Java code to get the weather data

Requirement 6

Show the date in UTC and temperature in Fahrenheit as a table.

We are using the single-line prompt technique. We write the prompt as follows:

```
// Display temperature in Fahrenheit and date in UTC for each weather
forecast entry
```

We will then get the following output:

```
72      // Display temperature in Fahrenheit  and date in UTC for each weather forecast entry    ← Single-line prompt
73
74
75      for (int i = 0; i < forecastList.length(); i++) {    ← Auto generated multiple line of code
76          JSONObject forecast = forecastList.getJSONObject(i);
77          double temperature = forecast.getJSONObject(key:"main").getDouble(key:"temp");
78          double temperatureFahrenheit = (temperature * 9/5) + 32;
79          long timestamp = forecast.getLong(key:"dt") * 1000; |
80          Date date = new Date(timestamp);
81          SimpleDateFormat sdf = new SimpleDateFormat(pattern:"yyyy-MM-dd HH:mm:ss");
82
83          // Display temperature and date
84          System.out.println("Forecast " + (i + 1));
85          System.out.println("Temperature in " + city + ", " + country + ": " + temperatureFahrenheit + "°F");
86          System.out.println("Date and time: " + sdf.format(date));
87          System.out.println();
88      }
```

Figure 4.15 – The Java code to display weather data as a table

Requirement 7

Plot a graph for the specified city with `Temperature` (°F) on the *y* axis and `Date` on the *x* axis based on weather data.

We are using the single-line prompt technique. We write the prompt as follows:

```
// Plot a graph for the specified city with "Temperature (°F)" on the
Y-axis and "Date" on the X-axis for weather forecast entry
```

```
88
89      // Plot a graph for the specified city with "Temperature (°F)" on the Y-axis and "Date" on the X-axis for weather forecast entry
90
91      TimeSeriesCollection dataset = new TimeSeriesCollection(series);
92      JFreeChart chart = ChartFactory.createTimeSeriesChart(    ← Single-line prompt
93          "Weather Forecast",
94          "Date (UTC)",
95          "Temperature (°F)",    ← Auto generated multiple line of code
96          dataset
97      );
98
99      ChartPanel chartPanel = new ChartPanel(chart);
100     chartPanel.setPreferredSize(new java.awt.Dimension(width:800, height:600));
101     ch5_book_weather_data frame = new ch5_book_weather_data(title:"Weather Forecast Chart");
102     frame.setContentPane(chartPanel);
103     frame.pack();
104     RefineryUtilities.centerFrameOnScreen(frame);
105     frame.setVisible(b:true);
106
```

Figure 4.16 – The Java code to display weather data as a chart

Now let's ensure that the script is running as expected. Run the code and enter the country and city to get the weather data.

For testing, I will use the following values, but feel free to choose your own:

- Country name: US
- City name: Atlanta

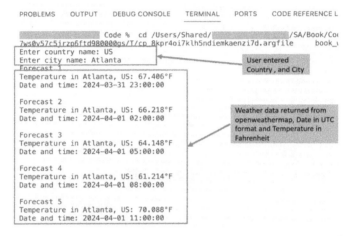

Figure 4.17 – The Java code output for user input country, city, and weather information

Now let's review the second part of the output related to the chart plotted for the date and corresponding temperature.

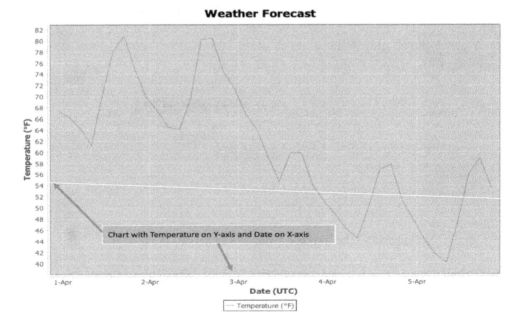

Figure 4.18 – The Java output for a weather information chart

Requirement 8

Generate the documentation for the script.

Let's use the Amazon Q Developer chat with code assistant technique to generate documentation. Remember that Amazon Q Developer supports the Explain prompt or command (see *Chapter 3* for reference).

Alternatively, you can highlight the entire code, then right-click and select the **Amazon Q** option from the pop-up menu followed by **Explain**.

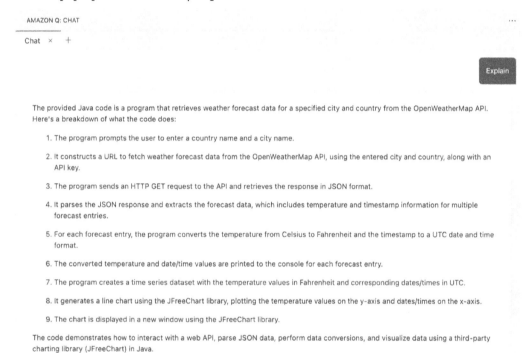

AMAZON Q: CHAT

Chat × +

Explain

The provided Java code is a program that retrieves weather forecast data for a specified city and country from the OpenWeatherMap API. Here's a breakdown of what the code does:

1. The program prompts the user to enter a country name and a city name.

2. It constructs a URL to fetch weather forecast data from the OpenWeatherMap API, using the entered city and country, along with an API key.

3. The program sends an HTTP GET request to the API and retrieves the response in JSON format.

4. It parses the JSON response and extracts the forecast data, which includes temperature and timestamp information for multiple forecast entries.

5. For each forecast entry, the program converts the temperature from Celsius to Fahrenheit and the timestamp to a UTC date and time format.

6. The converted temperature and date/time values are printed to the console for each forecast entry.

7. The program creates a time series dataset with the temperature values in Fahrenheit and corresponding dates/times in UTC.

8. It generates a line chart using the JFreeChart library, plotting the temperature values on the y-axis and dates/times on the x-axis.

9. The chart is displayed in a new window using the JFreeChart library.

The code demonstrates how to interact with a web API, parse JSON data, perform data conversions, and visualize data using a third-party charting library (JFreeChart) in Java.

Figure 4.19 – Amazon Q Developer in Java for overview documentation

As you can see, Amazon Q Developer analyzed the script that was open in the editor window. It then attempted to comprehend the code to derive its logic. Finally, it consolidated all its findings to generate documentation aimed at helping users understand the code better.

Now let's use the suggested prompts for additional documentation as seen in the following screenshot.

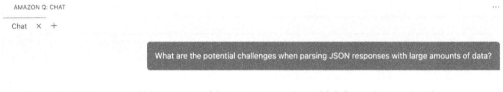

AMAZON Q: CHAT

Chat   ✕   +

> What are the potential challenges when parsing JSON responses with large amounts of data?

When parsing JSON responses with large amounts of data, there are several potential challenges that can arise: [1]

1. **Memory Constraints** : Large JSON responses can consume a significant amount of memory when loaded into memory all at once. This can lead to out-of-memory errors or performance issues, especially on resource-constrained devices or systems. [2]

2. **Processing Time** : Parsing and processing large JSON responses can be computationally intensive and time-consuming, leading to slower response times and potential performance bottlenecks in your application.

3. **Network Bandwidth** : Transferring large JSON responses over the network can consume significant bandwidth, which may be a concern in scenarios with limited or expensive network connectivity.

4. **Data Extraction** : As the size of the JSON data increases, it becomes more challenging to efficiently identify and extract the relevant data you need, especially if the data is nested deeply within the JSON structure.

5. **Streaming and Buffering** : When dealing with large JSON responses, it may be necessary to process the data in a streaming or buffered manner, rather than loading the entire response into memory at once. This can introduce additional complexity in terms of data handling and state management.

6. **Error Handling** : With larger JSON responses, there is a higher likelihood of encountering errors or inconsistencies in the data, which can be more difficult to detect and handle effectively.

7. **Scalability** : As the volume of data grows, it becomes increasingly important to design and implement scalable solutions that can handle the increased load and data processing requirements.

To mitigate these challenges, you may need to consider techniques such as:

Figure 4.20 – Amazon Q Developer suggested prompts in Java

Note that Amazon Q Developer shares additional suggestions to obtain further in-depth documentation. Please feel free to explore them further.

Application developers can use Amazon Q Developer for code improvements simply by using the chat with code assistant technique. Let's ask Amazon Q Developer to provide updated or improved code for the previously generated 'Converting temperature to Fahrenheit and date to UTC, and adding data to a time series' requirement.

We write the prompt as follows:

```
Improve following code :
for (int i = 0; i < forecastList.length(); i++) {
    JSONObject forecast = forecastList.getJSONObject(i);
    double temperatureCelsius = \
        forecast.getJSONObject("main").getDouble("temp");
    double temperatureFahrenheit = (temperatureCelsius * 9/5) + 32;
    long timestamp = forecast.getLong("dt") * 1000; // Convert
timestamp to milliseconds
```

```
Date date = new Date(timestamp);
SimpleDateFormat sdf = \
    new SimpleDateFormat("yyyy-MM-dd HH:mm:ss");
sdf.setTimeZone(java.util.TimeZone.getTimeZone("UTC"));
series.add(new org.jfree.data.time.Second(date),
    temperatureFahrenheit);
}
```

Observe in the following screenshot that Amazon Q Developer has recommended multiple code changes to improve the code, making it easier to understand. This includes creating new methods and standardizing the code to use object-oriented approaches such as class encapsulation instead of plain variables.

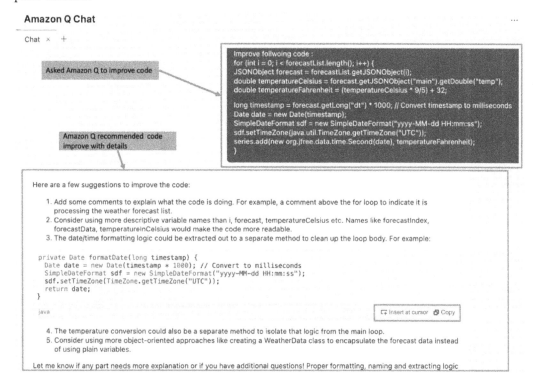

Figure 4.21 – Amazon Q Developer for Java code improvement

Applications can use these recommendations to obtain the exact code using Amazon Q Developer. Additionally, Amazon Q Developer provided code fragments. You can update the script simply by using either **Insert at cursor** or **Copy**.

I have illustrated a simple use case with one of the code fragments, but by following the preceding steps, you can get recommended updated code from Amazon Q Developer for any use case. Additionally, you can adjust the prompts to instruct Amazon Q Developer to generate code with error and exception handling.

Use-case summary

As illustrated, we employed a variety of prompt types – chain-of-thought, single-line, and multi-line – to develop a JavaScript script for a weather data application. We utilized VS Code IDE, which was configured to integrate seamlessly with Amazon Q Developer. Using Amazon Q Developer, we applied specific prompts to automatically generate functions for retrieving weather data from OpenWeatherMap, converting temperature from Celsius to Fahrenheit, converting dates to UTC format, and plotting charts. Moreover, we took advantage of Amazon Q Developer's chat feature to generate comprehensive documentation and receive suggestions for enhancing our weather data application's code. This approach highlights the versatility of using different prompt styles to tailor code generation to specific requirements within the JavaScript ecosystem. By leveraging advanced IDEs such as VS Code alongside intelligent code assistants such as Amazon Q Developer, developers can streamline the development process and enhance productivity. The ability to receive detailed documentation and actionable insights through chat interactions further demonstrates how integrating these tools can foster efficient and effective software development practices.

By adjusting prompts or simply instructing Amazon Q Developer in a chat-style interface, readers can further direct Amazon Q Developer to generate code with enhanced error and exception handling, making the solution robust and versatile for different use cases. This example demonstrates the power of combining advanced IDEs and intelligent code assistants to build sophisticated applications efficiently.

Summary

In this chapter, we covered how AI-powered code assistants can help Python and Java developers generate application code from the local IDE of their choice. To illustrate the functionality, we addressed the simple yet versatile application of weather data analysis.

Here are the key features covered throughout the application's development.

We walked through the prerequisites of generating the API key for OpenWeatherMap, allowing us to obtain weather data for a user-entered country and city combination. We used the API key to retrieve the weather data provided by OpenWeatherMap, which offers a rich set of APIs. We gathered forecast weather data from `http://api.openweathermap.org/data/2.5/forecast`.

For coding, we used VS Code IDE for Java, which has been set up to work with Amazon Q Developer. For Python, we utilized JetBrains' PyCharm IDE, which was also configured to work with Amazon Q Developer.

To get the code recommendations, we employed a combination of chain-of-thought, single-line, and multi-line prompts to create a JavaScript script for the weather data application. We used Amazon Q Developer with specific prompts to auto-generate functions to retrieve weather data from OpenWeatherMap, convert temperature from Celsius to Fahrenheit, convert dates to UTC format, and then plot the chart. Additionally, for documentation and code improvement, we used the chat with code assistant technique to interact with Amazon Q Developer. This allowed us to generate detailed documentation and receive code improvement recommendations for our weather data application.

By leveraging advanced IDEs such as VS Code and PyCharm alongside Amazon Q Developer, we demonstrated how various prompt styles can streamline the development process across multiple programming languages. This approach not only enhances productivity but also ensures that the generated code is robust and well-documented, making it easier for developers to understand and maintain.

In the next chapter, we will look at how Amazon Q Developer can generate code for multiple other programming languages such as JavaScript, C#, Go, PHP, Shell, and so on.

References

- OpenWeatherMap: `https://openweathermap.org/`

- Getting started with Amazon Q Developer in VS Code and JetBrains: `https://docs.aws.amazon.com/amazonq/latest/qdeveloper-ug/q-in-IDE-setup.html`

- Getting started with Java in VS Code: `https://code.visualstudio.com/docs/java/java-tutorial`

- Installing missing packages in JetBrains' PyCharm IDE: `https://www.jetbrains.com/help/pycharm/installing-uninstalling-and-upgrading-packages.html`

- Installing missing packages in VS Code IDE: `https://code.visualstudio.com/docs/java/java-project`

5

Boost Coding Efficiency for C and C++ with Auto-Code Generation

In this chapter we will look at the following key topics:

- Overview of programming language categories and foundational languages
- C auto-code generation using Amazon Q Developer
- C++ auto-code generation using Amazon Q Developer
- C and C++ code explainability and optimization Using Amazon Q Developer

In the previous chapter, we looked at how you can use Amazon Q Developer to suggest code in VS Code and PyCharm IDEs. We used the two most prominent programming languages used by developers, Python and Java, to demonstrate how auto-code can be generated using prompt techniques and leveraging a chat-with-code-assistant technique.

In this chapter, we will first explore some of the categories of programming languages. We will then discuss an overview of foundational programming languages before focusing on two dominant ones used in the software industry: C and C++. We will introduce how auto-code generation, especially in Amazon Q Developer, is integrated and can be used by C and C++ developers to understand, learn the syntax of, and automatically generate code. As a side benefit, by standardizing code across different programming languages, these tools can enhance code readability, maintainability, and interoperability across large-scale projects.

Overview of programming language categories and foundational languages

The landscape of software development has undergone a dramatic transformation over the past few decades. As the demand for robust, efficient, and scalable applications has grown, the complexity of the coding process has increased. Consequently, various categories of coding languages have emerged to meet these diverse needs.

Foundational languages, such as C and C++, provide the essential constructs for many systems. Web development languages, including HTML, CSS, and JavaScript, power dynamic and interactive web experiences. Database management languages, such as SQL, are crucial for handling large datasets. System administration and automation languages, such as Python and Bash, streamline repetitive tasks and enhance operational efficiency. **Infrastructure as Code** (**IaC**) languages, such as Terraform and AWS CloudFormation, allow developers to manage and provision computing infrastructure through code.

Each category addresses specific challenges, reflecting the multifaceted nature of modern technology ecosystems. Understanding these categories helps developers choose the right tools for their projects, leading to more effective solutions.

Foundational languages C and C++, and auto-code generation support

C and C++ are foundational programming languages that have shaped the landscape of software development. Known for their performance and control over system resources, these languages are crucial in various domains, including system programming, game development, and real-time processing applications. However, the very strengths that make C and C++ powerful also contribute to their complexity. Manual memory management, low-level operations, and intricate syntax can make coding in these languages error-prone and time-consuming.

To mitigate these challenges and enhance productivity, auto-code generation has emerged as a transformative solution. Auto-code generation involves using advanced tools to automatically produce source code based on predefined templates or specifications. This automation can significantly reduce the manual effort required in coding, minimize errors, and standardize code quality. Among the leading tools in this domain is Amazon Q Developer, which brings powerful capabilities to the table for C and C++ code generation. As discussed in previous chapters, Amazon Q Developer is an innovative tool that leverages artificial intelligence and machine learning to assist developers in writing code. By integrating Amazon Q Developer into the development workflow, programmers can benefit from its sophisticated code generation features tailored specifically for C and C++.

One of the primary advantages of using Amazon Q Developer for C and C++ code generation is the significant reduction in development time. Writing boilerplate code, such as memory management routines, data structures, and error handling, can be tedious and repetitive. Amazon Q Developer can provide recommendations to automate these tasks, allowing developers to focus on learning and implementing the more complex and creative aspects of their projects. This not only accelerates the development process but also enhances the overall code quality by ensuring that repetitive tasks are handled consistently and correctly.

Furthermore, Amazon Q aids in bridging the knowledge gap for developers who are new to C and C++. The learning curve for these languages can be steep, given their complexity and need for meticulous attention to detail. By providing auto-generated code examples and templates, Amazon Q serves as an educational tool, helping developers understand and adopt the proper syntax and coding conventions more quickly. This feature is particularly beneficial in educational settings and for onboarding new team members, as it provides concrete examples of how theoretical concepts are applied in practice.

Moreover, Amazon Q Developer is equipped with advanced debugging and optimization capabilities. It can identify potential issues in the code and suggest optimizations, ensuring that the generated code is not only correct but also efficient. This feature is crucial in performance-critical applications where even minor inefficiencies can have significant impacts.

Now, let's dive deep to see how Amazon Q Developer can help a programmer. To illustrate this, I will be using the Amazon Q Developer interface with VS Code (refer to *Chapter 2* for the configuration steps for Amazon Q Developer with VS Code and the *References* section at the end of this chapter for C/C++ for VS Code).

> **Note**
>
> Amazon Q Developer uses LLMs, which, by nature, are non-deterministic, so you may not get exactly the same answers or code blocks shown in the code snapshots. Try to update your prompts to get the desired recommendations. However, logically, the generated answer or code should meet the requirements.

C auto-code generation using Amazon Q Developer

First, assume a new programmer is planning to work with C and does not have a background in this language. Let's use Amazon Q Developer's chat-style interface to educate the programmer.

C language basics

Let's learn about C programming language:

```
Q: What is programming language C and what are use cases best suitable
for C ?
```

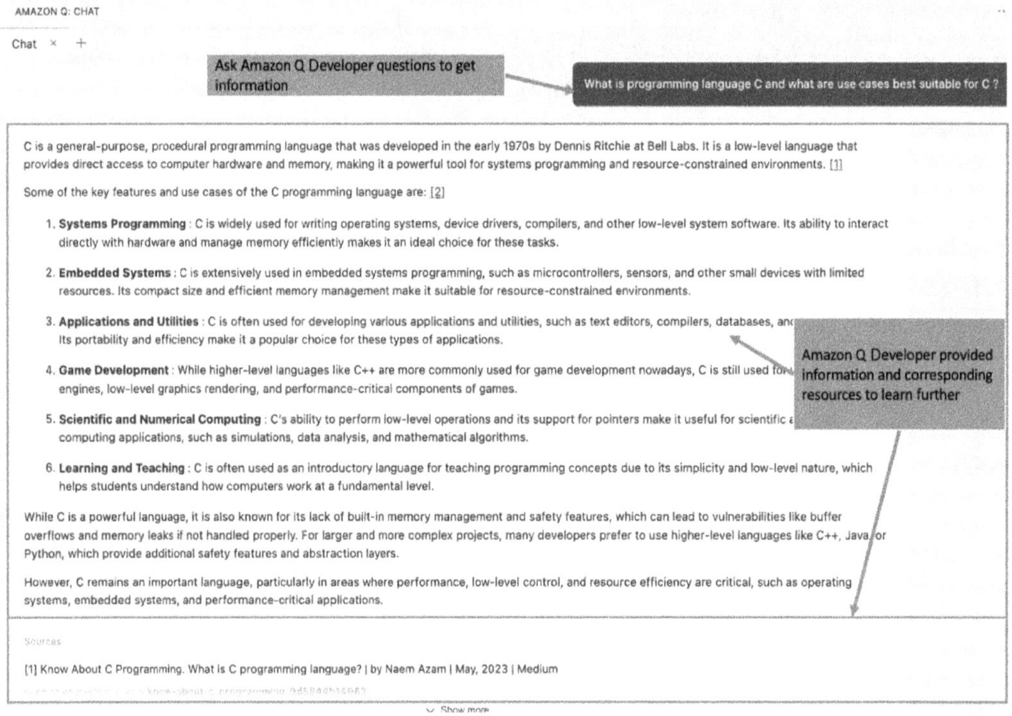

Figure 5.1 – Learning about the C language in chat style

As you can see in the preceding screenshot, Amazon Q Developer provided historical information about the original developers of the C language. It then summarized the common use cases by categorizing where the C language excels. Additionally, it offered sources for further reference and learning.

Now let's ask for commands that are commonly used by C programmers:

```
Q: What are the common commands in programming language C ?
```

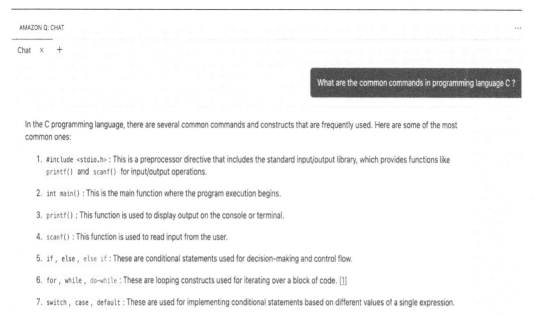

Figure 5.2 – Learning about common C programmer commands

We will not be going through each command in detail, but for illustration, as shown in the preceding partial screenshots, Amazon Q Developer provided several examples of common commands and constructs in the C programming language. To make it easy to follow, it categorized them into groups such as conditional constructs, looping constructs, functions, and more. Additionally, you can see that it provided references at the end for users to obtain further detailed information on the topics.

Now let's ask Amazon Q Developer for the syntax and an example of a `for` loop:

```
Q: Provide syntax and example of "for loop" in C
```

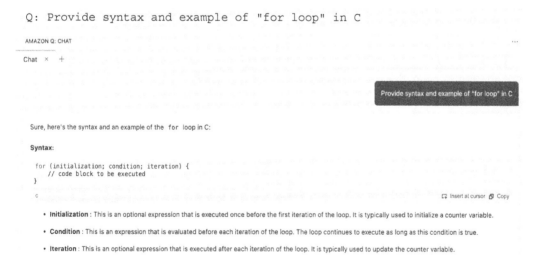

Figure 5.3 – Learning about the for loop

Observe, in the preceding and following screenshots, that Amazon Q Developer provided the syntax, details, and an example of a `for` loop.

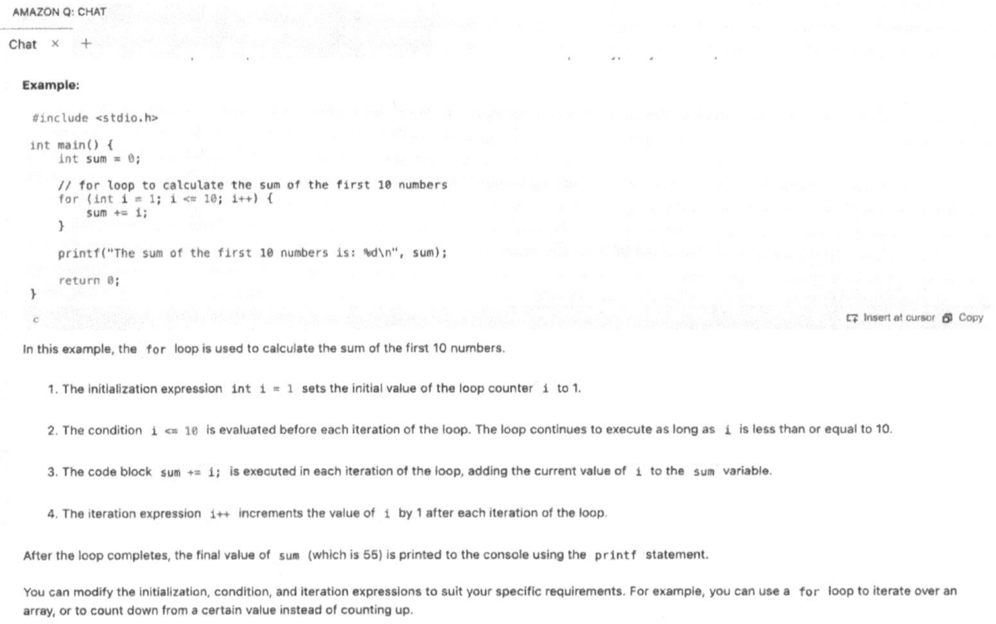

AMAZON Q: CHAT

Chat   ×   +

Example:

```
#include <stdio.h>

int main() {
    int sum = 0;

    // for loop to calculate the sum of the first 10 numbers
    for (int i = 1; i <= 10; i++) {
        sum += i;
    }

    printf("The sum of the first 10 numbers is: %d\n", sum);

    return 0;
}
```

c   ⌨ Insert at cursor 🗐 Copy

In this example, the `for` loop is used to calculate the sum of the first 10 numbers.

1. The initialization expression `int i = 1` sets the initial value of the loop counter `i` to 1.

2. The condition `i <= 10` is evaluated before each iteration of the loop. The loop continues to execute as long as `i` is less than or equal to 10.

3. The code block `sum += i;` is executed in each iteration of the loop, adding the current value of `i` to the `sum` variable.

4. The iteration expression `i++` increments the value of `i` by 1 after each iteration of the loop.

After the loop completes, the final value of `sum` (which is 55) is printed to the console using the `printf` statement.

You can modify the initialization, condition, and iteration expressions to suit your specific requirements. For example, you can use a `for` loop to iterate over an array, or to count down from a certain value instead of counting up.

Figure 5.4 – A for loop example

End-to-end C language use case development

Now let's move to the sample use case to show the power of Amazon Q Developer's ability to recommend code. To illustrate this, I am going to use the VS Code script editor window with single and multi-line prompts.

Let's say that we want to write a C program to read a list of integers from a file, sort them, and write the sorted list to another file. This is a very common use case to see multiple functionalities of C language.

Let's use the multi-line prompt technique to achieve the preceding use case.

```
Prompt:
/* Write the end-to-end C language program to read a list of integers
from a file "input.txt" then sorts that list and then writes the
sorted list to output file "output.txt"
*/
```

```
2    /*
3    Write the end-to-end C language program
4    to read a list of integers from a file "input.txt"
5    then sorts that list
6    and then writes the sorted list to output file "output.txt"
7    */
8
9    #include <stdio.h>
10   #include <stdlib.h>
11
12   #define MAX_SIZE 100
13   #define FILE_NAME "input.txt"
14   #define OUT_FILE "output.txt"
15
16   void readIntegersFromFile(int *arr, int *size, FILE *fp) {
17       *size = 0;
18       while (fscanf(fp, "%d", &arr[*size]) == 1) {
19           (*size)++;
20       }
21   }
22   void sortIntegers(int *arr, int size) {
23       for (int i = 0; i < size - 1; i++) {
```

Figure 5.5 – A C program sample snapshot

The preceding screenshot shows a partial code suggested by Amazon Q Developer. The end-to-end generated C code by Amazon Q Developer demonstrates several fundamental and advanced concepts in C programming, including functions, arrays, pointers, and file I/O operations. It also added error handling related to "Error opening file".

Now let's use an interactive inline prompt with the chain-of-thought technique to enhance the code. As you can see in the readIntegersFromFile() function, Amazon Q Developer already added

`"Error opening file"`. Now, let's instruct it to check for zero-byte files as well. Just go inside the function and use the single-line prompt technique as follows:

```
Prompt :
/* check if file is zero byte*/
```

```
16    void readIntegersFromFile(int *arr, int *size, FILE *fp) {
17        /* check if file is zero byte*/
18        fseek(fp, 0, SEEK_END);
19        if (ftell(fp) == 0) {
20            printf("File is empty\n");
21            exit(1);
22        }
23        rewind(fp);
24        *size = 0;
25        while (fscanf(fp, "%d", &arr[*size]) == 1) {
26            (*size)++;
27        }
28    }
```

Figure 5.6 – A C program inline function enhancement

As shown in the screenshot, based on our prompt, Amazon Q Developer added an `if` condition to check whether the input file is zero bytes. If the file is empty, the code gracefully exits with a `File is empty` message and a non-zero exit code.

In the next section, let's explore the functionality of Amazon Q Developer with C++.

C++ auto-code generation using Amazon Q Developer

Similar to the preceding section, let's first assume that a new programmer is planning to work with C++ and does not have a background in this language. We will use similar chat-style interface prompts from the C language section to learn the basics of the C++ language with the help of Amazon Q Developer.

C++ language basics

Let's learn about C++ programming language:

```
Q: What is programming language C++ and what are use cases best
suitable for C++ ?
```

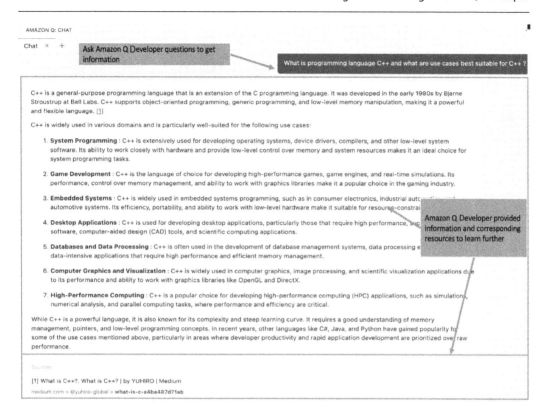

Figure 5.7 – Learning about C++ programming

As you can see in the preceding screenshot, Amazon Q Developer provided historical information about the original developers. It then summarized the common use cases by categorizing where the C++ language excels. Additionally, it offered sources for further reference and learning.

Now let's ask for commands that are commonly used by C++ programmers:

```
Q: What are the common commands in programming language C++?
```

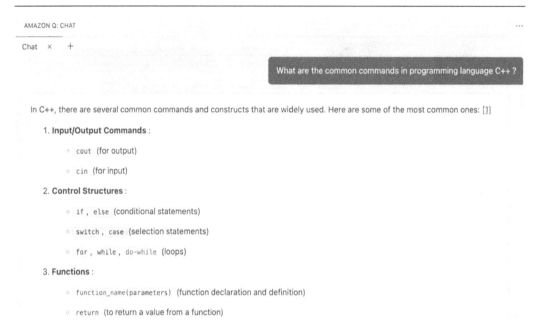

Figure 5.8 – Learning about common programmer C++ commands

We will not be going through each command but just for illustration, as you can see in the following partial screenshots, Amazon Q Developer provided several categories of the common commands and constructs in the C++ programming language. To make it easy to follow, it also included some of the common key commands for each sample command. Additionally, you can see that it provided references at the end for users to obtain further detailed information on the topics.

Based on the information provided you can observe that C++ shines in the area of **Object-Oriented Programming** (**OOP**), so let's ask Amazon Q Developer to provide details and examples related to C++ support for OOP:

```
Q: What is Object-Oriented Programming and how C++ supports it ?
```

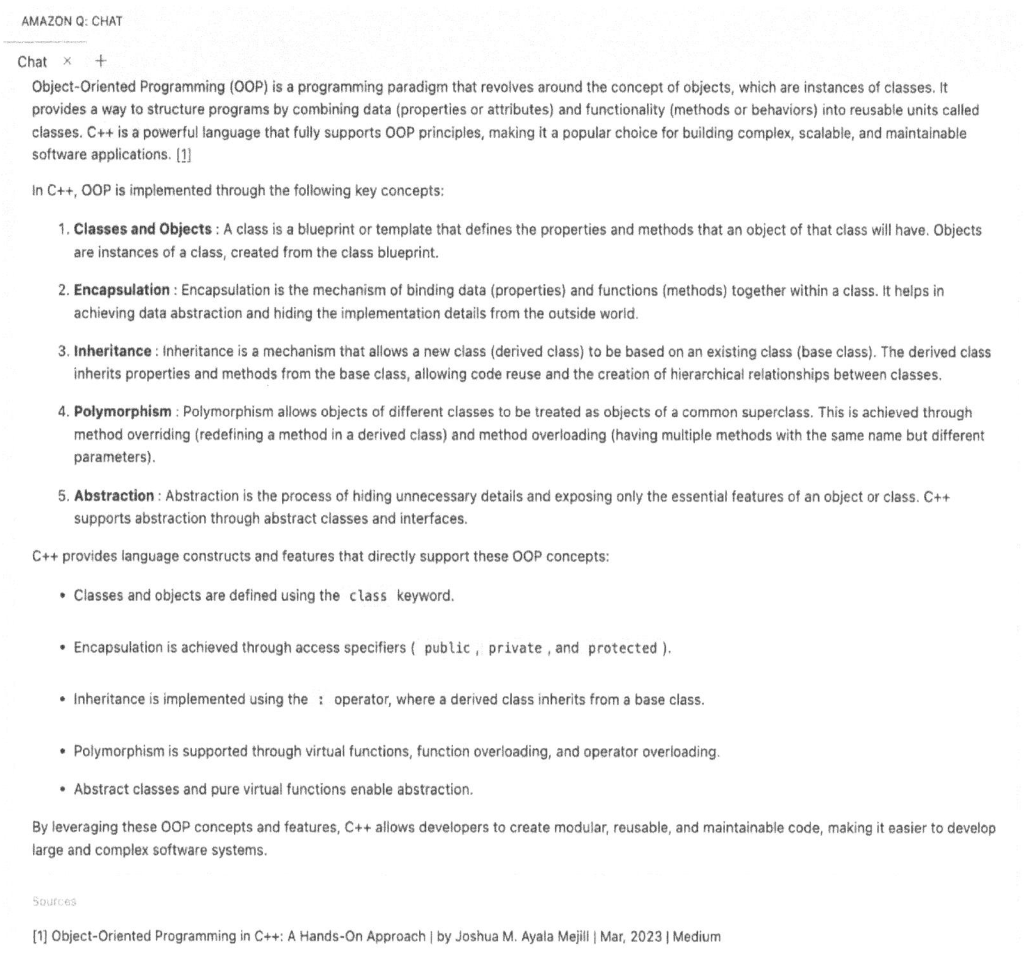

Chat × +

Object-Oriented Programming (OOP) is a programming paradigm that revolves around the concept of objects, which are instances of classes. It provides a way to structure programs by combining data (properties or attributes) and functionality (methods or behaviors) into reusable units called classes. C++ is a powerful language that fully supports OOP principles, making it a popular choice for building complex, scalable, and maintainable software applications. [1]

In C++, OOP is implemented through the following key concepts:

1. **Classes and Objects** : A class is a blueprint or template that defines the properties and methods that an object of that class will have. Objects are instances of a class, created from the class blueprint.

2. **Encapsulation** : Encapsulation is the mechanism of binding data (properties) and functions (methods) together within a class. It helps in achieving data abstraction and hiding the implementation details from the outside world.

3. **Inheritance** : Inheritance is a mechanism that allows a new class (derived class) to be based on an existing class (base class). The derived class inherits properties and methods from the base class, allowing code reuse and the creation of hierarchical relationships between classes.

4. **Polymorphism** : Polymorphism allows objects of different classes to be treated as objects of a common superclass. This is achieved through method overriding (redefining a method in a derived class) and method overloading (having multiple methods with the same name but different parameters).

5. **Abstraction** : Abstraction is the process of hiding unnecessary details and exposing only the essential features of an object or class. C++ supports abstraction through abstract classes and interfaces.

C++ provides language constructs and features that directly support these OOP concepts:

- Classes and objects are defined using the `class` keyword.

- Encapsulation is achieved through access specifiers (`public` , `private` , and `protected`).

- Inheritance is implemented using the `:` operator, where a derived class inherits from a base class.

- Polymorphism is supported through virtual functions, function overloading, and operator overloading.

- Abstract classes and pure virtual functions enable abstraction.

By leveraging these OOP concepts and features, C++ allows developers to create modular, reusable, and maintainable code, making it easier to develop large and complex software systems.

Sources

[1] Object-Oriented Programming in C++: A Hands-On Approach | by Joshua M. Ayala Mejill | Mar, 2023 | Medium

Figure 5.9 – Learning about C++ OOP

As seen in the preceding screenshot, Amazon Q provided an overview of OOP basics and key concepts such as classes, objects, inheritance, encapsulation, and abstraction. It also included corresponding C++ constructs such as `class`, `public`, `private`, and `protected`.

Now let's ask Amazon Q Developer for the syntax and example of the `do... while` loop:

```
Q: Provide syntax and example of "do... while loop" in C++
```

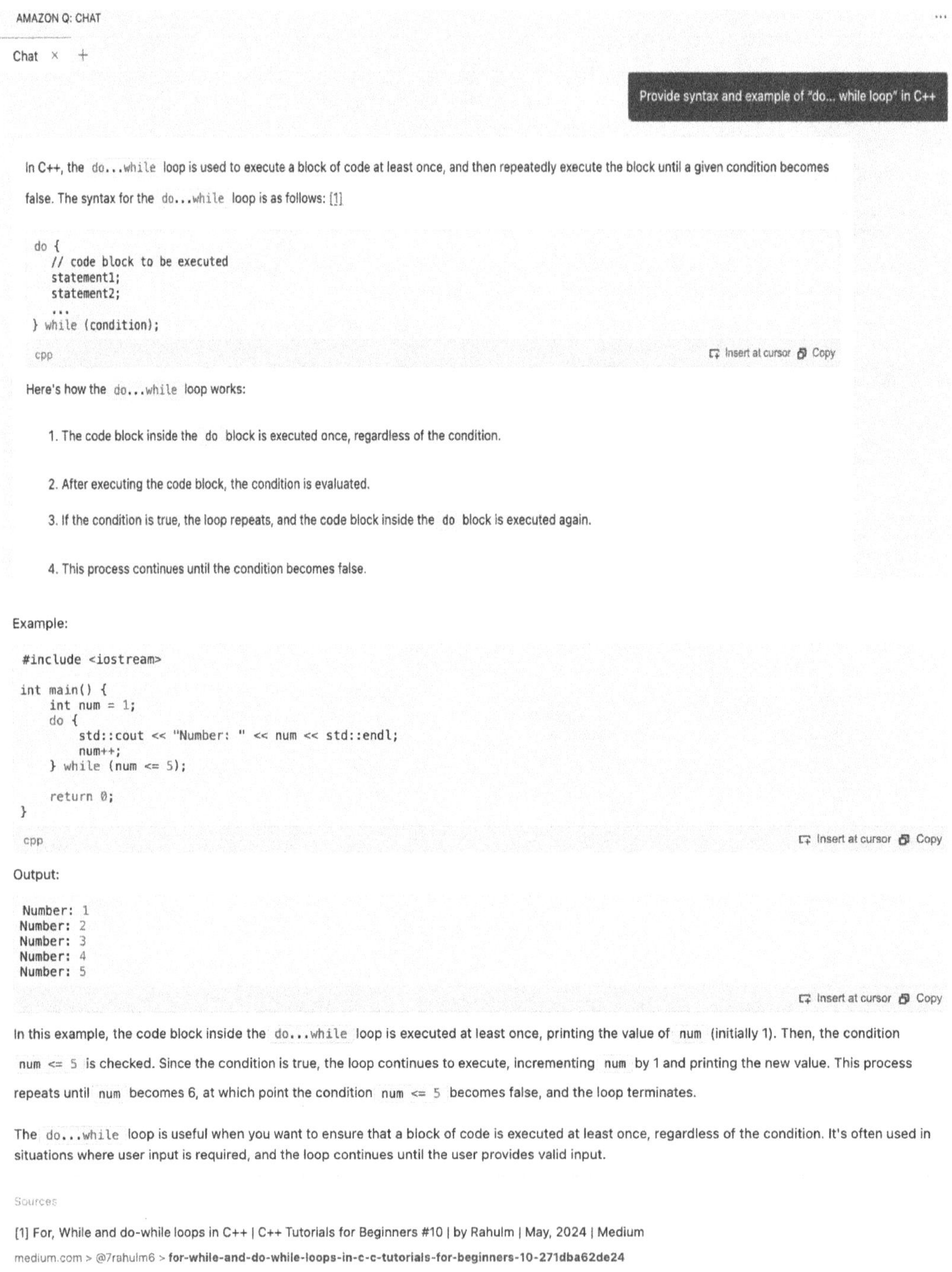

AMAZON Q: CHAT   ...

Chat   ×   +

> Provide syntax and example of "do... while loop" in C++

In C++, the `do...while` loop is used to execute a block of code at least once, and then repeatedly execute the block until a given condition becomes false. The syntax for the `do...while` loop is as follows: [1]

```cpp
do {
    // code block to be executed
    statement1;
    statement2;
    ...
} while (condition);
```

cpp   ⌁ Insert at cursor 📋 Copy

Here's how the `do...while` loop works:

1. The code block inside the `do` block is executed once, regardless of the condition.

2. After executing the code block, the condition is evaluated.

3. If the condition is true, the loop repeats, and the code block inside the `do` block is executed again.

4. This process continues until the condition becomes false.

Example:

```cpp
#include <iostream>

int main() {
    int num = 1;
    do {
        std::cout << "Number: " << num << std::endl;
        num++;
    } while (num <= 5);

    return 0;
}
```

cpp   ⌁ Insert at cursor 📋 Copy

Output:

```
Number: 1
Number: 2
Number: 3
Number: 4
Number: 5
```

⌁ Insert at cursor 📋 Copy

In this example, the code block inside the `do...while` loop is executed at least once, printing the value of `num` (initially 1). Then, the condition `num <= 5` is checked. Since the condition is true, the loop continues to execute, incrementing `num` by 1 and printing the new value. This process repeats until `num` becomes 6, at which point the condition `num <= 5` becomes false, and the loop terminates.

The `do...while` loop is useful when you want to ensure that a block of code is executed at least once, regardless of the condition. It's often used in situations where user input is required, and the loop continues until the user provides valid input.

Sources

[1] For, While and do-while loops in C++ | C++ Tutorials for Beginners #10 | by Rahulm | May, 2024 | Medium
medium.com > @7rahulm6 > **for-while-and-do-while-loops-in-c-c-tutorials-for-beginners-10-271dba62de24**

Figure 5.10 – Learning about the do…while loop

Observe, in the preceding and following screenshots, that Amazon Q Developer provided the syntax, details, and an example for the do... while loop.

End-to-end C++ language use case development

Now let's move to the sample use case to show the power of Amazon Q Developer's ability to recommend code. To illustrate, I am going to use the VS Code script editor window with single and multi-line prompts.

Let's say that we want to write a C++ program to develop a new operating system kernel, which can control system resources and has the ability to handle complex memory management tasks. The kernel must interact with hardware, manage processes, and ensure stability across different hardware architectures.

Let's use the multi-line prompt technique to achieve the preceding use case.

```
Prompt:
/* Write an end to end executable C++ script
to develop a new operating system kernel, that can control over system
resources and ability to handle complex memory management tasks. The
kernel must interact with hardware, manage processes, and ensure
stability across different hardware architectures.
*/
```

```
1   /*
2   Write an end to end executable C++ script
3   to develop a new operating system kernel,
4   that can control over system resources and ability to handle complex memory management tasks.
5   The kernel must interact with hardware, manage processes,
6   and ensure stability across different hardware architectures.
7   */
8
9   #include <iostream>
10  #include <string>
11  #include <vector>
12
13  // Represents a process
14  class Process {
15  public:
16      Process(std::string name, int id) : name(name), id(id) {}
17      std::string name;
18      int id;
19  };
20  std::vector<Process*> processes;
21  // Simulates loading a process into memory
22  void loadProcess(Process* process) {
23      processes.push_back(process);
24      std::cout << "Loaded process: " << process->name << " (ID: " << process->id << ")" << std::endl;
25  }
```

Figure 5.11 – Writing a C++ program with Amazon Q Developer

The preceding screenshot shows partial code suggested by Amazon Q Developer. You can observe that end-to-end code demonstrates several fundamental and advanced concepts in C++ programming, including OOP, classes, vectors, functions, error handling, and more.

Now let's use an interactive inline prompt with the chain-of-thought technique to enhance the code. As you can see, the loadProcess() function does not have exception handling, so let's instruct it to add an exception handle for the null process with the single-line prompt technique:

```
Prompt :
/* add exception handling for loadProcess */
21    // Simulates loading a process into memory
22    void loadProcess(Process* process) {
23    /* add exception handling for loadProcess */
24        if (process == nullptr) {
25            std::cout << "Error: process is null" << std::endl;
26            return;
27        }
28        processes.push_back(process);
29        std::cout << "Loaded process: " << process->name << " (ID: " << process->id << ")" << std::endl;
30    }
```

Figure 5.12 – Prompt – C++ program sample snapshot

As shown in the screenshot, based on our prompt, Amazon Q Developer added a nullptr condition to check the process load exception. If it encounters a null pointer, then the code gracefully exits with a Error: process is null message.

C and C++ code explainability and optimization using Amazon Q Developer

Amazon Q provides options for achieving code explainability and optimization and is applicable to both C and C++. For more details on explainability and optimization, please refer to *Chapter 12*.

To demonstrate code explainability and optimization, I will use the end-to-end code generated for the C language. As shown in the following screenshot, simply highlight the entire code, right-click to open the pop-up menu, then select **Amazon Q** and choose **Explain** for code explainability or **Optimize** for optimization recommendations.

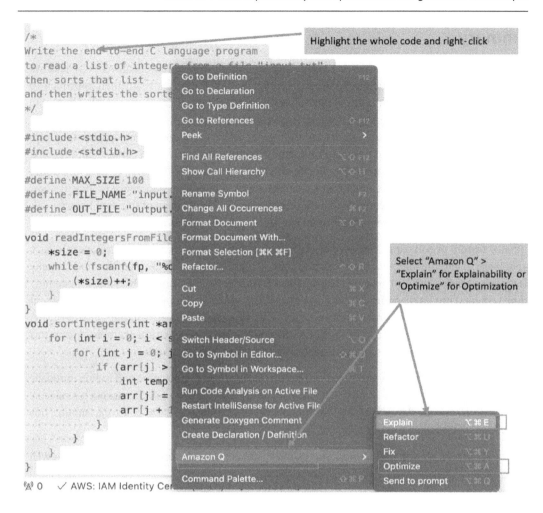

Figure 5.13 –C and C++ program explainability and optimization

This will bring up Amazon Q Developer's chat-style interface and move the full code for analysis. For explainability, Amazon Q Developer offers detailed information about each function and the overall code logic. Regarding optimization, Amazon Q Developer provides end-to-end optimized code that can be directly inserted into your editor by clicking the **Insert at cursor** button.

In addition to C and C++, Amazon Q Developer supports multiple other programming languages such as C#, Go, Rust, and more. There are many enhancements happening in this area, and we anticipate further enhancements in support for additional languages (check the *References* section at the end of the chapter).

Summary

In this chapter, we learned that Amazon Q Developer stands at the forefront of auto-code generation for foundational languages such as C and C++, offering developers a streamlined approach to software development. Through intuitive chat-style interfaces, programmers engage with Amazon Q Developer to explore these languages comprehensively. Furthermore, we learned that the platform leverages advanced prompting techniques to assist in generating code fragments or complete solutions, automating repetitive tasks, and ensuring uniformity in code structure. With robust support for C and C++ programming, Amazon Q Developer effectively handles a wide range of tasks including functions, loops, data structures, and memory management. It maintains adherence to industry best practices in error handling, indentation, naming conventions, and utilization of standard libraries. We learned that this meticulous approach enhances overall code quality and facilitates easier maintenance.

A notable feature of Amazon Q Developer is its Explain functionality, enabling developers to gain deeper insights into the functionality of existing code, as we learned in this chapter. This tool not only aids in understanding but also promotes learning and proficiency enhancement. By automating routine coding tasks and providing optimization suggestions, Amazon Q Developer accelerates development cycles while fostering code consistency and quality, as we explored in this chapter. Furthermore, the platform serves as an educational resource, offering tutorials and practical examples to support developers in honing their skills and staying updated with industry standards.

Overall, Amazon Q Developer empowers developers to focus more on innovation and less on mundane coding tasks, as mentioned in this chapter. It enhances productivity across projects, making it a valuable tool in modern software development environments where efficiency and agility are paramount; this was the main point learned in this chapter.

In the next chapter, we will look at how you can use Amazon Q Developer to suggest code in web development languages such as JavaScript and PHP.

References

- C/C++ for Visual Studio Code: `https://code.visualstudio.com/docs/languages/cpp`
- Supported languages for Amazon Q Developer in the IDE: `https://docs.aws.amazon.com/amazonq/latest/qdeveloper-ug/q-language-ide-support.html`

6

Boost Coding Efficiency for JavaScript and PHP with Auto-Code Generation

In this chapter, we will look at the following key topics:

- Overview of web development programming languages
- JavaScript auto-code generation using Amazon Q Developer
- PHP auto-code generation using Amazon Q Developer
- JavaScript and PHP code explainability and optimization using Amazon Q Developer

In the previous chapter, we looked at some of the categories of programming languages. We then discussed an overview of foundational programming languages before focusing on two dominant ones used in the software industry: C and C++. We introduced how auto-code generation, especially using Amazon Q Developer, is integrated and can be used by C and C++ developers to understand, learn the syntax, and automatically generate code.

Similar to the previous chapter, in this chapter, we will focus on two programming languages. We will start with an overview of web programming languages and then focus on two that are commonly used in the software industry: **JavaScript** (**JS**) and **PHP**. We will introduce how auto-code generation, particularly through Amazon Q Developer, can be used by JS and PHP developers for understanding the basics, learning the syntax, automatically generating code, code explainability, and optimization through code recommendations.

Overview of web development programming languages

In earlier chapters, we reviewed different programming language categories and delved deep into foundational languages such as C and C++, which power many system constructs. Now, let's dive into web development languages such as JS and PHP, which enable dynamic and interactive web experiences. These are some of the most popular languages for server-side scripting, client-side scripting, **Document Object Model (DOM)** programming, web APIs, mobile applications, and more.

A few decades ago, with the growth in web application development, many languages were introduced in the software industry. Web development languages play a crucial role in shaping the interactive and functional aspects of modern web applications. Among these, JS and PHP are prominent for their respective roles in client-side and server-side scripting. JS powers dynamic and interactive user experiences in web browsers, while PHP facilitates server-side logic for dynamic content generation and database interactions. In the realm of web development, efficiency and productivity are key factors for success. Developers often face challenges such as repetitive coding tasks, maintaining code consistency across different modules, and keeping up with evolving best practices. Auto-code generation has emerged as a solution to these challenges, enabling developers to automate routine coding tasks and streamline their workflow.

Amazon Q Developer, an advanced tool leveraging artificial intelligence and machine learning, offers significant capabilities in auto-code generation for JavaScript and PHP. By integrating Amazon Q Developer into the development process, developers can leverage its features to enhance productivity, improve code quality, and expedite the development cycle.

JS, as a frontend scripting language, is instrumental in creating interactive user interfaces and dynamic web content. However, JS development can become complex due to the need for cross-browser compatibility, asynchronous programming patterns, and managing event-driven interactions. Amazon Q Developer simplifies JS development by generating code snippets and functions tailored to specific requirements instructed in the prompts. Web developers can leverage different prompting techniques and chat style integrations (refer to *Chapter 3*) to get recommendations for the automated creation of event handlers, DOM manipulation functions, and asynchronous operations, thereby reducing development time and ensuring code consistency across different browser environments.

On the server side, PHP remains a popular choice for dynamic web applications and backend services. PHP's versatility in handling form submissions, session management, and database interactions makes it indispensable for web developers. However, writing efficient and secure PHP code requires adherence to coding standards and best practices. Amazon Q Developer assists developers by generating optimized PHP code snippets for common tasks such as database queries, form processing, and error handling. This not only accelerates development but also enhances code reliability and security.

Moreover, Amazon Q Developer supports the learning and understanding of JS and PHP syntax. For developers new to these languages, Amazon Q Developer provides interactive tutorials and code examples through chat-style interactions. Developers can use prompts to request explanations of language features, obtain code snippets for specific functionalities, or generate complete scripts based on predefined templates. This educational aspect of Amazon Q Developer helps shorten the learning curve and empowers developers to quickly grasp essential concepts and techniques.

Now, let's dive deep to see how Amazon Q Developer can help a web programmer. To illustrate, I am using the Amazon Q Developer interface with VS Code (refer to *Chapter 2* for the configuration steps for Amazon Q Developer with VS Code and check the *References* section at the end of this chapter for the URLs for JS and PHP in VS Code).

> **Note**
> Amazon Q Developer uses LLMs, which, by nature, are non-deterministic, so you may not get exactly the same answers/code blocks shown in the code snapshots; try to update prompts to get the desired recommendations. However, logically, the generated answer/code should meet the requirements.

JS auto-code generation using Amazon Q Developer

First, let's explore how Amazon Q Developer can assist a new web programmer planning to work with JS but lacking a background in the language. We will use Amazon Q Developer's chat-style interface to educate the programmer.

JS language basics

Let's learn about the JS programming language by asking about the very basics of JS:

```
Q: What is programming language JavaScript(JS) and what are use cases
best suitable for JS ?
```

As you can see in the following screenshot, Amazon Q Developer provided historical information about the original developers. It then summarized the common use cases by categorizing where the JS language excels. Additionally, it offered sources for further reference and learning.

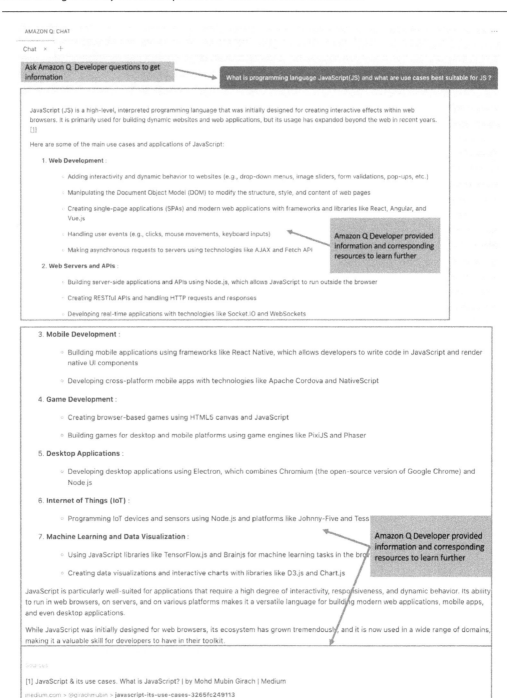

Figure 6.1 – Chat style – JS basics

You can observe that, in addition to web development, Amazon Q Developer pointed out the versatility of JS applications, such as game development, desktop applications, **Internet of Things (IoT)** applications, and machine learning applications. Amazon Q also provided guidelines about JS use in **"applications that require a high degree of interactivity, responsiveness, and dynamic behavior"**.

Based on the preceding information, you can see that JS is so versatile and can be used in various software development projects across industries. Let's look at the commonly used commands by JS web programmers:

```
Q: What are the common commands in programming language JS?
```

We will not be going through each command, but just for illustration, as you can see in the following partial screenshot, Amazon Q Developer provided several categories of the common commands and constructs in the JS programming language. To make it easy to follow, it also included some of the common key commands for each of the categories.

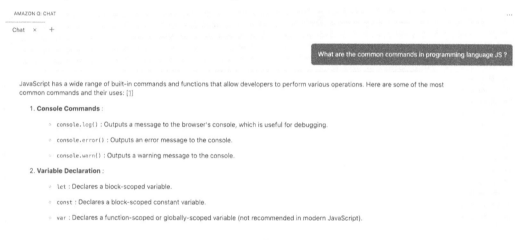

Figure 6.2 – Chat style – JS commands

Amazon Q Developer highlighted multiple areas where JS has various supported command sets such as console commands, DOM manipulation commands, array manipulation, and more.

Let's use one of the areas to get information and ask Amazon Q Developer to provide details and examples related to JS support for DOM manipulation:

```
Q: What is DOM Manipulation and how JS supports it ?
```

AMAZON Q: CHAT   ...

Chat × +

                                                    What is DOM Manipulation and how JS supports it ?

DOM (Document Object Model) manipulation refers to the ability to dynamically access and modify the content, structure, and style of a web page using JavaScript. The DOM represents the hierarchical structure of an HTML document, where each element, attribute, and text node is an object that can be accessed and manipulated.

JavaScript provides various methods and properties to interact with the DOM, allowing developers to create dynamic and interactive web applications. Here are some common ways JavaScript supports DOM manipulation:

1. **Accessing Elements** :

    ○ `document.getElementById()` : Retrieves an element by its unique ID.

    ○ `document.getElementsByTagName()` : Returns a collection of elements with the specified tag name.

    ○ `document.getElementsByClassName()` : Returns a collection of elements with the specified class name.

    ○ `document.querySelector()` : Returns the first element that matches the specified CSS selector.

    ○ `document.querySelectorAll()` : Returns a collection of elements that match the specified CSS selector.

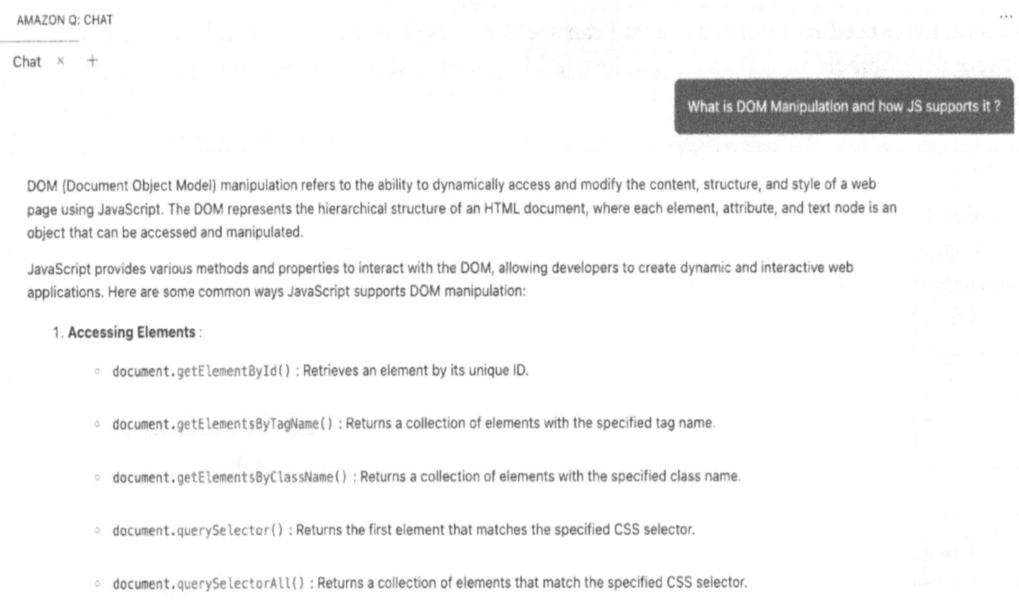

Figure 6.3 – Chat style – JS DOM manipulation

As shown in the preceding partial screenshot, Amazon Q Developer begins with the basics of DOM manipulation, which refers to the ability to dynamically access and modify the content, structure, and style of a web page using JS. The DOM represents the hierarchical structure of an HTML document, where each element, attribute, and text node is an object that can be accessed and manipulated. It then continues by providing specific details related to categories such as accessing DOM elements, creating and modifying elements, manipulating attributes, handling events, modifying styles, and DOM manipulation libraries and frameworks. At the end, it provides a summary highlighting the importance of JS for DOM manipulation: "**DOM manipulation is a crucial aspect of building dynamic and interactive web applications. By leveraging JavaScript's capabilities to access and modify the DOM, developers can create rich user experiences, update content on-the-fly, validate user input, and respond to user interactions without requiring a full-page refresh**", with reference sources included.

Now, let's ask Amazon Q Developer for the syntax of one of the most useful loops used while coding complex scripts, the `for...in` loop in JS:

```
Q: Provide syntax and example of "for...in loop" in JS
```

AMAZON Q: CHAT

Chat   ×   +

Provide syntax and example of "for...in loop" in JS

The `for...in` loop in JavaScript is used to iterate over the enumerable properties of an object. The syntax for the `for...in` loop is as follows. [1]

```javascript
for (variable in object) {
  // code block to be executed
}
```

Here's how the syntax breaks down:

- `variable` : This variable will hold the key (property name) of the current property being iterated over in each loop iteration.

- `object` : The object whose enumerable properties are to be iterated over.

Example:

```javascript
const person = {
  name: "John Doe",
  age: 30,
  city: "New York"
};

for (let property in person) {
  console.log(`${property}: ${person[property]}`);
}
```

Output:

```
name: John Doe
age: 30
city: New York
```

In this example, the `for...in` loop iterates over the `person` object's enumerable properties (`name` , `age` , and `city`). In each iteration, the `property` variable holds the current property name, and `person[property]` accesses the corresponding value.

It's important to note that the `for...in` loop iterates over all enumerable properties of an object, including those inherited from its prototype chain. If you want to iterate only over the object's own properties, you can use the `hasOwnProperty()` method or the `Object.keys()` method in combination with a regular `for...of` loop.

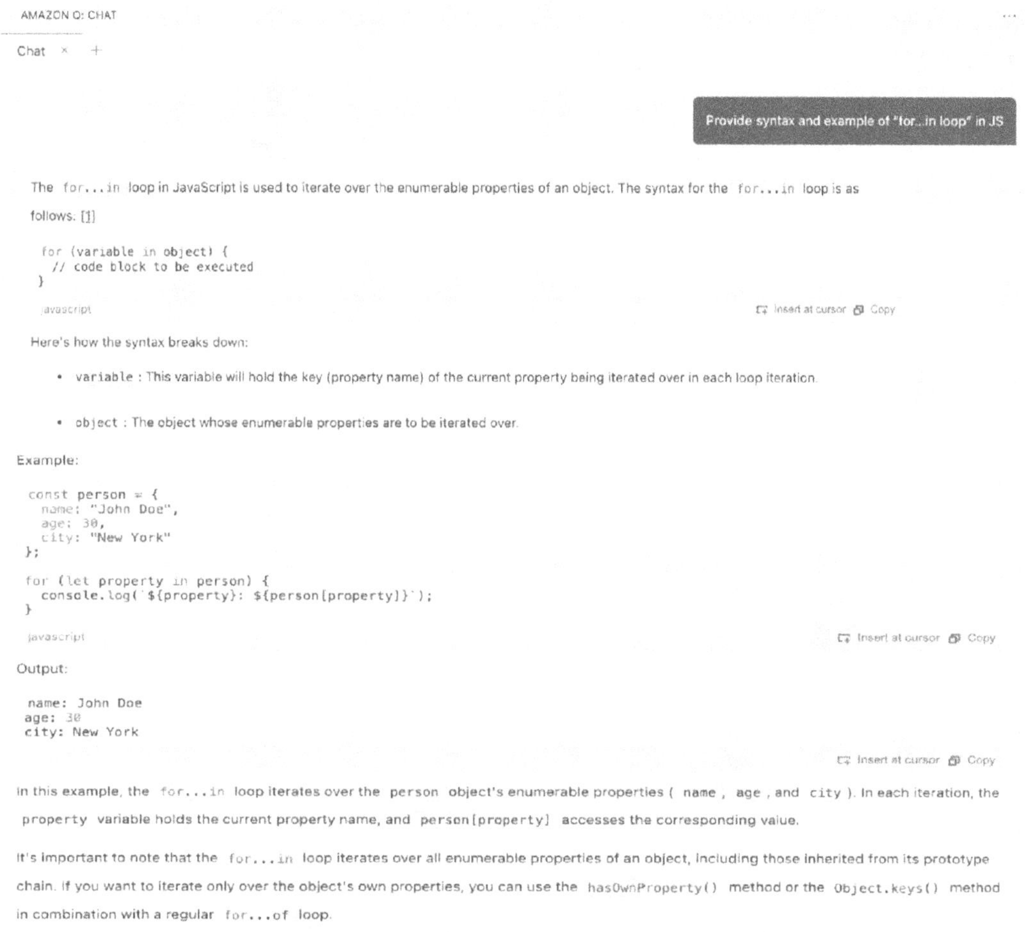

Figure 6.4 – Chat style – JS "for…in" loop example

Observe in the preceding screenshot that Amazon Q Developer provided the syntax, details, and an example of a `for...in` loop. Now, let's move to the sample use case to show the power of Amazon Q Developer's ability to recommend the code.

End-to-end JS use case development

To illustrate, I am going to use the VS Code script editor window with single and multi-line prompts.

Here is the use case: Create JS end-to-end code to build a to-do list application. Create tasks such as adding new items, marking items as completed, and deleting items. Use JS to manipulate the DOM and respond to user events. Here is the prompt:

```
Prompt:
/* Create JS end to end code to building a to-do list application.
Tasks to adding new items, mark items as completed, and delete items.
Use JavaScript to manipulate the DOM and responding to user events. */
```

```
1    /*
2    Create JS end to end code to building a to-do list application.
3    Tasks to adding new items, mark items as completed, and delete items.
4    Use JavaScript tomanipulate the DOM and responding to user events.
5    */
6
7
8
9    <!DOCTYPE html>
10   <html lang="en">
11   <head>
12       <meta charset="UTF-8">
13       <meta name="viewport" content="width=device-width, initial-scale=1.0">
14       <title>To-Do List</title>
15       <style>
16           body {
17               font-family: Arial, sans-serif;
18               background-color: #f4f4f4;
19               margin: 0;
20               padding: 0;
21               display: flex;
22               justify-content: center;
23               align-items: center;
24               height: 100vh;
```

Figure 6.5 – Prompt – JS program partial snapshot

The preceding screenshot shows a partial code snippet suggested by Amazon Q Developer. The full end-to-end code generated in response to our prompt demonstrates several fundamental and advanced concepts in JS and DOM manipulation. It effectively utilizes key JS functionalities related to the DOM, such as getting elements, creating elements, modifying elements, handling events, and applying styles.

It leverages multiple key out-of-box functions of JS, including `document.getElementById`, `document.createElement`, `taskItem.appendChild`, `taskList.removeChild`, `Button.onclick`, and more. It also included a custom `addTask()` function to support coding best practices. The end-to-end code meets the specifics of your prompts, showcasing Amazon Q Developer's capability to generate comprehensive and functional code for web development tasks.

In the current version of the code, Amazon Q Developer added two buttons: `Delete` and `Complete`. If we use the basic chain of thought for actions supported by the application, it appears that a `Cancel` button is missing. Let's use a single-line prompt technique to suggest to Amazon Q Developer to add a `Cancel` button:

```
Prompt:
/* add Cancel button */
 98
 99              /* add Cancel button */
100              const cancelButton = document.createElement('button');
101              cancelButton.textContent = 'Cancel';
102              cancelButton.onclick = () => {
103                  taskList.removeChild(taskItem);
104              };
```

Figure 6.6 – Prompt – JS program Cancel button

As you can see in the preceding screenshot, Amazon Q Developer added multiple lines of code to add the `Cancel` button. It effectively used JS functions such as `document.createElement`, `cancelButton.onclick`, and `taskList.removeChild`.

In the next section, let's see how Amazon Q Developer supports PHP developers.

PHP auto-code generation using Amazon Q Developer

Just like previously, first, let's assume that a new web programmer is planning to work with **PHP** (which is a recursive acronym for **PHP: Hypertext Preprocessor**) and does not have much experience coding with a PHP background. Let's use Amazon Q Developer's chat-style interface to educate the programmer.

PHP language basics

Let's learn about the PHP programming language:

Q: What is programming language PHP and what are use cases best
suitable for PHP ?

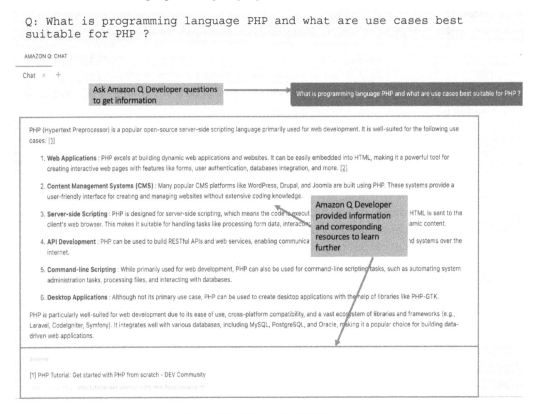

Figure 6.7 – Chat style – PHP basics

As you can see in the preceding screenshot, Amazon Q Developer provided basic information, such as informing us that PHP is a popular, open source, server-side scripting language widely used for web development. It is designed primarily for creating dynamic web pages and web applications. PHP's versatility and the availability of numerous libraries and frameworks have expanded its use cases beyond the web. However, it is still predominantly used for building dynamic web applications, content management systems, and web services.

Additionally, it outlined common applications where PHP plays an important role, such as web development, **Content Management Systems (CMSs)**, server-side scripting, command-line scripting, desktop applications, data processing and analysis, and embedded systems.

Since PHP is a popular, open source, server-side scripting language, let's ask Amazon Q Developer to provide details and examples of PHP's support for server-side scripting, which has contributed to its widespread use:

```
Q: What is server-side scripting and how PHP supports it?
```

AMAZON Q: CHAT

Chat × +

Ask Amazon Q Developer questions to get information

What is server-side scripting and how PHP supports it ?

Server-side scripting refers to the process of executing code on the server before sending the resulting output (typically HTML) to the client's web browser. This approach allows for dynamic content generation, data processing, and interaction with databases or other server resources. [1]

PHP is a server-side scripting language, which means it is designed to run on the server and generate dynamic web pages. Here's how PHP supports server-side scripting: [2]

1. **Embedding PHP in HTML** : PHP code can be embedded directly into HTML files using the opening and closing PHP tags (`<?php` and `?>`). This allows PHP to generate dynamic HTML content on the server before sending it to the client's browser.

Amazon Q Developer provided information

2. **Processing User Input** : PHP can handle user input from various sources, such as form submissions, URL parameters, a processed on the server, allowing for tasks like form validation, user authentication, and data manipulation.

3. **Database Integration** : PHP has built-in support for connecting to and interacting with various database management systems (DBMS) like MySQL, PostgreSQL, Oracle, and SQLite. This enables PHP to retrieve, store, and manipulate data from databases, making it suitable for building data-driven web applications.

4. **File System Operations** : PHP provides functions for reading, writing, uploading, and managing files on the server. This allows for tasks like file uploads, log management, and content generation based on server-side files.

5. **Session Management** : PHP includes built-in support for session management, which allows for maintaining user-specific data across multiple page requests. This is essential for building applications that require user authentication and personalization.

6. **Third-party Libraries and Frameworks** : PHP has a vast ecosystem of third-party libraries and frameworks that extend its functionality. These libraries and frameworks provide additional features and tools for tasks like routing, templating, authentication, and more, making it easier to build complex web applications.

When a client requests a PHP file from a web server, the server executes the PHP code and generates the resulting HTML output. This output is then sent back to the client's browser for rendering. This server-side execution allows for dynamic content generation, data processing, and secure handling of sensitive information (like passwords) without exposing the logic to the client.

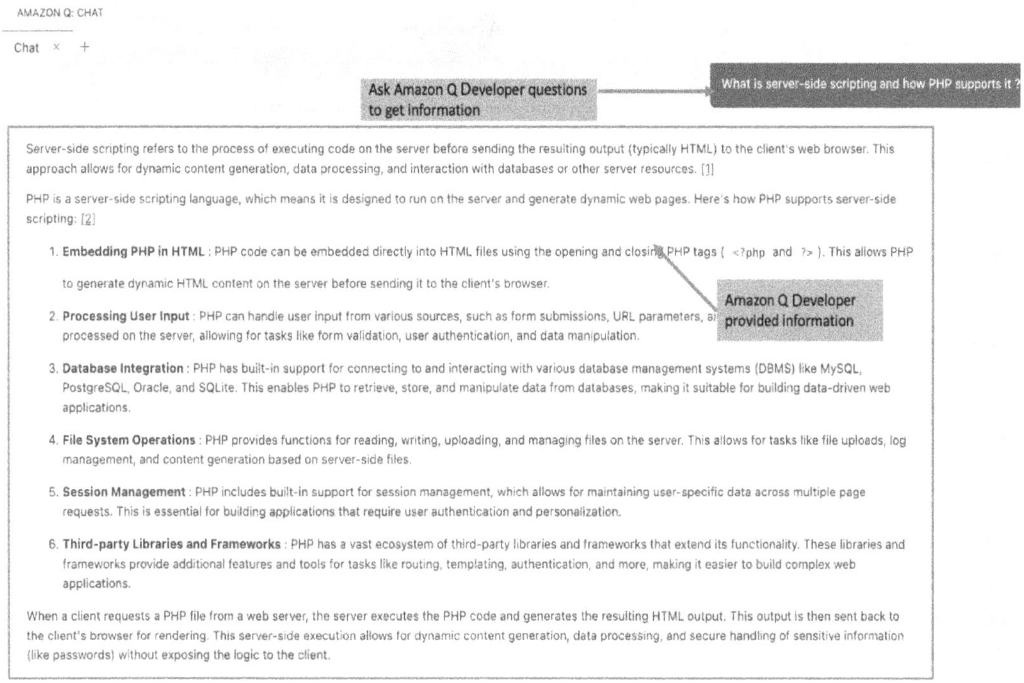

Figure 6.8 – Chat style – PHP server-side scripting support

Amazon Q Developer began with the basics of server-side scripting, explaining that "**Server-side scripting refers to the process of executing code on the server before sending the resulting output (typically HTML) to the client's web browser**". It then provided context for PHP, highlighting that PHP is a server-side scripting language designed to run on the server and generate dynamic web pages. Additionally, it listed key functionality supported by PHP, such as embedding PHP code in HTML, processing user input, filesystem operations, database integration, session management, web services and APIs, and server configuration and environment.

Now, let's ask for some commonly used commands by PHP web programmers:

```
Q: What are the common commands in programming language PHP?
```

AMAZON Q: CHAT

Chat   ×   +

What are the common commands in programming language PHP?

PHP (Hypertext Preprocessor) is a server-side scripting language widely used for web development. Here are some common commands and constructs in PHP:

1. **Output Commands** :

   ○ `echo` : Outputs one or more strings.

   ○ `print` : Outputs a string and returns 1 (true).

   ○ `print_r()` : Prints human-readable information about a variable.

   ○ `var_dump()` : Outputs information about a variable, including its data type and value.

2. **Variable Declaration** :

   ○ `$variable_name` : Declares a variable.

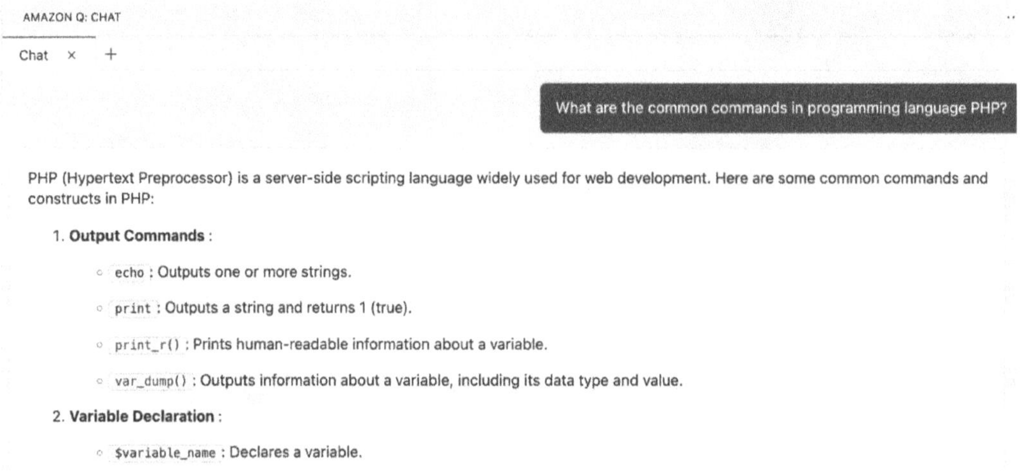

Figure 6.9 – Chat style – PHP common commands partial screenshot

We won't cover every command, but as illustrated in the preceding partial screenshot, Amazon Q Developer provided various command categories such as output commands, variable commands, conditional statements, database commands, form commands, error handling, array commands, and more. To make it easier to follow, it included some common key commands for each category and provided references for users to obtain further detailed information on these topics.

Database commands is one of the important categories that make PHP useful for server-side scripting. Now, let's ask Amazon Q Developer for the syntax and an example of mysqli_connect() in PHP:

```
Q: Provide syntax and example of mysqli_connect() in PHP
```

AMAZON Q: CHAT   ...

Chat   ×   +

Provide syntax and example of mysqli_connect() in PHP

The `mysqli_connect()` function in PHP is used to establish a connection to a MySQL database server. Here's the syntax: [1]

```php
mysqli_connect(host, username, password, database, port, socket)
```
php   ⌐ Insert at cursor 🗐 Copy

- `host` (required): The hostname or IP address of the MySQL server.

- `username` (required): The username to use for the MySQL connection.

- `password` (required): The password to use for the MySQL connection.

- `database` (optional): The name of the database to select after establishing the connection. [2]

- `port` (optional): The port number to use for the connection. If not specified, the default port (3306) is used.

- `socket` (optional): The path to the MySQL socket file, which is used for local connections.

The function returns a `mysqli` object on success or `false` on failure.

Here's an example of how to use `mysqli_connect()` :

```php
<?php
// MySQL server credentials
$servername = "localhost";
$username = "your_username";
$password = "your_password";
$database = "your_database";

// Create connection
$conn = mysqli_connect($servername, $username, $password, $database);

// Check connection
if (!$conn) {
    die("Connection failed: " . mysqli_connect_error());
}
echo "Connected successfully";

// Close connection
mysqli_close($conn);
?>
```
php   ⌐ Insert at cursor 🗐 Copy

In this example:

1. We define the MySQL server credentials: `$servername` , `$username` , `$password` , and `$database` .

2. We call `mysqli_connect()` with the provided credentials and store the connection object in `$conn` .

Figure 6.10 – Chat style –PHP mysqli_connect()

As shown in the preceding screenshot, Amazon Q Developer provided an example for `mysqli_ connect()` and explained the details of each line of the code to make it easier to understand.

End-to-end PHP language use case development

Now, let's move on to the sample use case to show the power of Amazon Q Developer's ability to recommend the code. To illustrate, I am going to use the same chat-style interface to generate the code and then move to the VS Code script editor window.

A common use case for PHP in server-side programming is handling user authentication and managing sessions. This involves verifying user credentials, starting sessions, and managing session data to maintain user state across different pages of a web application. Here is the prompt:

```
Prompt:
/* Write PHP server-side end to end programming to handle user
authentication and managing sessions. This involves verifying user
credentials, starting sessions, and managing session data to maintain
user state across different pages of a web application. */
```

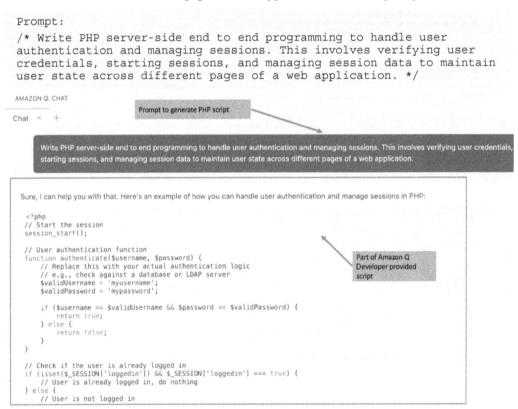

Figure 6.11 – Prompt – PHP program partial snapshot

The preceding screenshot displays a partial code snippet suggested by Amazon Q Developer. You can use the **Insert at cursor** button to port the script into the editor. The end-to-end script utilizes many key functionalities such as a form to accept usernames and passwords, an `authenticate` (`$username`, `$password`) function with error handling, and checks for valid users with session handling using `isset($_SESSION['loggedin']) && $_SESSION['loggedin']`. The code meets the specifics of your prompts, demonstrating Amazon Q Developer's ability to generate comprehensive and functional code for handling user authentication and managing sessions.

The script is already echoing the error message, but let's spice things up with some customization. Instead of just a boring error message, let's display a random error image from a folder full of random images every time an error occurs. Let's use the single-line prompt technique, to add a custom twist:

```
Prompt:
/* show random error image from folder '/user/error/images/'*/
62    if (isset($error)) {
63        echo $error;
64    /* show random error image from folder '/user/error/images/'*/
65        $dir = 'user/error/images/';
66        $files = scandir($dir);
67        $randomIndex = array_rand($files, 1);
68        $randomImage = $files[$randomIndex];
69        echo '<img src="user/error/images/'.$randomImage.'" alt="Random Error Image">';
70    }
71 }
```

Figure 6.12 – Prompt – PHP program custom error image

As you can see in the preceding screenshot, Amazon Q Developer added multiple lines of code to randomly pick one image from our `error/images` directory and added it as part of the display.

Now, let's explore how Amazon Q Developer supports code explainability and optimization for JS and PHP.

JS and PHP code explainability and optimization using Amazon Q Developer

Amazon Q Developer offers a simple interface for achieving code explainability and optimization for both JS and PHP. For more details on explainability and optimization, please refer to *Chapter 12*.

To illustrate, I am going to use the previously auto-generated code during the *PHP server-side end-to-end code* creation task. As shown In the following screenshot, highlight the code, right-click to open the pop-up menu, select **Amazon Q**, and choose **Explain** for code explainability or **Optimize** for optimization recommendations.

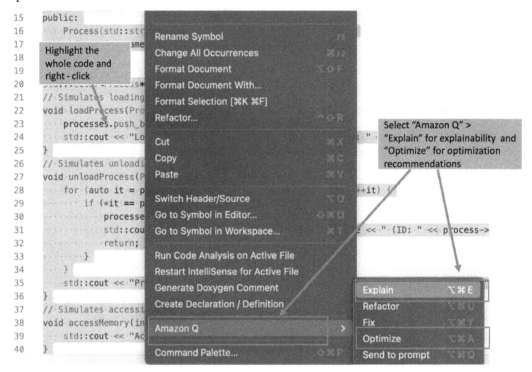

Figure 6.13 – Prompt – JS and PHP program explainability and optimization

This will open up Amazon Q Developer's chat-style interface to analyze the full code.

For explainability, Amazon Q Developer provides detailed information about each function and the overall code logic.

When it comes to optimization, Amazon Q Developer offers end-to-end optimized code that you can directly insert into your editor using the **Insert at cursor** button. It recommended using the ternary operator for the `authenticate()` function, combining conditions in `$isLoggedIn` for improved readability, and suggested output encoding to prevent potential XSS vulnerabilities. These optimizations enhance the code's security, readability, and performance without altering its core functionality. Additionally, it advised replacing the current authentication method with a more secure approach, such as employing a secure password hashing algorithm and storing user credentials in a database.

In addition to JS and PHP, Amazon Q Developer supports multiple other programming languages. There are many enhancements happening in this area, and we anticipate further enhancements in the support for additional languages (check the *References* section at the end of the chapter).

Summary

Amazon Q Developer represents a transformative tool for auto-code generation in web development languages such as JS and PHP. Programmers engage with Amazon Q Developer through chat-style interactions, making it an intuitive platform for learning and coding. By using various prompting techniques, developers can obtain code fragments or complete scripts from Amazon Q Developer, streamlining their workflow.

For JS, this capability is crucial for creating interactive user interfaces and dynamic web content. Developers can prompt Amazon Q Developer to generate JS code for common tasks such as form validation, event handling, and DOM manipulation.

Similarly, Amazon Q Developer's integration for PHP aids web developers in implementing dynamic web applications, server-side scripting, and backend services. As illustrated, it can generate PHP code for tasks such as handling user logins, sessions, authentication, and more. In addition to code generation, Amazon Q Developer offers explanations for the generated code, helping developers understand how each part of the script functions. This educational aspect is invaluable for both novice and experienced developers looking to enhance their coding skills.

Overall, Amazon Q Developer empowers web developers to be more productive by automating routine coding tasks and providing optimization insights. This enables developers to focus on the innovative aspects of their projects, improving development efficiency and code quality. The support for JS and PHP showcases the versatility and potential of Amazon Q Developer in modern web development.

In the next chapter, we will introduce how Amazon Q Developer with SQL provides benefits to DBAs and DEs.

References

- JavaScript in Visual Studio Code: `https://code.visualstudio.com/docs/languages/javascript`

- PHP in Visual Studio Code: `https://code.visualstudio.com/docs/languages/php`

- Supported languages for Amazon Q Developer in the IDE: `https://docs.aws.amazon.com/amazonq/latest/qdeveloper-ug/q-language-ide-support.html`

7

Boost Coding Efficiency for SQL with Auto-Code Generation

In this chapter, we will look at the following key topics:

- An overview of database management languages
- SQL auto-code generation for **database administrators** (**DBAs**) using Amazon Q Developer
- SQL auto-code generation for **data engineers** (**DEs**) using Amazon Q Developer
- SQL code explainability and optimization using Amazon Q Developer

In our previous chapter, we discussed an overview of web programming languages before focusing on two dominant ones used in the software industry – **JavaScript** (**JS**) and PHP. We demonstrated how auto-code generation, especially Amazon Q Developer, is integrated and can be used by JS and PHP developers to understand the basics, learn syntax, and automatically generate code.

In this chapter, we will focus on database management languages. We'll dive into one of the most dominant database management languages in the software industry, **Structured Query Language** (**SQL**). Using two user personas, DBAs and DEs, we will introduce how auto-code generation through Amazon Q Developer can help you understand the basics, learn syntax, and automatically generate code for common DBA and DE activities. Then, we will explore code explainability to support documentation and code optimization recommendations provided by Amazon Q Developer for SQL.

Overview of database management languages

Let's dive into another widely used area of systems across industries – database management – and explore SQL, a crucial language for handling large datasets, tables, views, users, and so on.

In the dynamic landscape of database management, SQL serves as the fundamental language for interacting with relational databases. SQL enables DBAs and DEs to retrieve, manipulate, and manage data efficiently across various **Database Management Systems (DBMs)**. SQL is vital when working on projects that interact with relational and/or columnar databases, such as MySQL, PostgreSQL, Oracle Database, SQL Server, Teradata, and Amazon RedShift. As databases become increasingly complex and critical to business operations, the ability to write efficient and reliable SQL code is paramount. Auto-code generation has emerged as a transformative approach to streamline SQL development processes, automate routine tasks, and enhance productivity for DBAs and DEs alike. Amazon Q Developer, leveraging advanced artificial intelligence and machine learning capabilities, offers robust support for SQL code generation. This section explores how Amazon Q Developer revolutionizes SQL development, empowers DBAs and DEs, and facilitates efficient database management and data manipulation.

For DBAs, SQL plays a pivotal role in database administration and application development by enabling tasks such as defining database structures, managing access control, monitoring log tables, and ensuring data integrity. For DEs, SQL is key for tasks such as loading data in tables, querying data by joining multiple tables, manipulating data using store procedures, and unloading data. However, writing SQL queries and maintaining database schemas can be time-consuming and prone to errors, particularly in environments with extensive data requirements and complex relationships. Amazon Q Developer simplifies SQL development by automating the generation of SQL queries, database schema definitions, and data manipulation scripts. Through its intuitive chat-style interface, DBAs and DEs can interact with Amazon Q Developer to request specific SQL queries or tasks using natural language prompts.

Furthermore, Amazon Q Developer serves as an educational resource for SQL learners and novice DE. It offers interactive tutorials, explanations of SQL concepts, and practical examples of SQL queries through its chat interface. This educational aspect helps DBAs and DEs grasp SQL fundamentals more effectively and apply them to real-world database management and data manipulation scenarios.

Experienced DBAs and DEs can use Amazon Q Developer's advanced features, such as query optimization and performance tuning suggestions. It analyzes SQL queries generated by users, identifies potential performance bottlenecks or inefficiencies, and offers recommendations to enhance query execution speed and resource utilization. These optimization capabilities are crucial for maintaining optimal database performance and scalability in production environments. Teams working on database-driven applications can benefit from standardized SQL code templates, shared best practices, and streamlined code review processes.

Now, let's dive deep to see how Amazon Q Developer can help a DBA and DE. To illustrate, I will use Amazon **Relational Database Service (RDS)** for MySQL, which is an offering from **Amazon Web Services (AWS)**, and the Amazon Q Developer interface with VS Code (refer to *Chapter 2* for configuration steps for Amazon Q Developer with VS Code).

> **Note**
>
> Amazon Q Developer uses **large language models** (**LLMs**), which by nature are non-deterministic, so you may not get exactly the same answers/code blocks shown in the code snapshots. Try to update prompts to get your desired recommendations. However, logically, the generated answer/code should meet the requirements.

SQL auto-code generation for DBAs using Amazon Q Developer

First, assume a new DBA is planning to work with Amazon RDS for MySQL and does not have any background in this service offering. Let's use Amazon Q Developer's chat-style interface to educate DBAs, where they can simply ask questions in natural language to learn about DBA activities and best practices when it comes to Amazon RDS for MySQL.

Amazon RDS for MySQL basics for DBAs

Let's first learn about Amazon RDS for MySQL:

```
Q: What is Amazon RDS for MySQL ?
```

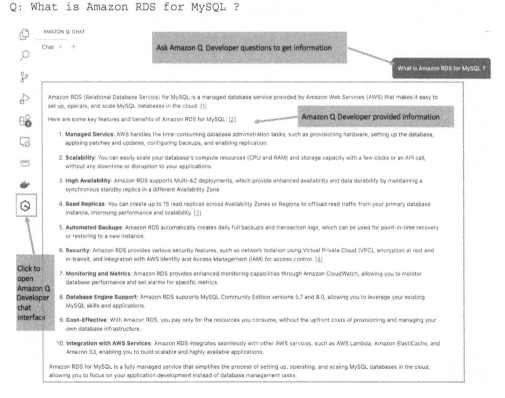

Figure 7.1 – Chat style – DBA Amazon RDS for MySQL basic

Note that Amazon Q Developer provided the key information about Amazon RDS for MySQL, related to important topics such as **Managed Service**, **Scalability**, **High Availability**, **Read Replicas**, and **Automated Backups**.

Now, let's dive deep to understand DBA-specific information related to Amazon RDS for MySQL:

```
Q: What are the common activities of Amazon RDS for MySQL DBA?
```

AMAZON Q: CHAT

Chat × +

Ask Amazon Q Developer questions to get information

what are the common activities of Amazon RDS for MySQL DBA ?

As an Amazon RDS for MySQL database administrator (DBA), you can perform various tasks to manage and maintain your databases effectively. However, it's important to note that Amazon RDS is a managed service, which means that some traditional DBA tasks are handled by AWS, while others are still the responsibility of the user. Here are some common activities that a DBA can perform with Amazon RDS for MySQL: [1]

Amazon Q Developer provided information and corresponding resources to learn further

1. **Creating and managing databases**: You can create new databases, configure database parameters, ... ols and commands.

2. **Monitoring and optimizing performance**: You can monitor the performance of your databases using Amazon CloudWatch metrics and enhanced monitoring. You can also analyze and optimize queries, indexes, and database configurations for better performance.

3. **Backup and recovery**: Amazon RDS automatically creates daily backups and transaction logs, which you can use to restore your databases to a specific point in time. You can also create manual snapshots and configure backup retention periods.

4. **High availability and replication**: You can configure Multi-AZ deployments for high availability and read replicas for read scaling. You can also set up replication between different AWS Regions for disaster recovery.

5. **Security management**: You can control access to your databases using AWS Identity and Access Management (IAM) policies, security groups, and SSL/TLS encryption for data in transit.

6. **Upgrading and patching**: AWS automatically applies minor version upgrades and patches to your DB instances, but you may need to initiate major version upgrades manually.

7. **Importing and exporting data**: You can import data from various sources, such as SQL scripts, flat files, or other databases, using tools like MySQL Workbench or AWS Database Migration Service (DMS).

8. **Configuring options and parameters**: You can enable or disable various MySQL options and modify database parameters to suit your application's needs.

9. **Monitoring and auditing**: You can enable and review database logs, such as error logs, slow query logs, and audit logs, to troubleshoot issues and monitor database activity.

10. **Scaling and resizing**: You can vertically scale your DB instances by modifying the instance type or storage allocation, or horizontally scale by adding read replicas.

It's important to note that while Amazon RDS handles many routine tasks, such as automated backups and software patching, DBAs still play a crucial role in ensuring the performance, security, and availability of their databases.

Sources

[1] Hosted MySQL - Amazon RDS for MySQL - AWS
aws.amazon.com > rds > **mysql**

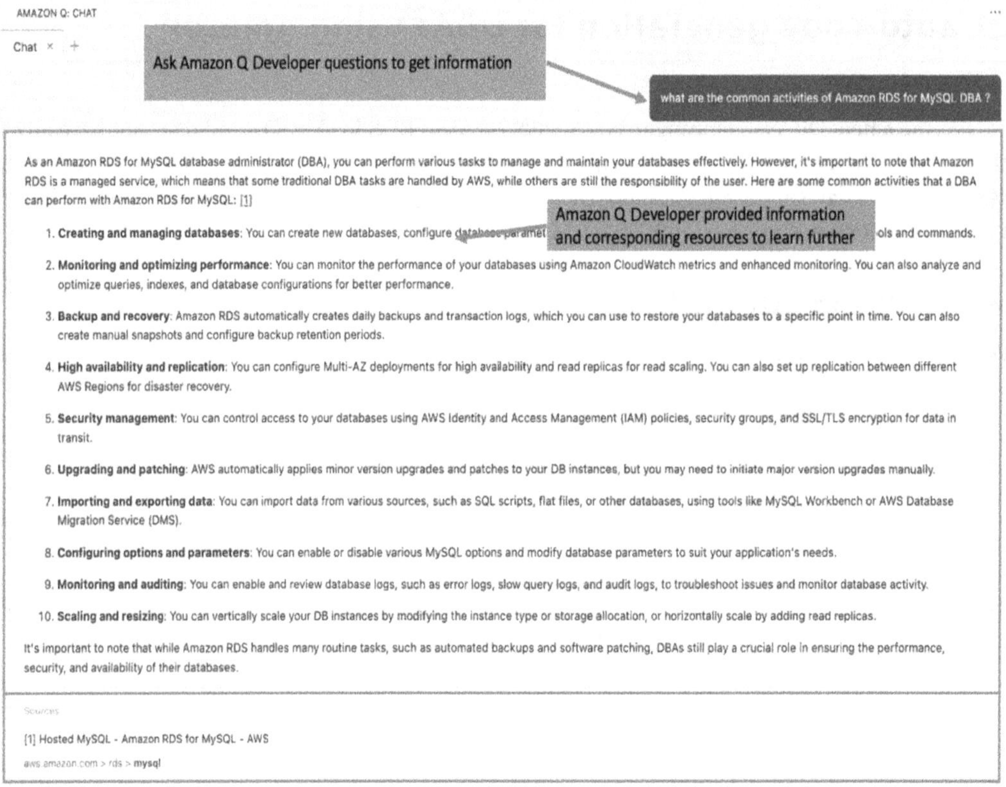

Figure 7.2 – Chat style – Amazon RDS for basic MySQL DBA activities

We will not dive deep into each DBA activity suggested by Amazon Q Developer; just note that Amazon Q Developer utilized the key areas of responsibilities of a DBA, such as creating and managing databases, monitoring and optimizing performance, backup and recovery, high availability and replication, security management, and upgrading and patching to provide Amazon RDS with MySQL-specific information. In addition, it also provided resources that can be used for further reading.

Security management is one of the most important responsibilities of DBAs. Managed AWS services such as Amazon RDS have mechanisms that differ from on-premises databases, so it's essential that DBAs understand these differences. Let's ask Amazon Q Developer to provide more information to help DBAs understand specific details and best practices.

To illustrate the best practice recommendations, we'll ask Amazon Q Developer about key security resource configurations, such as security groups for Amazon RDS for MySQL databases.

Q: What are the best practices for configuring security groups for Amazon RDS for MySQL databases?

Figure 7.3 – Chat style – Amazon RDS for MySQL security group best practices

As shown in the preceding screenshot, Amazon Q Developer provided the top 10 best practices that DBAs can follow to configure security groups for Amazon RDS for MySQL.

SQL generation for DBAs

Now, let's ask Amazon Q Developer about the common SQL commands used by Amazon RDS for MySQL DBAs:

Q: Which are the common SQL commands used by Amazon RDS for MySQL DBA?

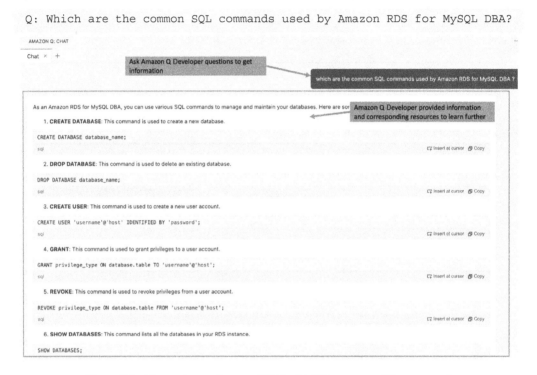

Figure 7.4 – Chat style – an Amazon RDS for MySQL common SQL command

As shown in the preceding screenshot, the Amazon Q Developer chat-style interface provides multiple commonly used SQL commands with syntax to make it easy to understand. New Amazon RDS for MySQL DBAs will realize that while some traditional DBA activities are managed by Amazon RDS, activities such as creating databases and schemas, simple tables, and database users with specific access permissions still require DBA attention.

So, based on the preceding information for DBAs, let's start writing SQL commands to carry out common DBA tasks with Amazon Q Developer's help. To illustrate, I am going to use the VS Code script editor window with single-line and multi-line prompts.

Let's ask Amazon Q Developer to provide a SQL command to create a database and schema:

```
Prompt :
/* Write a SQL command for Amazon RDS for MySQL to
create database "my_test_db", schema "my_test_db.my_test_schema"
*/
```

```
1    /* Write a SQL command for Amazon RDS for MySQL to
2    create database "my_test_db", schema "my_test_db.my_test_schema"
3    */
4    CREATE DATABASE `my_test_db` CHARACTER SET utf8 COLLATE utf8_general_ci;
5    CREATE SCHEMA `my_test_db`.`my_test_schema` ;
```

Figure 7.5 – A prompt for Amazon RDS for MySQL to create a database and schema

Note that Amazon Q Developer provided complete, error-free SQL code to create the database and schema, as per the names provided in the prompt following MySQL specifics.

Now, let's ask Amazon Q Developer to create a simple table:

```
Prompt :
/* Write a SQL command for Amazon RDS for MySQL
to create table "my_test_table" in schema "my_test_schema" with
columns sr_nbr as integer, name as varchar 50 where sr_nbr is a
primary key */
```

```
8    /* Write a SQL command for Amazon RDS for MySQL
9    to create table "my_test_table" in schema "my_test_schema" with
10   columns sr_nbr as integer, name as varchar 50 where sr_nbr is a primary key
11   */
12   CREATE TABLE `my_test_db`.`my_test_schema`.`my_test_table` (
13     `sr_nbr` INT NOT NULL,
14     `name` VARCHAR(50) NULL,
15     PRIMARY KEY (`sr_nbr`));
16
```

Figure 7.6 – A prompt for Amazon RDS for MySQL to create a table

Note that, as prompted, Amazon Q Developer provided table DDL with a primary key constraint.

Now, let's ask Amazon Q Developer to create a database user and grant specific access permission to the user in the preceding table:

```
Prompt 1:
/* Write a SQL command for Amazon RDS for MySQL
to create user "my_test_user" */
```

```
Prompt 2:
/* Write a SQL command for Amazon RDS for MySQL
to grant select, insert, and update access to user "my_test_user" on
table "my_test_db.my_test_schema.my_test_table"
*/
```

We get the SQL code for this as shown in *Figure 7.7*.

```
18    /* Write a SQL command for Amazon RDS for MySQL
19    to create user "my_test_user"
20    */
21     CREATE USER `my_test_user` IDENTIFIED WITH AWSAuthenticationPlugin AS 'RDS';
22
23
24    /* Write a SQL command for Amazon RDS for MySQL
25    to grant select, insert, and update  access to user "my_test_user" on
26    table "my_test_db.my_test_schema.my_test_table"
27    */
28     GRANT SELECT , INSERT , UPDATE ON `my_test_db`.`my_test_schema`.`my_test_table` TO `my_test_user`@`%`  ;
29     FLUSH PRIVILEGES;
30
```

Figure 7.7 – A prompt for Amazon RDS for MySQL to create a user and grant user access

Now, let's create SQL to remove the previously granted `update` privilege:

```
Prompt :
/* Write a SQL command for Amazon RDS for MySQL
to revoke update access to user "my_test_user" on
table "my_test_db.my_test_schema.my_test_table"
*/
```

```
31  /* Write a SQL command for Amazon RDS for MySQL
32  to revoke update  access to user "my_test_user" on
33  table "my_test_db.my_test_schema.my_test_table"
34  */
35   REVOKE UPDATE ON `my_test_db`.`my_test_schema`.`my_test_table` FROM `my_test_user`@`%`  ;
36   FLUSH PRIVILEGES;
```

Figure 7.8 – A prompt for Amazon RDS for MySQL to revoke access

Now, let's carry out the clean-up tasks to drop the user, table, schema, and database:

```
Prompt :
/* Write a SQL command for Amazon RDS for MySQL to
drop user "my_test_user"
drop table "my_test_db.my_test_schema.my_test_table"
drop schema "my_test_db.my_test_schema"
drop database "my_test_db"
*/
```

We get the SQL code for this as shown in *Figure 7.9*.

```
39    /* Write a SQL command for Amazon RDS for MySQL to
40    drop user "my_test_user"
41    drop table "my_test_db.my_test_schema.my_test_table"
42    drop schema  "my_test_db.my_test_schema"
43    drop database  "my_test_db"
44    */
45    DROP USER `my_test_user`@`%` ;
46    DROP TABLE `my_test_db`.`my_test_schema`.`my_test_table` ;
47    DROP SCHEMA `my_test_db`.`my_test_schema` ;
48    DROP DATABASE `my_test_db` ;
49    FLUSH PRIVILEGES;
```

Figure 7.9 – A prompt for Amazon RDS for MySQL to drop objects

As illustrated here, by using Amazon RDS for MySQL, a DBA can simply use a chat-style and/or prompt-style interface to interact with Amazon Q Developer, receiving error-free SQL code to carry out daily activities. To explore further, feel free to use additional prompts to get SQL code from Amazon Q Developer related to DBA activities.

In the next section, let's see how Amazon Q Developer can help a DE who is mainly responsible for writing SQL code to understand data, join multiple tables, manipulate data, and load data into tables.

SQL auto-code generation for DEs using Amazon Q Developer

First, let's assume that a new DE is planning to work with Amazon RDS for MySQL and does not have any background in the SQL command offerings. In the next sections, we will explore how Amazon Q Developer can support DEs with common **Data Manipulation Language** (**DML**) support, syntaxes, and basic SQL queries. Let's start with basic DML support.

Amazon RDS for MySQL basics for DEs

Let's use Amazon Q Developer's chat-style interface to educate DEs, where they can simply ask questions in a natural language to learn about the available commands and functions for Amazon RDS for MySQL. Since data manipulation is a crucial activity for DEs, which falls under a DML, let's ask Amazon Q Developer to provide information on the support offered by Amazon RDS for DML operations:

```
Q: What are supported SQL DML commands in Amazon RDS for MySQL?
```

As you can see in the following screenshot, Amazon Q Developer provides commonly used available SQL commands such as **SELECT**, **INSERT**, **UPDATE**, and **DELETE**.

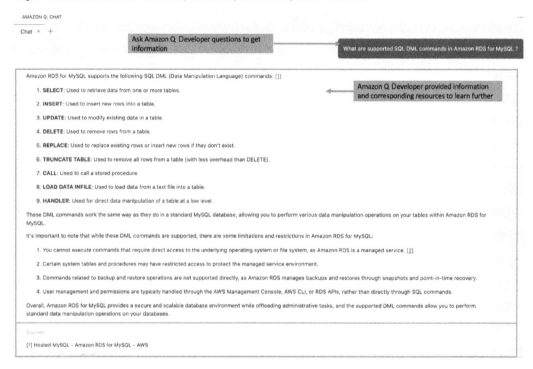

Figure 7.10 – Chat style – Amazon RDS for MySQL DMLs for a DE

SQL generation for DEs

Let's dive deeper and ask about the SQL commands, working with some of the common SQL operations to understand how Amazon Q Developer can help DEs learn more, as well as auto-generate SQL code using different prompting techniques.

Let's ask one more question about SQL syntax before we start writing SQL code:

```
Q: What is syntax of SELECT SQL commands in Amazon RDS for MySQL ?
```

AMAZON Q: CHAT

Chat × +

What is syntax of SELECT SQL commands in Amazon RDS for MySQL ?

In Amazon RDS for MySQL, the syntax for SELECT SQL commands follows the standard SQL syntax. Here's the basic structure: [1]

```
SELECT column1, column2, ...
FROM table_name
[WHERE condition]
[ORDER BY column1, column2, ... [ASC|DESC]]
[LIMIT n];
```

Amazon Q Developer provided syntax, details of each clause, and sample example

sql   ⌨ Insert at cursor 🗇 Copy

Here's a breakdown of the different clauses:

1. `SELECT column1, column2, ...` : Specifies the columns you want to retrieve from the table. You can use `*` to select all columns.

2. `FROM table_name` : Specifies the name of the table from which you want to retrieve data.

3. `WHERE condition` : An optional clause that filters the rows based on a specified condition.

4. `ORDER BY column1, column2, ... [ASC|DESC]` : An optional clause that sorts the result set by one or more columns in ascending (`ASC`) or descending (`DESC`) order.

5. `LIMIT n` : An optional clause that limits the number of rows returned by the query to `n` rows.

You can also use other SQL clauses like `GROUP BY` , `HAVING` , `JOIN` , etc., along with various SQL functions and operators, just like you would in a standard MySQL environment.

Here's an example query that selects the `name` and `email` columns from a `users` table, ordered by the `name` column in ascending order, and limits the result to the first 10 rows:

```
SELECT name, email
FROM users
ORDER BY name ASC
LIMIT 10;
```

Figure 7.11 – Chat style – an Amazon RDS for MySQL SELECT command for a DE

Note that Amazon Q Developer provided detailed clauses and a sample example at the end to help users understand the usage of the `select` command.

Now, let's start writing SQL commands to carry out common DE tasks with Amazon Q Developer's help. To illustrate, I am going to use the VS Code script editor window with single-line and multi-line prompts.

Let's start with simple `select` commands from tables, assuming that tables are already present and the user has access to select from them:

```
Prompt :
/* Write a SQL command for Amazon RDS for MySQL to
select sample 10 records from
Table "table_emp" and Table "table_dept"
*/
```

```
 2    /* Write a SQL command for Amazon RDS for MySQL to
 3    select sample 10 records from
 4    Table "table_emp"   and Table "table_dept"
 5    */
 6    SELECT *
 7    FROM table_emp
 8    LIMIT 10;
 9
10    SELECT *
11    FROM table_dept
12    LIMIT 10;
```

Figure 7.12 – A prompt for an Amazon RDS for MySQL select SQL command for a DE

Now, let's join these two tables to de-normalize the data, by using `join`:

```
Prompt :
/* Write a SQL command for Amazon RDS for MySQL to join
Table "table_emp" - columns emp_id as integer, emp_name as varchar
50  with Table "table_dept" - columns emp_id as integer, dept_id as
int, and dept_name as varchar 50
on emp_id to pull emp_id, emp_name, dept_name
and pull all records from table_emp
*/
```

```
16   /* Write a SQL command for Amazon RDS for MySQL to join
17   Table "table_emp" – columns emp_id as integer, emp_name as varchar 50  with
18   Table "table_dept" – columns emp_id as integer, dept_id as int, and dept_name as varchar 50
19   on emp_id to pull emp_id, emp_name, dept_name
20   and pull all records from table_emp
21   */
22   SELECT e.emp_id ,  e.emp_name ,   d.dept_name
23   FROM table_emp e
24   LEFT OUTER JOIN table_dept d
25   ON d.emp_id = e.emp_id
26   LIMIT 10;
```

Figure 7.13 – A prompt for an Amazon RDS for MySQL join SQL command for a DE

Note the `pull all records from table_emp` instruction in our prompt, which Amazon Q Developer correctly converted to `LEFT OUTER JOIN` between two tables.

Now, let's use prompts to create a "store procedure" with output from the preceding join to load into an existing table:

```
Prompt :
/* Write a SQL command for Amazon RDS for MySQL to create Store
Procedure  to
Join between Table "table_emp" and Table "table_dept"
on emp_id to pull uppercase(emp_id), emp_name, dept_name
and pull all records from table_emp
load data into table "derom_data"
*/
```

```
30    /* Write a SQL command for Amazon RDS for MySQL to create Store Procedure  to
31    Join between Table "table_emp" and Table "table_dept"
32    on emp_id to pull uppercase(emp_id), emp_name, dept_name
33    and pull all records from table_emp
34    load data into table "derom_data"
35    */
36
37    DELIMITER $$
38    CREATE PROCEDURE `join_emp_dept`()
39    BEGIN
40        insert into derom_data
41        SELECT UPPER(e.emp_id) ,  e.emp_name ,  d.dept_name
42        FROM table_emp e
43        LEFT OUTER JOIN table_dept d
44        ON d.emp_id = e.emp_id;
45    END$$
46    DELIMITER ;
```

Figure 7.14 – A prompt for an Amazon RDS for MySQL join SQL command for DE

Note that Amazon Q Developer created error-free code to create a "procedure" based on our instructions, and it also added the appropriate function to handle uppercase instructions for `emp_id`.

In the preceding examples, we just scratched the surface of Amazon Q Developer's integration for SQL code development. Feel free to try different questions in chat-style prompts and interactive coding to explore further.

Now, let's explore how the Amazon Q Developer supports code explainability and optimization for SQL.

SQL code explainability and optimization using Amazon Q Developer

Amazon Q Developer offers a simple interface to achieve code explainability and optimization. For more details on the explainability and optimization topic, refer to *Chapter 12*.

To illustrate, I am going to use the previously auto-generated code during the store procedure creation task. As shown in the following screenshot, highlight the code section generated during the store procedure, right-click to open the pop-up menu, select **Amazon Q**, and then choose **Explain** for code explainability or **Optimize** for optimization recommendations.

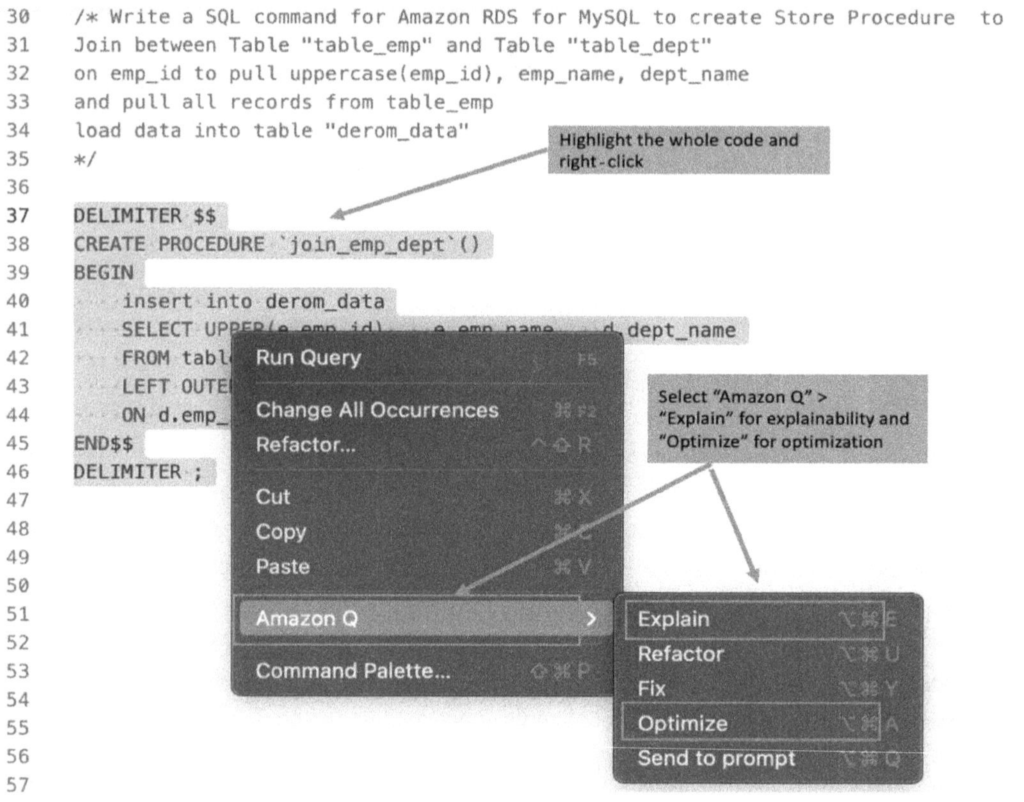

```
30    /* Write a SQL command for Amazon RDS for MySQL to create Store Procedure  to
31    Join between Table "table_emp" and Table "table_dept"
32    on emp_id to pull uppercase(emp_id), emp_name, dept_name
33    and pull all records from table_emp
34    load data into table "derom_data"
35    */
36
37    DELIMITER $$
38    CREATE PROCEDURE `join_emp_dept`()
39    BEGIN
40        insert into derom_data
41        SELECT UPPER(e.emp_id)    e.emp_name    d.dept_name
42        FROM tabl
43        LEFT OUTE
44        ON d.emp_
45    END$$
46    DELIMITER ;
47
48
49
50
51
52
53
54
55
56
57
```

Highlight the whole code and right-click

Run Query   F5

Change All Occurrences   ⌘F2

Refactor...   ⌃⇧R

Cut   ⌘X

Copy   ⌘C

Paste   ⌘V

Amazon Q   >

Command Palette...   ⇧⌘P

Select "Amazon Q" > "Explain" for explainability and "Optimize" for optimization

Explain   ⌥⌘E

Refactor   ⌥⌘U

Fix   ⌥⌘Y

Optimize   ⌥⌘A

Send to prompt   ⌥⌘Q

Figure 7.15 –SQL code explainability and optimization

This will pop up Amazon Q Developer's chat-style interface and use the full code for analysis.

For explainability, Amazon Q Developer provides details for almost every line of code and at the end provides a summary, such as the following: "**When the join_emp_dept stored procedure is called, it will execute the SELECT statement, which performs a left outer join between the table_emp and table_dept tables based on the emp_id column. The resulting data, including the employee ID in uppercase, employee name, and department name, will be inserted into the derom_data table**".

For optimization, even though we had a small code fragment with relatively simple logic, Amazon Q Developer still provided recommendations to improve the code and rewrite it, following the best practices on `join` implementation, added column listing to the `insert` statement, added `coalesce`, and so on.

As we can see in the preceding topics, Amazon Q Developer provides meaningful inputs for DBAs and Des, with auto-generated SQL for relational databases such as Amazon RDS MySQL. Additionally, Amazon Q Developer seamlessly integrates with columnar databases such as Amazon Redshift. AWS has integrated Amazon Redshift, a fully managed AI-powered data warehouse service, with a code assistant that allows users to use **natural language processing** (**NLP**) to generate complete SQL queries. We will explore this topic in *Chapter 14* in the *Code assistance integration with Amazon Redshift* section.

In addition to SQL, Amazon Q Developer supports several other programming languages. There are many developments happening in this area, and we anticipate several of these in support of additional languages (check the *References* section at the end of the chapter).

Summary

Amazon Q Developer represents a groundbreaking innovation in SQL code generation and database management for DBAs and DEs. Through various prompting techniques and chat-style interactions, professionals can obtain SQL code from Amazon Q Developer. This capability helps automate routine SQL tasks, improve code consistency, and offer advanced optimization capabilities. It empowers database professionals to focus on strategic initiatives, innovation, and delivering superior database-driven applications.

The integration of Amazon Q Developer with SQL also facilitates collaborative database development. Teams working on database-driven applications benefit from standardized SQL code templates, shared best practices, and streamlined code review processes. This consistency enhances team productivity, reduces the risk of errors during database schema modifications or data migrations, and accelerates the time-to-market for database-driven applications.

Additionally, Amazon Q Developer offers insights into SQL code optimization, ensuring efficient query performance and resource utilization. By continuously improving its recommendations, Amazon Q Developer adapts to the evolving needs of database professionals, providing cutting-edge solutions to complex database challenges.

Thus, Amazon Q Developer not only automates routine SQL tasks but also fosters collaboration and innovation within database teams, ultimately leading to higher quality and more efficient database-driven applications.

In this chapter, we scratched the surface of Amazon Q Developer's support for DBAs and DEs. Feel free to explore other SQL options related to creating common database objects, such as views, materialized views, functions, and complex SQL queries.

In the next chapter, we will introduce how Amazon Q Developer benefits system administration and automation using shell scripts.

References

Supported languages for Amazon Q Developer in the IDE: `https://docs.aws.amazon.com/amazonq/latest/qdeveloper-ug/q-language-ide-support.html`

8

Boost Coding Efficiency for Command-Line and Shell Script with Auto-Code Generation

In this chapter, we will look at the following key topics:

- Overview of command-line and shell script

- Command-line auto-code generation for system administrators using Amazon Q Developer

- Shell script auto-code generation for programmers using Amazon Q Developer

- Shell script explainability and optimization using Amazon Q Developer

In the previous chapter, we dived into one of the most dominant database management languages in the software industry: **Structured Query Language** (**SQL**). Using two user personas—**database administrators** (**DBAs**) and **data engineers** (**DEs**)—we introduced how auto-code generation through Amazon Q Developer can help understand the basics, learn the syntax, and automatically generate code for common DBA and DE activities.

In this chapter, we continue with a similar theme and focus on system administrators and shell script programmers. System administrators primarily use the **command-line interface** (**CLI**) for various system configuration tasks, and shell script programmers use **operating system** (**OS**) commands to automate repetitive tasks such as development processes, file transfers, and preprocessing tasks. We will introduce how Amazon Q Developer can help these professionals understand the basics, learn the syntax, and automatically generate code for common system administration and shell script programming activities. Then, we will explore code explainability to support documentation and code optimization recommendations provided by Amazon Q Developer for shell scripts.

Overview of command-line and shell script

OS code plays a pivotal role in managing and maintaining the stability, security, and efficiency of IT infrastructures. Shell scripting and the command line serve as fundamental tools for automating routine tasks, executing system commands, and orchestrating complex workflows across Unix-based operating systems such as Linux and macOS. As organizations strive to optimize operational processes and enhance scalability, the demand for efficient shell scripting and command-line solutions becomes increasingly significant.

Auto-code generation has emerged as a transformative approach to streamline shell script development using an IDE and/or command line, reduce manual effort, and improve productivity for system administrators, programmers, and engineers.

CLI and shell scripts enable system administrators, programmers, and engineers to automate repetitive tasks such as file management, system monitoring, user administration, and backup operations. It also facilitates chaining multiple commands together for the execution of system commands and complex workflows, ensuring consistency and reliability in IT operations. However, writing and maintaining complex chained commands and shell scripts can be time-consuming, prone to errors, and needs expertise, especially as scripts become more intricate and critical to business operations. System administrators, programmers, and engineers can leverage different prompting techniques and chat-style integrations (refer to *Chapter 3*) to get recommendations for the automated creation of shell scripts tailored to specific system tasks. This interaction model accelerates script development cycles, reduces human error, and ensures uniformity in script structure and execution across different OS environments.

Moreover, Amazon Q Developer serves as an educational resource for learning shell scripting techniques, available command-line syntax, and best practices. It offers interactive tutorials, explanations of shell script concepts, and practical examples of automation scripts through its chat interface. This educational aspect helps system administrators, programmers, and engineers enhance their scripting skills, adopt industry-standard practices, and leverage advanced automation capabilities effectively.

For experienced users, Amazon Q Developer provides advanced features such as script optimization and error-handling suggestions. It analyzes generated shell scripts, identifies potential inefficiencies or pitfalls, and offers recommendations to enhance script performance and reliability. These optimization capabilities are crucial for maintaining operational efficiency and scalability in dynamic IT environments. The integration of Amazon Q Developer with shell scripting also facilitates collaborative practices. Teams working on system automation initiatives benefit from standardized shell script templates, shared automation workflows, and streamlined code review processes. This consistency enhances team productivity, fosters collaboration between development and operations teams, and accelerates time-to-deployment for critical IT infrastructure changes.

In the next sections, we'll explore how Amazon Q Developer can help generate OS code using the Unix CLI and shell scripting. To illustrate, I will use an overall chain-of-thought prompt technique to mimic a sequence of activities, breaking them into single-line and multi-line prompts (refer to *Chapter 3* for additional details). For the platform, I'll use the Amazon Q Developer interface with VS Code for shell scripts and macOS Terminal for the CLI. For configuration steps to integrate Amazon Q Developer with VS Code and the CLI, refer to *Chapter 2* and check the *References* section at the end of the chapter for the URLs for *Terminal shell integration*, and *Installing Amazon Q for the command line*.

> **Note**
>
> Amazon Q Developer uses **large language models** (**LLMs**), which, by nature, are non-deterministic, so you may not get exactly the same answers/code blocks shown in the code snapshots; try to update prompts to get desired recommendations. However, logically, the generated answer/code should meet the requirements.

Command-line auto-code generation for system administrators using Amazon Q Developer

I will assume the role of a system administrator who is familiar with responsibilities such as disk space management, user management, process management, and library management, but requires assistance with writing commands with the correct syntax. Using macOS, I will demonstrate how Amazon Q Developer interacts with the CLI. This example showcases how Amazon Q Developer can simplify and automate complex tasks, even for those with limited command-line experience.

By following these steps, you will be able to leverage Amazon Q Developer's powerful features to streamline your command-line tasks and enhance your overall productivity as a system administrator.

To start the interaction with Amazon Q Developer, open the terminal and type q ai:

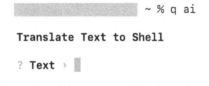

Figure 8.1 – CLI – Amazon Q Developer basics

Once the integration window is open, let's ask Amazon Q Developer to provide the commands related to common tasks of a system administrator in the following sections. These tasks encompass various areas, including disk space management, user management, library management, process management, network configuration, and system monitoring. By leveraging Amazon Q Developer, system administrators can receive detailed, step-by-step commands for a wide range of tasks, ensuring efficient and accurate execution. The CLI of Amazon Q Developer allows users to execute commands without explicitly typing them in the Terminal prompt.

Disk space management CLI auto-code generation

Disk space management is one of the key responsibilities of system administrators. Effective space management helps maintain system stability, performance, and security, making it a crucial task for system administrators. Let's start with some of the disk space management-related tasks. Let's use a chain of thoughts to mimic common tasks and break them into simple single-line prompts.

Let's start with finding out the available disk space:

```
Prompt: command to find out the disk space available
```

```
[  Text · command to find out the disk space available
   Shell   df -h

>  ⚡ Execute command
   Edit command
   Regenerate answer
?  Ask another question
✖  Cancel
```

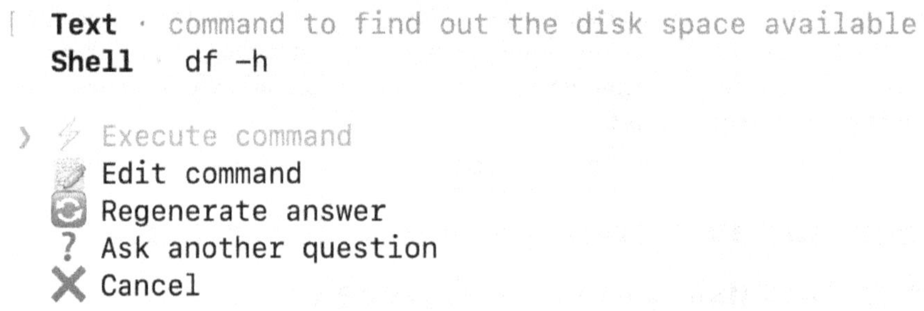

Figure 8.2 – CLI – Amazon Q Developer available disk space

Now, let's determine the utilized disk space and sort the results from highest to lowest based on the amount of space occupied on the disk:

```
Prompt: command to find out the disk space utilized sort by high to
low
```

```
[  Text · command to find out the disk space utilized sort by high to low
   Shell   du -h | sort -nr

>  ⚡ Execute command
   Edit command
   Regenerate answer
?  Ask another question
✖  Cancel
```

Figure 8.3 – CLI – Amazon Q Developer disk usage

Now, let's find the sizes of files and sort them from highest to lowest size, so system administrators can identify which files contribute most to disk space utilization:

```
Prompt: command to find out the largest file on the disk
```

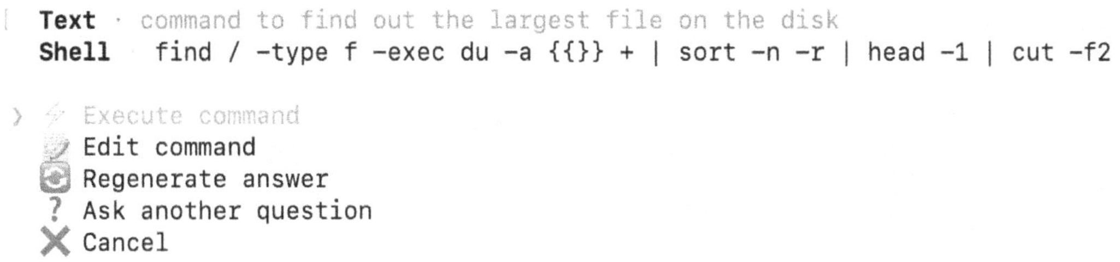

```
Text · command to find out the largest file on the disk
Shell   find / -type f -exec du -a {{}} + | sort -n -r | head -1 | cut -f2

> ⚡ Execute command
  Edit command
  Regenerate answer
? Ask another question
✖ Cancel
```

Figure 8.4 – CLI – Amazon Q Developer large file

Observe in the preceding screenshot that Amazon Q Developer provided a single command by chaining multiple commands to achieve the instruction in the prompt.

Now, let's free up some space by tasking Amazon Q Developer with finding the top 10 largest files and archiving them. This may involve a slightly complex process:

```
Prompt: command to find out top 10 largest file on the disk and
archive them
```

```
Text · command to find out top 10 largest file on the disk and archive them
Shell   find . -type f -exec du -Sh {} + | sort -rh | head -10 | xargs -I{} tar -zcvf {}.tar.gz {}

> ⚡ Execute command
  Edit command
  Regenerate answer
? Ask another question
✖ Cancel
```

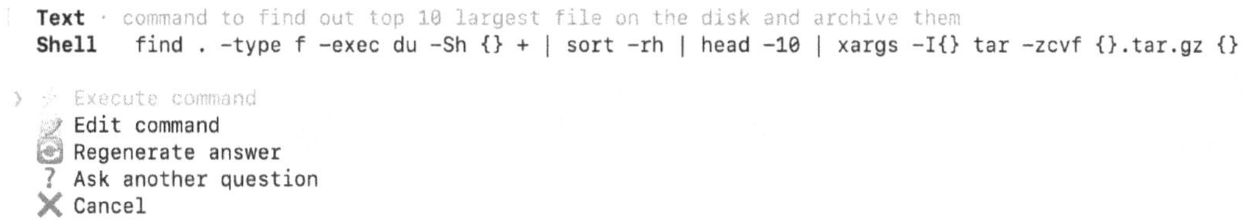

Figure 8.5 – CLI – Amazon Q Developer archive files

As you can see in the preceding screenshot, Amazon Q Developer provided a single command by chaining multiple commands to achieve the instruction in the prompt. It also used the number of files we specified and passed only those files to archive as `tar.gz`.

Now, let's look at user management, another very important area of responsibility for a system administrator.

User management CLI auto-code generation

User management for system administrators involves tasks related to creating, configuring, maintaining, and securing user accounts on a computer system or network. Effective user management is crucial for maintaining system security, ensuring data integrity, and supporting efficient operations within an organization. Let's use a chain of thoughts to mimic common tasks and break them into simple single-line prompts.

Let's begin with basic user management tasks, such as identifying a list of all users who have access to the system:

```
Prompt: command to print list of all the user logins

         | Text · command to print list of all the user logins
           Shell   ls /Users

         ⟩  ⚡ Execute command
            📝 Edit command
            🔄 Regenerate answer
            ?  Ask another question
            ✖ Cancel
```

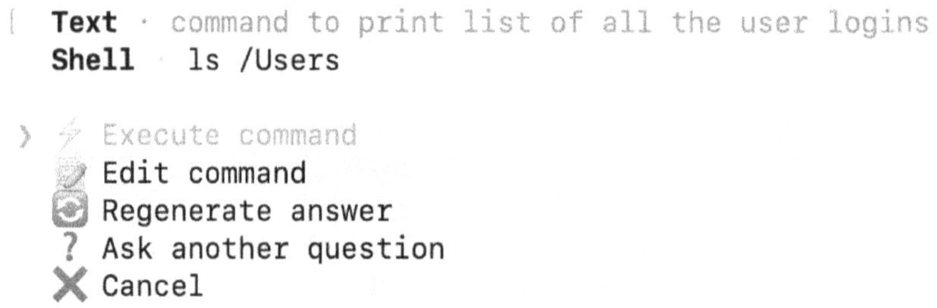

Figure 8.6 – CLI – Amazon Q Developer user list

Now, let's identify active users on the system and determine when they last logged in:

```
Prompt: command to print user list with last login date and time

| Text · command to print user list with last login date and time
  Shell   last | cut -d " " -f1 | sort | uniq | xargs -I{} bash -c 'echo -n "{}: "; last -F | grep "{}"'

⟩ ⚡ Execute command
   📝 Edit command
   🔄 Regenerate answer
   ? Ask another question
   ✖ Cancel
```

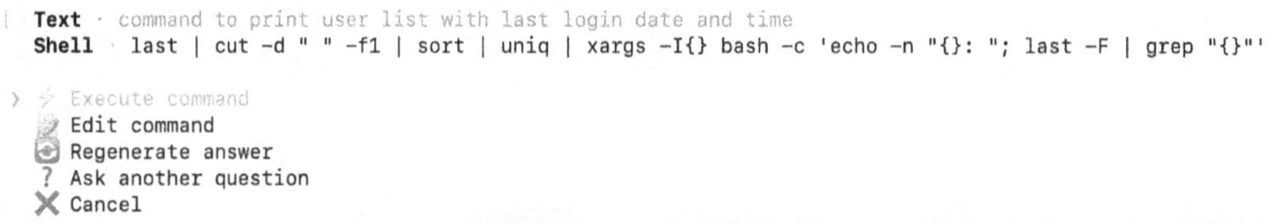

Figure 8.7 – CLI – Amazon Q Developer users' last login

Now, let's identify dormant users on the system who have not logged in recently:

```
Prompt: command to print user name those do not have last log-in date

     Text · command to print user name those do not have last log-in date
     Shell   last | grep never | awk '{print $1}' | sort

     ⟩  ⚡ Execute command
        📝 Edit command
        🔄 Regenerate answer
        ?  Ask another question
        ✖ Cancel
```

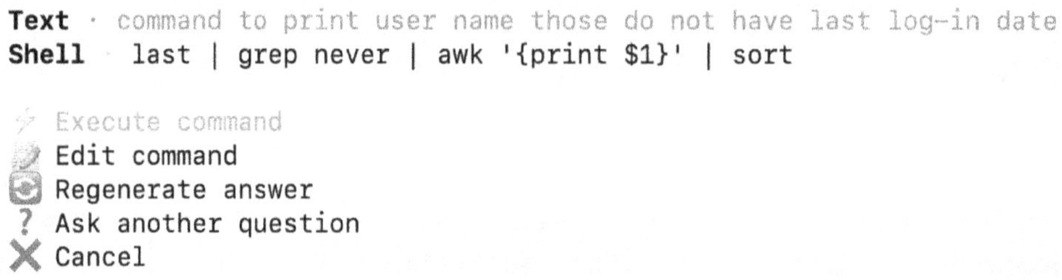

Figure 8.8 – CLI – Amazon Q Developer user never logged in

To enhance security and reduce vulnerabilities, let's remove dormant users from the system who do not have a recent login date associated with them. This process may be somewhat complex so let's see how Amazon Q Developer can help:

```
Prompt: command to drop user those do not have login date

    Text · command to drop user those do not have login date
    Shell    awk -F: '($3==""){print $1}' /etc/passwd | xargs -n1 userdel

>  ⚡ Execute command
   ✎ Edit command
   ⟳ Regenerate answer
   ? Ask another question
   ✗ Cancel
```

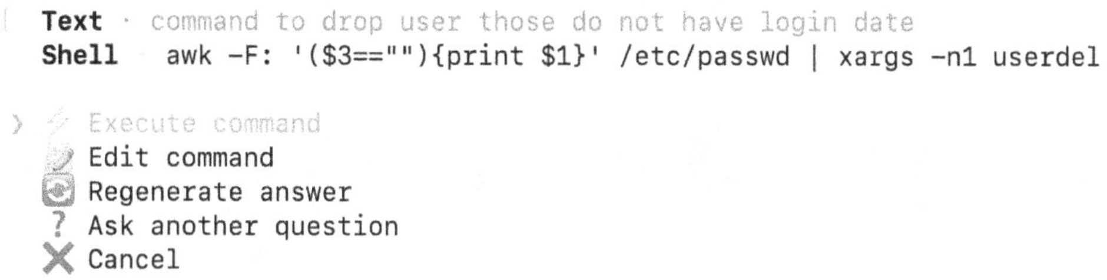

Figure 8.9 – CLI – Amazon Q Developer dropping users who never logged in

Observe in the preceding screenshots that as the complexity of requirements increases, Amazon Q Developer chains multiple commands to achieve the instruction in the prompt.

Now, let's look at some of the process management-related tasks and corresponding support by Amazon Q Developer.

Process management CLI auto-code generation

Process management is a critical responsibility for system administrators, involving tasks such as monitoring, controlling, and optimizing running processes on a computer system to ensure efficient resource utilization and performance. Maintaining the health of the servers is one of the important responsibilities of a system administrator.

Let's begin by asking Amazon Q Developer to generate commands for finding all active processes:

```
Prompt: command to list all the active processes

        Text · command to list all the active processes
        Shell   ps -ef

    >  ⚡ Execute command
       ✎ Edit command
       ⟳ Regenerate answer
       ? Ask another question
       ✗ Cancel
```

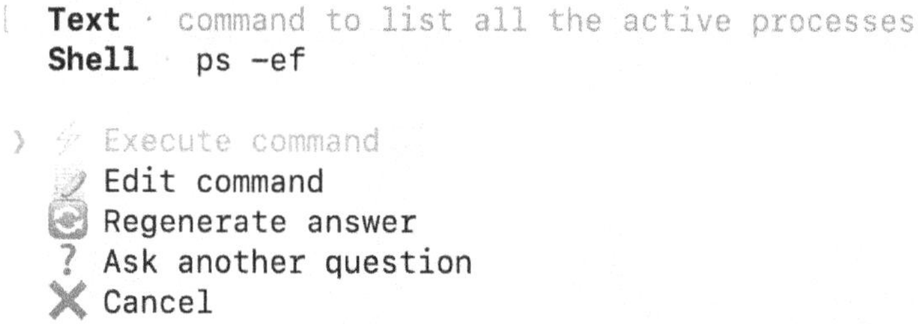

Figure 8.10 – CLI – Amazon Q Developer active processes

In process management, the CPU plays a vital role, so let's find the top 10 CPU-intensive processes:

```
Prompt: command to list top 10 CPU intense processes

    [ Text · command to list top 10 CPU intense processes
      Shell    ps -axch -o cmd,%cpu --sort=-%cpu | head -n 10

    )  ⚡ Execute command
       ✏ Edit command
       🔄 Regenerate answer
       ? Ask another question
       ✗ Cancel
```

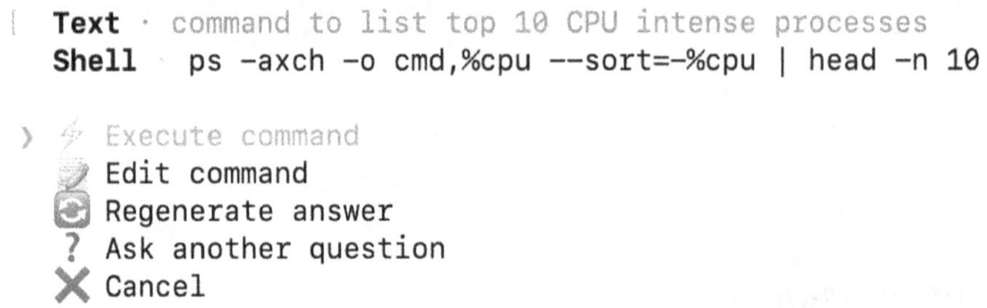

Figure 8.11 – CLI – Amazon Q Developer top CPU processes

Similar to the CPU, memory plays a crucial role in effective multiple-process execution across the system, so let's identify the top 10 memory-intensive processes:

```
Prompt: command to list top 10 memory intense processes

    [ Text · command to list top 10 memory intense processes
      Shell    ps -axo %mem,pid | sort -nr | head -10

    )  ⚡ Execute command
       ✏ Edit command
       🔄 Regenerate answer
       ? Ask another question
       ✗ Cancel
```

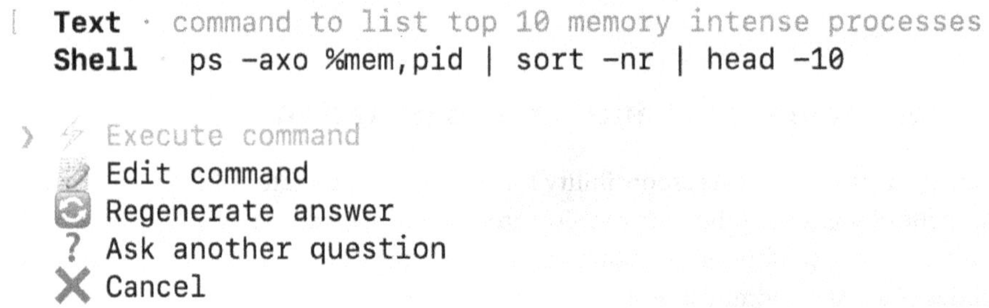

Figure 8.12 – CLI – Amazon Q Developer top memory processes

System administrators often need to terminate processes that are consuming high memory or CPU resources. To simulate this scenario, let's use Amazon Q Developer to create a command to kill the top two processes consuming the most memory:

```
Prompt: command to list kill top two memory intense processes
```

```
Text · command to kill top two memory intense processes
Shell   ps −ax −o %mem | sort −nr | head −2 | awk '{print $1}' | xargs kill −9
```

```
> ; Execute command
  Edit command
  Regenerate answer
? Ask another question
X Cancel
```

Figure 8.13 – CLI – Amazon Q Developer kill top memory processes

As you can see in the preceding screenshot, Amazon Q Developer efficiently generated the precise command required to free up memory by terminating the top two memory-intensive processes. This involved a slightly complex command where multiple commands were linked together end to end.

Now, let's look at one last area related to library management, which includes installing new libraries and upgrading existing ones.

Library management CLI auto-code generation

There are relatively fewer activities in library management compared to other sections, but many of these tasks are crucial for ensuring that all applications run smoothly on the server. Proper library management, including the installation of new libraries and the upgrading of existing ones, is essential for maintaining the functionality and performance of applications. These tasks, while less frequent, play a significant role in preventing software conflicts, enhancing security, and ensuring that applications have access to the latest features and optimizations. Therefore, even though library management activities might seem less numerous, their impact on the overall health and efficiency of the server environment is substantial.

Let's start with the basics to get the list of all the libraries and save them in a file:

```
Prompt: command to list all the installed libraries and their version
and save them in a file
```

```
Text · command to list all the installed libraries and their version and save them in a file
Shell   ldd --version > ldd_version.txt
```

```
> ; Execute command
  Edit command
  Regenerate answer
? Ask another question
X Cancel
```

Figure 8.14 – CLI – Amazon Q Developer list of all libraries

It's standard practice for system administrators to verify the version of an installed library. For example, let's utilize Python and request Amazon Q Developer to generate the command that identifies the associated version:

```
Prompt: command to find python version
```

```
[ Text  ·  command to find python version
  Shell    python --version

>  ⚡ Execute command
   📝 Edit command
   🔄 Regenerate answer
   ?  Ask another question
   ✖  Cancel
```

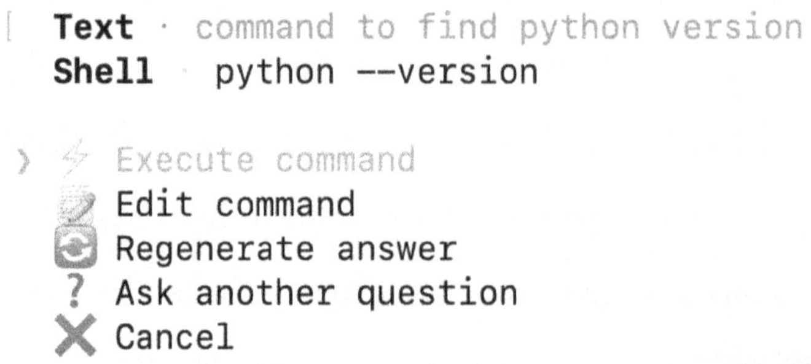

Figure 8.15 – CLI – Amazon Q Developer finding the version

Similarly, system administrators often need to upgrade installed libraries. For example, let's use Python and ask Amazon Q Developer to generate the command for upgrading the Python version:

```
Prompt: command to upgrade python version
```

Translate Text to Shell

```
[ Text  ·  command to upgrade python version
  Shell    python3 -m pip install --upgrade pip

>  ⚡ Execute command
   📝 Edit command
   🔄 Regenerate answer
   ?  Ask another question
   ✖  Cancel
```

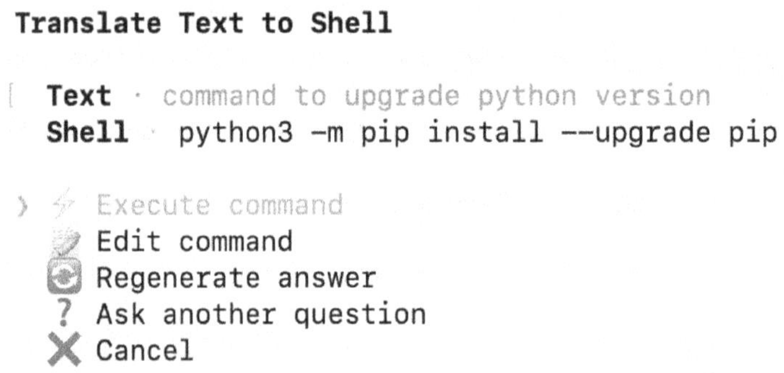

Figure 8.16 – CLI – Amazon Q Developer upgrading the version

I have illustrated a few commonly used CLI commands that system administrators use, demonstrating Amazon Q Developer's capability to provide commands ranging from single commands to chaining multiple ones to achieve the requirements. Additionally, try using Amazon Q Developer for other user personas who could benefit from generating CLI commands using simple single-line prompts.

In the next section, we'll explore how Amazon Q Developer can help programmers write shell scripts.

Shell script auto-code generation for programmers using Amazon Q Developer

Continuing with the theme of previous chapters, I am going to start with the persona of a programmer or an engineer who wishes to learn and develop code using shell scripts.

Shell script basics

Let's use Amazon Q Developer's chat-style interface to educate programmers, where they can simply ask questions in natural language to learn about the available commands and functions. I'll use chat-style interaction and single-line and multi-line prompt techniques:

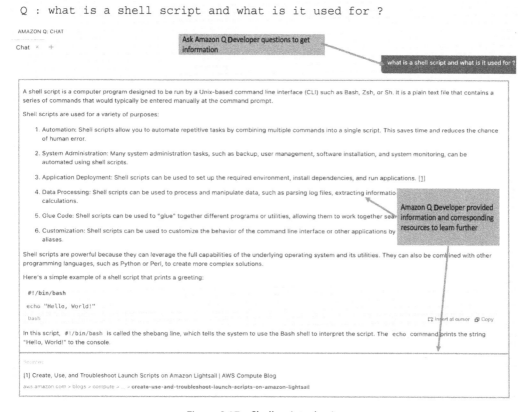

Figure 8.17 – Shell script – basics

As you can see in the preceding screenshot, for learners, Amazon Q Developer pointed out important areas in which shell scripts are effective, such as automation, system administration, application deployment, and more. It also provided a `"Hello, World !"` example of the shell script with a declaration of the bash shell. In addition, it provided sources that can be used for additional information.

Now, let's get more information and ask about commonly used shell script commands and corresponding syntax with examples:

```
Q : what are the  available bash shell script commands ?
```

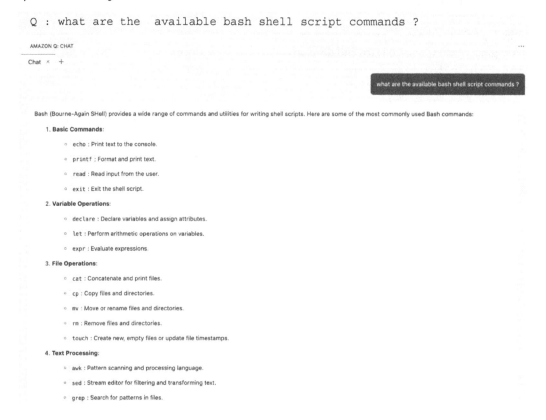

Figure 8.18 – Shell – list of commands, partial screenshot

As you can see in the preceding partial screenshot, Amazon Q Developer offers a catalog of shell script commands categorized into file and directory commands, text processing commands, system information commands, process management commands, network commands, and user and permission commands. Additionally, Bash supports control structures, functions, and other features, making it a robust scripting language for automating tasks and system administration on Unix-like operating systems.

Let's now request Amazon Q Developer to provide the syntax and an example for the "cat" command:

```
Q : provide details and example for shell command "cat"
```

AMAZON Q: CHAT

Chat × +

provide deatils and example for shell command "cat"

The `cat` command in Bash is used to concatenate and display the contents of one or more files. It stands for "concatenate" and is one of the most commonly used commands in Unix-based operating systems.

Syntax:

```
cat [OPTIONS] [FILE(S)]
```

Insert at cursor Copy

Options:

- `-n` or `--number` : Displays line numbers along with the contents of the file(s).
- `-b` or `--number-nonblank` : Displays line numbers for non-blank lines.
- `-s` or `--squeeze-blank` : Suppresses repeated empty output lines.
- `-v` or `--show-nonprinting` : Displays non-printing characters using the `^` and `M-` notation.
- `-E` or `--show-ends` : Displays a `$` character at the end of each line.
- `-T` or `--show-tabs` : Displays tab characters as `^I` .

Examples:

1. **Display the contents of a file:**

```
cat file.txt
```
bash

Insert at cursor Copy

This command will display the entire contents of the `file.txt` file in the terminal.

2. **Concatenate multiple files:**

```
cat file1.txt file2.txt file3.txt
```
bash

Insert at cursor Copy

This command will concatenate the contents of `file1.txt` , `file2.txt` , and `file3.txt` and display them in the terminal.

Figure 8.19 – Shell – the "cat" command

As you can see in the preceding screenshots, Amazon Q Developer provided basic usage of the command. Then, it provided the syntax, cat [OPTIONS] [FILE(S)], and different options such as -n, -b, -s, -v, -E, and -T with examples.

Shell script generation for programmers

Now that we understand the basics of shell scripting, let's use the preceding information for a use case.

Here is the use case: Create a shell script to accept a filename as a parameter. Check that the file is available and readable. If it is, then copy the file in the /user/input/files directory and start the initial_data_load.sh script. Then, check for successful completion of the process.

Let's use the chain-of-thought prompt technique to break this use case down into smaller tasks that, when combined, will achieve the end-to-end scenario:

1. Create a shell script to accept a filename as a parameter.
2. Check that the file is available and readable, then copy the file in the /user/input/files directory.
3. Assume that the programmer has already created a script called initial_data_load.sh. Let's start the initial_data_load.sh script and check for its successful completion.

To illustrate, I am going to use the VS Code script editor window with single and multi-line prompts.

Requirement 1

This is how we handle the first requirement and create a shell script to accept a filename as a parameter:

```
Prompt 1 :
/* Create a shell script */

Prompt 2 :
/* get file name from as parameter */
```

```
1    /* Create a shell script */
2    #!/bin/bash
3
4
5    /* get file name from as parameter */
6    echo "File name: $1"
7
```

Figure 8.20 – Shell – input parameters

Observe that Amazon Q Developer initialized the shell and accepted a parameter. Now, let's handle the next requirements.

Requirements 2 and 3

heck that the file is available and readable and then copy the file in the `/user/input/files` directory. We'll then start the `initial_data_load.sh` script and check for successful completion:

```
Prompt 1 :
/* check if a file exists and readable in the current directory then
copy in "/user/input/files/" directory */


Prompt 2 :
/* run "/user/scripts/initial_data_load.sh"  and check it's status  */
    8    /* check if a file exists and readable in the current directory
    9    then copy in "/user/input/files/" directory
   10    */
   11
   12    if [ -f "$1" ] && [ -r "$1" ]; then
   13        cp "$1" /user/input/files/
   14        echo "File $1 copied to target directory."
   15
   16
   17    /*  run "/user/scripts/initial_data_load.sh"  and check it's status */
   18    sh initial_data_load.sh
   19    if [ $? -eq 0 ]; then
   20        echo "Data load successful."
   21        exit 0
```

Figure 8.21 – Shell – copy file and run script

As illustrated, you can utilize multiple single-line prompts to auto-generate code with the help of Amazon Q Developer. Since this is a relatively straightforward use case, feel free to try the single and multi-line prompt techniques to observe the final code generated by Amazon Q Developer.

Shell script explainability and optimization using Amazon Q Developer

Amazon Q Developer offers a simple interface for achieving code explainability and optimization. For more details on explainability and optimization, please refer to *Chapter 12*.

To illustrate, I am going to use the previously auto-generated code during the shell script creation task. As shown in the following screenshot, highlight the code, right-click to open the pop-up menu, select **Amazon Q**, and choose **Explain** for code explainability or **Optimize** for optimization recommendations:

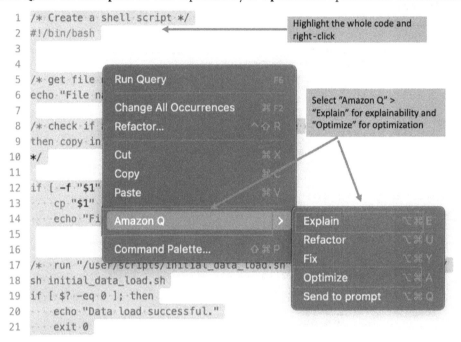

Figure 8.22 – Shell – Amazon Q explainability and optimization

This will pop up Amazon Q Developer's chat-style interface and move the full code for analysis.

For explainability, Amazon Q Developer provided details for each line of code and, at the end, provided a summary: "**In summary, this shell script takes a file name as an argument, checks if the file exists and is readable in the current directory, copies the file to a specific location (/user/input/files/), runs another script (initial_data_load.sh), and prints a success or failure message based on the exit status of the second script**".

Since we constructed this script using multiple single-line prompts, for optimization, Amazon Q Developer thoroughly analyzed the end-to-end script. It provided several recommendations to optimize and enhance the script's readability, maintainability, and error handling. These improvements were made without altering its core functionality. Recommendations spanned variable assignment, conditional optimization, error handling, path assignment, command substitution, error redirection, exit status, and code formatting. Additionally, Amazon Q Developer generated an updated end-to-end shell script incorporating these best practices, which you can add directly by clicking the **Insert at cursor** button.

In addition to command-line and shell scripts, Amazon Q Developer supports multiple other programming languages. There are many enhancements happening in this area, and we anticipate further enhancements to support additional languages (check the *References* section at the end of the chapter).

Summary

As illustrated, Amazon Q Developer serves as a valuable assistant for system administrators, programmers, and engineers. Amazon Q Developer can be used with a CLI to generate complex commands just by using simple prompts. The output commands can be simple single commands or a combination of a chain of commands to achieve administrative tasks such as disk space management, user management, process management, library management, and more.

Shell script programmers and engineers can use chain-of-thought prompts, single-line prompts, and multi-line prompt techniques to auto-generate end-to-end error-free scripts. Moreover, they can use the **Explain** and **Optimize** options to easily get the code details and customized code recommendations. Using these options encourage adherence to shell scripting best practices and command-line syntaxes, promoting code quality and standardization. The tool also verifies script syntax and command-line options and recommends secure scripting practices to mitigate potential vulnerabilities. This ensures that shell scripts and command-line integrations produced by system administrators and DevOps engineers are robust, secure, and compliant with organizational standards and regulatory requirements.

In the next chapter, we will look at how you can use Amazon Q Developer to suggest code in command-line and shell scripts.

References

- Installing Amazon Q for the command line: `https://docs.aws.amazon.com/amazonq/latest/qdeveloper-ug/command-line-getting-started-installing.html`

- Terminal shell Integration: `https://code.visualstudio.com/docs/terminal/shell-integration`

- Supported languages for Amazon Q Developer in the IDE: `https://docs.aws.amazon.com/amazonq/latest/qdeveloper-ug/q-language-ide-support.html`

Boost Coding Efficiency for JSON, YAML, and HCL with Auto-Code Generation

In this chapter, we will look at the following key topics:

- An overview of **Infrastructure as Code (IaC)** languages
- JSON and YAML auto-code generation for AWS CloudFormation using Amazon Q Developer
- HCL auto-code generation for Terraform using Amazon Q Developer
- JSON, YAML, and HCL code explainability and optimization using Amazon Q Developer

In the previous chapter, we explored the **command-line interface** (**CLI**) and shell scripting using two user personas: system administrators and shell script programmers. We introduced how auto-code generation through Amazon Q Developer can assist these professionals. Using the Amazon Q Developer CLI, we demonstrated how to achieve various administrative tasks such as disk space management, user management, process management, library management, and more. Additionally, we generated an end-to-end shell script utilizing chain-of-thought prompts, single-line prompts, and multi-line prompt techniques.

In this chapter, we will explore different categories of programming languages associated with IaC. In IaC, the programming languages are tightly coupled with the tools or services used. Firstly, we will focus on JSON and YAML, which are used by the AWS CloudFormation service. Then, we will explore HCL, which is used by Terraform. We will introduce how auto-code generation through Amazon Q Developer can help understand the basics, learn the syntax, and automatically generate code using JSON, YAML, and HCL. Then, we will explore code explainability to support documentation and code optimization recommendations provided by Amazon Q Developer for IaC programmers.

Overview of IaC languages

In today's cloud computing landscape, IaC has emerged as a fundamental paradigm shift in managing and provisioning infrastructure. AWS CloudFormation and Terraform are two leading IaC tools that enable developers and system administrators to define and manage cloud resources programmatically, treating infrastructure as software.

AWS CloudFormation simplifies the provisioning and management of AWS resources using declarative JSON or YAML templates. These templates define the configuration and interdependencies of AWS services, enabling users to provision multiple resources predictably and repeatably. CloudFormation automates the creation, updating, and deletion of infrastructure stacks, ensuring consistency and reducing the risk of manual errors in resource provisioning.

Terraform, developed by HashiCorp, takes a broader approach by supporting multiple cloud providers (including AWS, Azure, and Google Cloud Platform) and on-premises infrastructure. It uses a declarative language called **HashiCorp Configuration Language** (HCL) to define infrastructure as code. Terraform's state management capabilities allow it to plan and execute changes to infrastructure with minimal downtime and manage complex dependencies between resources across different providers.

Both AWS CloudFormation and Terraform play critical roles in enabling organizations to achieve IaC principles, including versioning, automation, and scalability. They facilitate collaborative development practices, integrate with CI/CD pipelines, and enable the management of infrastructure configurations as code repositories. IaC developers can leverage various prompt techniques and chat-style integration with Amazon Q Developer to enhance AWS CloudFormation and Terraform capabilities in several key areas. Amazon Q Developer automates the generation of AWS CloudFormation templates and Terraform configurations based on specific infrastructure requirements. Using natural language processing and chat-style interactions, developers can specify infrastructure needs and receive generated code that adheres to best practices and architectural guidelines. Beyond basic template generation, Amazon Q Developer offers optimization suggestions and enforces best practices for AWS CloudFormation and Terraform configurations. It analyzes generated code, identifies potential performance bottlenecks or security vulnerabilities, and provides recommendations to optimize infrastructure deployments.

Moreover, Amazon Q Developer serves as an educational resource for learning IaC principles and mastering AWS CloudFormation and Terraform. It offers interactive tutorials, explanations of IaC concepts, and practical examples through its intuitive interface. This educational aspect helps developers and system administrators improve their proficiency in efficiently and securely managing cloud resources. Additionally, by integrating Amazon Q Developer with AWS CloudFormation and Terraform, teams benefit from standardized templates, shared infrastructure configurations, and streamlined code review processes. This integration enhances collaboration between development, operations, and security teams, ensuring that infrastructure changes are implemented consistently and securely. As cloud environments evolve, Amazon Q Developer continuously updates its recommendations and capabilities to align with the latest best practices and architectural patterns in AWS CloudFormation and Terraform. This ensures that organizations can leverage cutting-edge infrastructure management techniques and maintain operational excellence in cloud deployments.

Let's dive deep to see how Amazon Q Developer can help an IaC developer. To illustrate, I am going to use the Amazon Q Developer interface with VS Code (refer to *Chapter 2* for configuration steps for Amazon Q Developer with VS Code). Additionally, for information on the Terraform Extension for Visual Studio Code, please refer to the *References* section at the end of the chapter.

> **Note**
>
> Amazon Q Developer uses LLMs, which, by nature, are non-deterministic, so you may not get exactly the same answers/code blocks shown in the code snapshots; try to update prompts to get desired recommendations. However, logically, the generated answer/code should meet the requirements.

JSON and YAML auto-code generation for AWS CloudFormation using Amazon Q Developer

Similar to the previous chapter, let's assume that a new IaC developer plans to work with JSON and YAML coding to build infrastructure on the AWS cloud using AWS CloudFormation but lacks background knowledge about this service. We will use Amazon Q Developer's chat-style interface to educate IaC developers. They can simply ask questions in natural language to learn about AWS CloudFormation and its support for JSON and YAML.

AWS CloudFormation basics

Let's first learn about AWS CloudFormation:

```
Q: What is AWS CloudFormation and what is it use for?
```

What is AWS CloudFormation and what is it use for?

AWS CloudFormation is a service provided by Amazon Web Services (AWS) that allows you to model and provision a collection of related AWS and third-party resources using templates. It enables you to define and manage your infrastructure as code, making it easier to create, update, and delete resources in a consistent and repeatable manner. [1]

The primary use of AWS CloudFormation is to simplify the process of setting up and managing AWS resources, such as Amazon EC2 instances, Amazon RDS databases, Amazon S3 buckets, and many others. Instead of manually provisioning and configuring resources one by one, you can define them in a CloudFormation template and let CloudFormation handle the deployment and management of those resources. [2]

Here are some key benefits and use cases of AWS CloudFormation:

1. **Infrastructure as Code (IaC)** : CloudFormation allows you to treat your infrastructure as code, which means you can version control, review, and audit your infrastructure definitions just like you would with application code. [3]

2. **Automated Provisioning** : With CloudFormation, you can automate the provisioning of AWS resources, ensuring consistent and repeatable deployments across different environments (e.g., development, staging, production). [4]

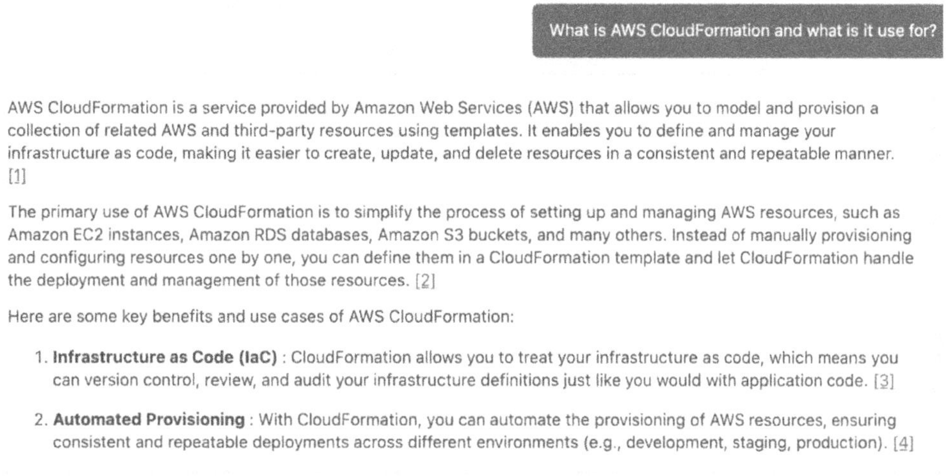

Figure 9.1 – Prompt – IaC AWS CloudFormation basic

Amazon Q Developer provided an overview of the AWS CloudFormation service: "**AWS CloudFormation is a service provided by Amazon Web Services (AWS) that allows you to model and provision a collection of related AWS and third-party resources using templates. It enables you to define and manage your infrastructure as code, making it easier to create, update, and delete resources in a consistent and repeatable manner**".

Additionally, Amazon Q Developer listed some AWS resources that can be provisioned using AWS CloudFormation. Then, it provided the key features of AWS CloudFormation, including IaC, automated provisioning, resource orchestration, reusability and modularity, change management, rollback, and deletion. Furthermore, it highlighted the use of JSON and YAML templates as common programming mechanisms.

So, now, let's ask Amazon Q Developer to provide more information about the AWS resources supported by AWS CloudFormation using JSON and YAML templates:

```
Q: Which AWS resources AWS CloudFormation support using JSON and YAML
templates?
```

Which AWS resources AWS CloudFormation support using JSON and YAML templates ?

AWS CloudFormation supports creating and managing a wide range of AWS resources using JSON or YAML templates. Here are some of the key AWS resource types that can be provisioned with CloudFormation templates: [1]

- Compute: EC2 instances, Auto Scaling groups, ECS/EKS clusters, Lambda functions

- Networking: VPCs, subnets, internet gateways, NAT gateways, route tables

- Storage: EBS volumes, EFS file systems, S3 buckets

- Databases: RDS instances, Aurora clusters, DynamoDB tables

- Analytics: Kinesis streams, EMR clusters, Athena queries

- Security: IAM roles, security groups, ACLs

- Application Services: ELB load balancers, API Gateways, SNS topics, SQS queues

- Management: CloudWatch alarms, CloudTrail trails, Config rules

CloudFormation supports creating resources across almost all AWS services. The templates define the desired state of the resources using JSON or YAML syntax. During stack creation or updates, CloudFormation provisions and configures the specified resources based on the template. [2]

Some higher-level AWS services like Amazon Location Service, AWS Transfer Family, AWS AppSync also support defining their resources via CloudFormation templates. This allows you to declaratively manage your entire AWS infrastructure using infrastructure as code principles.

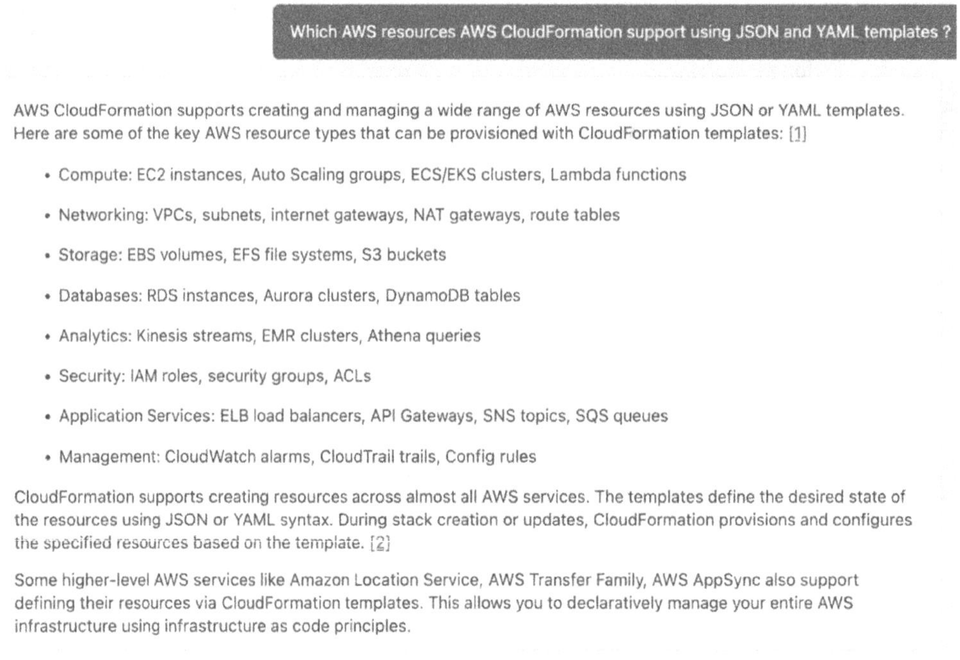

Figure 9.2 – Prompt – IaC AWS CloudFormation AWS resources

As you can see in the preceding screenshot, Amazon Q Developer gave a well-formatted answer with specifications related to categories of resources such as compute, networking, storage, databases, analytics, security, application services, and management, and sample resources from each of those categories that are supported by AWS CloudFormation using JSON and YAML.

JSON auto-code generation

Based on the previous information, an IaC developer now understands the support provided by AWS CloudFormation for provisioning resources using JSON. As the next step, let's begin writing a JSON template for AWS CloudFormation to create a simple AWS resource.

Let's ask Amazon Q Developer to create JSON code for a commonly requested AWS resource such as an S3 bucket from a storage resource and a networking component such as VPC. For customization, use the specific S3 bucket name, VPC name, and the CIDR block as 10.0.0.0/16. We will use prompts to do the following:

1. Create an AWS CloudFormation JSON template to create an S3 bucket named "MyTestBucket".

2. Create an AWS CloudFormation JSON template to create a "MyTestVPC" VPC with a CIDR block of 10.0.0.0/16.

To illustrate, I will use the VS Code script editor integrated with Amazon Q Developer. As these are relatively easy requirements, let's use single-line prompts for each of the preceding resources:

```
Prompt 1:
/* Write AWS CloudFormation JSON code to
create s3 bucket "MyTestBucket" */

Prompt 2:
/* Write AWS CloudFormation JSON code to
create Amazon VPC "MyTestVPC" with CIDR block "10.0.0.0/16" */
```

```
 1   /*
 2   Write AWS CloudFormation JSON code to
 3   create s3 bucket "MyTestBucket"
 4   */
 5
 6   {
 7     "Resources": {
 8       "MyTestBucket": {
 9         "Type": "AWS::S3::Bucket",
10         "Properties": {
11           "BucketName": "MyTestBucket"
12         }
13       }
14     }
15   }
16
17   /*
18   Write AWS CloudFormation JSON code to
19   create Amazon VPC "MyTestVPC" with CIDR block "10.0.0.0/16"
20   */
21   {
22     "Resources": {
23       "MyTestVPC": {
24         "Type": "AWS::EC2::VPC",
25         "Properties": {
26           "CidrBlock": "10.0.0.0/16"
27         }
28       }
29     }
30   }
```

Figure 9.3 – Prompt – IaC AWS CloudFormation JSON, S3 and VPC

As you can see in the preceding screenshot, Amazon Q Developer provided the correct end-to-end JSON template syntax for S3 and VPC. It used specific keywords supported by AWS CloudFormation to provision these resources, such as `"AWS::S3::Bucket"` and `"AWS::EC2::VPC"`, which combine the resource type and the corresponding parent service name. Additionally, it used the required properties to meet our requirements, such as `"BucketName"` set to `"MyTestBucket"` and `"CidrBlock"` with the value `"10.0.0.0/16"`.

YAML auto-code generation

YAML is another language supported by AWS CloudFormation for provisioning resources. As the next step, let's begin writing a YAML template for AWS CloudFormation to create a simple AWS resource. We'll use the same requirements for the Amazon S3 bucket and VPC creation as we did for the JSON template. Now, let's modify our prompts to instruct Amazon Q Developer to generate a YAML template.

Similar to the previous example, I will use the VS Code script editor integrated with Amazon Q Developer and utilize single-line prompts for each of these resources:

```
Prompt 1:
/* Write AWS CloudFormation YAML code to create s3 bucket
"MyTestBucket" */


Prompt 2:
/* Write AWS CloudFormation YAML code to create Amazon VPC "MyTestVPC"
with CIDR block "10.0.0.0/16" */
```

```
 1  /*
 2  Write AWS CloudFormation YAML code to
 3  create s3 bucket "MyTestBucket"
 4  */
 5  ---
 6  Resources:
 7    MyTestBucket:
 8      Type: AWS::S3::Bucket
 9      Properties:
10        BucketName: mytestbucket
11        AccessControl: PublicRead
12        AvailabilityZone: us-east-1a
13        BucketEncryption:
14          ServerSideEncryptionConfiguration:
15            - ServerSideEncryptionByDefault:
16                SSEAlgorithm: AES256
17                KMSMasterKeyID: alias/aws/s3
18
19  /*
20  Write AWS CloudFormation YAML code to
21  create Amazon VPC "MyTestVPC" with CIDR block "10.0.0.0/16"
22  */
23  ---
24  Resources:
25    MyTestVPC:
26      Type: AWS::EC2::VPC
27      Properties:
28        CidrBlock: 10.0.0.0/16
29        EnableDnsSupport: true
30        EnableDnsHostnames: true
31        InstanceTenancy: default
```

Figure 9.4 – Prompt – IaC AWS CloudFormation YAML, S3 and VPC

Notice in the preceding screenshot that Amazon Q Developer created an end-to-end YAML script for S3 and the VPC. It used specific keywords supported by AWS CloudFormation to provision these resources, such as AWS::S3::Bucket and AWS::EC2::VPC, which combine the resource type and the corresponding parent service name. Additionally, it used the required properties to meet our requirements, such as BucketName set to mytestbucket and CidrBlock with the value 10.0.0.0/16.

If you compare the YAML template screenshot (*Figure 9.4*) with the JSON template screenshot (*Figure 9.3*), you will notice that Amazon Q Developer included additional properties in the YAML version. Due to the non-deterministic nature of LLMs, Amazon Q Developer may add extra properties. However, it always includes the necessary properties to meet your requirements. If desired, you can adjust the prompts to add further requirements and receive recommendations with additional properties.

We will revisit the JSON template in the *JSON, YAML, and HCL code explainability and optimization using Amazon Q Developer* section to review additional recommendations.

HCL auto-code generation for Terraform using Amazon Q Developer

We'll follow the theme to start with a basic knowledge of the language in this section, too. I'll first assume the persona of a new IaC developer who is planning to work with HCL by Terraform and does not have any background with the tool.

Terraform basics

Let's use Amazon Q Developer's chat-style interface to educate IaC developers, where they can simply ask questions in natural language to learn about Terraform.

Let's first learn about Terraform from Amazon Q Developer:

```
Q: What is Terraform and what is it use for?
```

AMAZON Q: CHAT

Chat   ×   +

What is Terraform and what is it use for?

Terraform is an open-source infrastructure as code (IaC) tool created by HashiCorp. It is used for provisioning and managing cloud infrastructure resources across multiple cloud providers. [1]

The main use cases for Terraform are:

1. **Infrastructure Provisioning** : Terraform allows you to define your desired infrastructure state in configuration files using a declarative coding syntax. It can then provision that infrastructure across various cloud providers like AWS, Azure, GCP, as well as on-premises environments.

2. **Infrastructure as Code** : Terraform configurations are written in HashiCorp Configuration Language (HCL) or JSON, allowing infrastructure to be versioned and treated as code. This enables collaborative editing, code review, and automated testing of infrastructure definitions.

3. **Multi-Cloud Support** : Terraform supports a wide range of cloud providers and services, making it easy to provision and manage resources across multiple clouds using a consistent workflow and syntax.

4. **Resource Lifecycle Management** : Terraform can create, update, and delete resources based on the

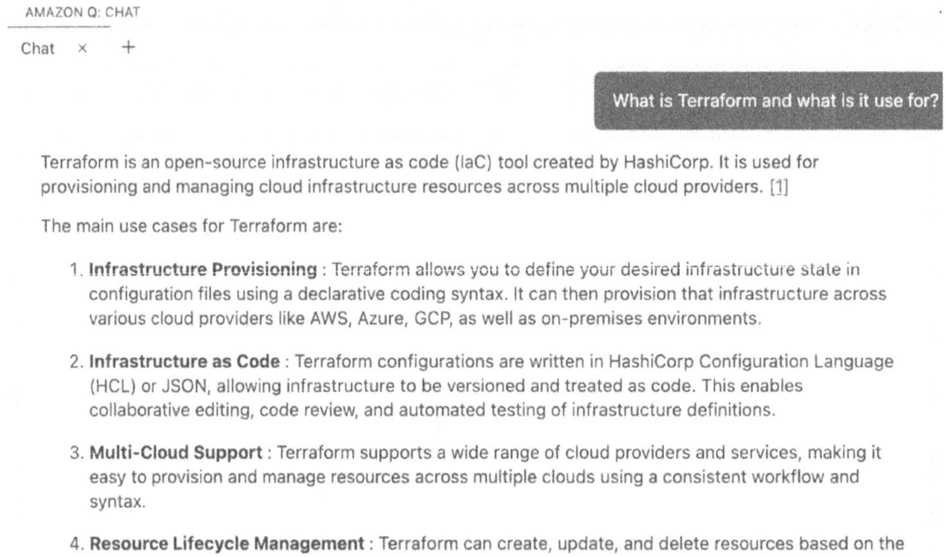

Figure 9.5 – Prompt – IaC Terraform basic partial screenshot

As you can see in the preceding partial screenshot, Amazon Q Developer provided basic information: **"Terraform is an open-source infrastructure as code (IaC) tool created by HashiCorp. It is used for provisioning and managing cloud infrastructure resources across multiple cloud providers".** Then, it listed the use cases and features supported by Terraform such as infrastructure provisioning, IaC, multi-cloud support, resource lifecycle management, state management, dependency handling, and more.

Now, based on the information from Amazon Q Developer's answer, as an IaC developer, we understand that it used HCL to create resources across multiple cloud providers and on-premises, including AWS resources such as EC2 instances, VPCs, RDS databases, and more.

Let's now ask Amazon Q Developer to list the different AWS resources that are supported by HCL:

```
Q: Which AWS resources are supported by Terraform HCL?
```

Which AWS resources are supported by Terraform HCL ?

Terraform supports a wide range of AWS resources through its AWS provider. The AWS provider allows you to manage resources across almost all AWS services using HashiCorp Configuration Language (HCL) or JSON configuration files. [1]

Here are some of the key AWS resource types supported by Terraform:

- **Compute** : EC2 instances, Auto Scaling groups, ECS/EKS clusters, Lambda functions, Batch jobs

- **Networking** : VPCs, subnets, internet gateways, NAT gateways, route tables, security groups, network ACLs

- **Storage** : EBS volumes, EFS file systems, S3 buckets, FSx file systems

- **Databases** : RDS instances, Aurora clusters, DynamoDB tables, ElastiCache clusters

- **Analytics** : Kinesis streams, Athena queries, EMR clusters, Glue jobs

- **Security and Identity** : IAM users, groups, roles, policies, security hub

- **Application Services** : ELB load balancers, API Gateways, SNS topics, SQS queues, AppSync

- **Monitoring and Logging** : CloudWatch logs, metrics, alarms, CloudTrail trails

- **Management and Governance** : CloudFormation stacks, Config rules, Systems Manager resources

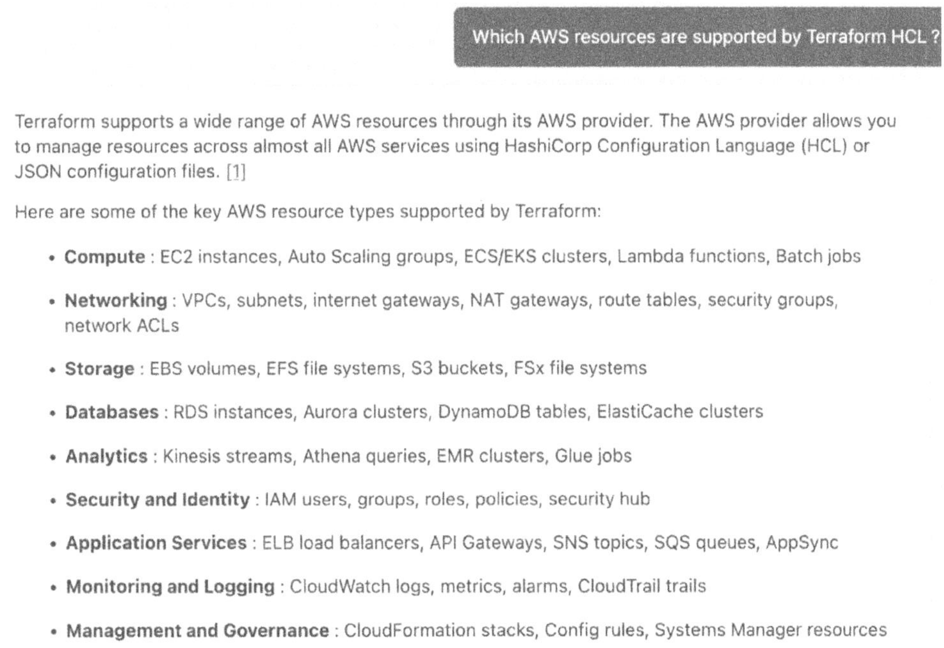

Figure 9.6 – Prompt – IaC Terraform AWS resources

Notice that Amazon Q Developer provided a categorized list including compute, networking, storage, databases, analytics, security, application services, and management, and sample resources within each category supported by Terraform HCL.

HCL auto-code generation

Based on the preceding basic information from Amazon Q Developer, HCL is the language supported by Terraform to provision multiple AWS cloud resources including Amazon S3 buckets and VPCs. As the next step, let's use the same requirements for the S3 bucket and VPC creation as we used in the previous section on AWS CloudFormation with JSON.

Now, let's modify our prompts to instruct Amazon Q Developer to generate an HCL template. Similar to the previous example, I will use the VS Code script editor integrated with Amazon Q Developer and utilize single-line prompts for each of these resources:

```
Prompt 1:
/* Write Terraform HCL to create s3 bucket "MyTestBucket" */


Prompt 2:
/* Write Terraform HCL code to create Amazon VPC "MyTestVPC" with CIDR
block "10.0.0.0/16" */
```

```
 2    /* Write Terraform HCL to
 3    create s3 bucket "MyTestBucket"
 4    */
 5    resource "aws_s3_bucket" "MyTestBucket" {
 6      bucket = "MyTestBucket"
 7      acl    = "public-read"
 8      region = "us-east-1"
 9      force_destroy = true
10      tags = {
11        Name = "MyTestBucket"
12      }
13    }
14
15
16    /* Write Terraform HCL code to
17    create Amazon VPC "MyTestVPC" with CIDR block "10.0.0.0/16"
18    */
19    resource "aws_vpc" "MyTestVPC" {
20      cidr_block           = "10.0.0.0/16"
21      enable_dns_support   = true
22      enable_dns_hostnames = true
23      tags = {
24        Name = "MyTestVPC"
25      }
26    }
27
```

Figure 9.7 – Prompt – IaC Terraform HCL, S3 and VPC

Observe that the HCL end-to-end code provided by Amazon Q Developer used Terraform modules and corresponding required properties. It used `resource "aws_s3_bucket"`, `resource "aws_vpc"`, `bucket = "MyTestBucket"`, and `cidr_block = "10.0.0.0/16"` to meet the exact specification provided in the prompt. Feel free to update prompts to get specific HCL code recommendations from Amazon Q Developer.

JSON, YAML, and HCL code explainability and optimization using Amazon Q Developer

Amazon Q Developer offers a simple interface for achieving code explainability and optimization and supports AWS CloudFormation scripts with JSON and YAML templates and Terraform HCL. For more details on explainability and optimization, please refer to *Chapter 12*.

To illustrate, I am going to use the code that was auto-generated during the AWS CloudFormation JSON creation task. As shown in the following screenshot, highlight the code, right-click to open the pop-up menu, select **Amazon Q**, and choose **Explain** for code explainability or **Optimize** for optimization recommendations.

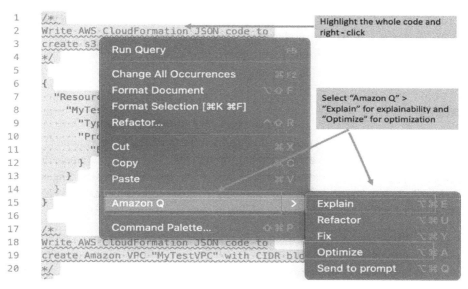

Figure 9.8 – JSON, YAML, and HCL – Amazon Q explainability and optimization

This will pop up Amazon Q Developer's chat-style interface and move the full code for analysis.

For explainability, Amazon Q Developer recognizes two distinct templates and segregates each block accordingly, offering detailed information that can be utilized for documentation and understanding resource specifications. For instance, Amazon Q Developer provided the following details for the S3 bucket section: "**This template defines a single resource named MyTestBucket of type AWS::S3::Bucket. The Properties section specifies the properties for the S3 bucket resource. In this case, it sets the BucketName property to "MyTestBucket", which will be the name of the S3 bucket created**".

For optimization, Amazon Q Developer recommended several additional properties to enhance the security, functionality, and organizational structure of the CloudFormation templates, while maintaining their core purpose of creating an S3 bucket and a VPC. Here are some key enhancements suggested: "**Enable S3 Bucket Encryption: Implement server-side encryption with AES256 for enhanced data security, Configure VPC DNS Support: Enable EnableDnsHostnames and EnableDnsSupport to ensure proper DNS resolution within the VPC, Set VPC Instance Tenancy: Set the tenancy to default, which is recommended for most use cases, and Implement Tagging: Add tags to resources for improved identification and organization purposes**".

Additionally, Amazon Q Developer generated an updated end-to-end script incorporating all the preceding changes, which you can add directly by clicking the **Insert at cursor** button.

In addition to JSON, YAML, and HCL for IaC, Amazon Q Developer supports **AWS Cloud Development Kit (CDK)** (TypeScript and Python), and multiple other programming languages. There are many enhancements happening in this area, and we anticipate further enhancements in support for additional languages (check the *References* section at the end of the chapter).

Summary

JSON, YAML, and HCL with AWS CloudFormation and Terraform, enhanced by Amazon Q Developer, revolutionize modern infrastructure management through IaC. These tools automate deployment workflows, ensuring consistency and adherence to best practices, while also providing educational resources to facilitate learning and adoption. This comprehensive approach empowers organizations to achieve enhanced agility, reliability, and scalability in their cloud environments.

As illustrated in this chapter, the integration of Amazon Q Developer with AWS CloudFormation and Terraform to generate JSON, YAML, and HCL code can significantly enhance the efficiency and innovation of infrastructure provisioning and management. By leveraging Amazon Q Developer's capabilities, teams can streamline the development of IaC templates, optimize resource utilization, and expedite deployment cycles. This not only reduces manual effort but also minimizes the risk of errors during infrastructure updates and scaling operations. Furthermore, Amazon Q Developer provides insights into advanced configurations and offers recommendations for optimizing infrastructure designs. This enables organizations to implement robust, scalable architectures that meet evolving business needs effectively.

Ultimately, the synergy between Amazon Q Developer, JSON, YAML, and HCL with AWS CloudFormation and Terraform empowers enterprises to achieve operational excellence in cloud operations, driving continuous improvement and innovation in their IT environments.

In the next chapter, we will look at how you can use the Amazon Q Developer Customizations feature to allow code suggestions that align with the team's internal libraries, proprietary algorithmic techniques, and enterprise code style.

References

- Terraform Extension for Visual Studio Code: `https://marketplace.visualstudio.com/items?itemName=HashiCorp.terraform`

- Supported languages for Amazon Q Developer in the IDE: `https://docs.aws.amazon.com/amazonq/latest/qdeveloper-ug/q-language-ide-support.html`

Part 3:
Advanced Assistant Features

In this part, we will explore some of the advanced features of Amazon Q Developer that enhance developer productivity by providing capabilities beyond the basic auto-code generation.

This part contains the following chapters:

- *Chapter 10, Customizing Code Recommendations*
- *Chapter 11, Understanding Code References*
- *Chapter 12, Simplifying Code Explanation, Optimization, Transformation, and Feature Development*
- *Chapter 13, Simplifying Scanning and Fixing Security Vulnerabilities in Code*

10

Customizing Code Recommendations

In this chapter, we will look at the following key topics:

- Prerequisites for Amazon Q customizations
- Creating customization in Amazon Q
- Evaluating and optimizing the customization
- Creating different versions of customization
- Adding users and groups
- Using customization in integrated development environments (IDEs)

With customizations, **Amazon Q Developer** can support software development tailored to your team's internal libraries, proprietary algorithms, and enterprise code style. By connecting a data source containing your code, Amazon Q leverages your content to provide assistance that aligns with your organization's development style.

Let's get started by looking at some of the prerequisites before you can leverage customization with Amazon Q.

Prerequisites for Amazon Q customizations

Since customization is mostly used by organizations, it is only available with the Pro tier of Amazon Q Developer. However, you can also try it out on your own by getting the Pro subscription and adding the customization feature from the Amazon Q administrative console.

When using Amazon Q customizations, ensure that your Amazon Q administrator is authorized to access your code base, which can be stored either on **Amazon S3** or through **AWS CodeConnections**. Notably, in the standard setup procedure for Amazon Q Developer Pro, your **AWS Organizations** administrator doesn't grant the Amazon Q administrator access to these services.

Hence, before utilizing Amazon Q customizations, it's imperative to include the following permissions to the role of your Amazon Q administrator. Note the legacy name of **CodeWhisperer** still shows in the policy, even though it pertains to Amazon Q Developer now:

```
{
    "Version": "2012-10-17",
    "Statement": [{
            "Effect": "Allow",
            "Action": [
                "sso-directory:DescribeUsers"
            ],
            "Resource": [
                "*"
            ]
        },
        {
            "Effect": "Allow",
            "Action": [
                "codewhisperer:CreateCustomization",
                "codewhisperer:DeleteCustomization",
                "codewhisperer:ListCustomizations",
                "codewhisperer:UpdateCustomization",
                "codewhisperer:GetCustomization",
                "codewhisperer:ListCustomizationPermissions",
                "codewhisperer:AssociateCustomizationPermission",
                "codewhisperer:DisassociateCustomizationPermission"
            ],
            "Resource": [
                "*"
            ]
        },
        {
            "Effect": "Allow",
            "Action": [
                "codeconnections:ListConnections",
                "codeconnections:ListOwners",
                "codeconnections:ListRepositories",
                "codeconnections:GetConnection"
            ],
            "Resource": [
                "*"
            ]
        },
        {
```

```
            "Effect": "Allow",
            "Action": "codeconnections:UseConnection",
            "Resource": "*",
            "Condition": {
                "ForAnyValue:StringEquals": {
                    "codeconnections:ProviderAction": [
                        "GitPull",
                        "ListRepositories",
                        "ListOwners"
                    ]
                }
            }
        },
        {
            "Effect": "Allow",
            "Action": [
                "s3:GetObject*",
                "s3:GetBucket*",
                "s3:ListBucket*"
            ],
            "Resource": [
                "*"
            ]
        }
    ]
}
```

The preceding policy grants the role the permissions to leverage the customization APIs of Amazon Q Developer. It also allows the role to establish connections and access the code repositories either via Amazon S3 or AWS CodeConnections. If you are planning to try out customizations, feel free to copy the **Identity and Access Management** (**IAM**) policy code from the AWS documentation, a link to which has been provided in the *References* section at the end of this chapter.

Amazon Q also retains data regarding the creation of your customization within **Amazon CloudWatch Logs**. Grant your Amazon Q administrator permission to access these logs with the following authorization set. The following permissions allow the Amazon Q Developer administrator to view these logs. We will see in the following sections how these logs are helpful for tracking as well as debugging purposes:

```
{
    "Version": "2012-10-17",
    "Statement": [
        {
            "Sid": "AllowLogDeliveryActions",
```

```
            "Effect": "Allow",
            "Action": [
                "logs:PutDeliverySource",
                "logs:GetDeliverySource",
                "logs:DeleteDeliverySource",
                "logs:DescribeDeliverySources",
                "logs:PutDeliveryDestination",
                "logs:GetDeliveryDestination",
                "logs:DeleteDeliveryDestination",
                "logs:DescribeDeliveryDestinations",
                "logs:CreateDelivery",
                "logs:GetDelivery",
                "logs:DeleteDelivery",
                "logs:DescribeDeliveries",
                "firehose:ListDeliveryStreams",
                "firehose:DescribeDeliveryStream",
                "s3:ListAllMyBuckets",
                "s3:ListBucket",
                "s3:GetBucketLocation"
            ],
            "Resource": [
                "arn:aws:logs:us-east-1:account number:log-group:*",
                "arn:aws:firehose:us-east-1:account
 number:deliverystream/*",
                "arn:aws:s3:::*"
            ]
        }
    ]
}
```

Ensuring the quality of your customization begins with selecting optimal source material. When preparing your data source, incorporate code that features team-endorsed patterns while avoiding code that contains anti-patterns, bugs, security vulnerabilities, performance issues, and similar concerns. Organizations establish best-of-breed coding standards and patterns by creating comprehensive guidelines and best practices that are thoroughly documented and easily accessible to all developers. They enforce these standards through code reviews and security standards in the development projects. All organizations can decide for themselves what project code repositories are gold standards for them so that they can use those for customizations in Amazon Q Developer.

Your data source should include between 2 MB and 20 GB of source code files in supported languages. The Amazon CloudWatch logs will indicate the total size of all the code bases used during the customization training process. In our next few sections, we will highlight this aspect in the logs.

There is no restriction on the number of files, but ensure each language you wish to support includes at least 10 files. If using Amazon S3 as the data source, organize all source code within a directory, avoiding placement at the root level, as files at the root level will be disregarded.

> **Supported languages**
>
> At this time, Amazon Q Developer supports customization for Java, Python, JavaScript, and TypeScript programming languages, and it can be used from the VS Code and JetBrains IDEs only. Keep an eye on new releases that may include more options.

Now, let's learn how to create the customization in Amazon Q Developer.

Creating a customization in Amazon Q

To get started with customizations, first set them up in the Amazon Q Developer console. We assume you have already followed the steps to set up your Pro tier.

The following screenshot shows the customization settings inside the Amazon Q Developer console. To access the Amazon Q Developer console, search for the service name within the AWS console and click the **Settings** button to reach the following screen.

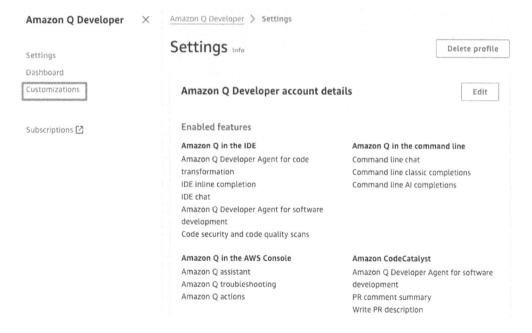

Figure 10.1 – Amazon Q customizations setup

After that, click on the **Customization** link and follow a simple three-step process inside the **Customizations** page: connecting to a repository, creating customization, and activating it. The following screenshot shows the process.

Amazon Q Developer > Customizations

Customizations Info

Receive code recommendations that match the style and rules of your codebase.

▼ **How it works**

Step 1: Connect to a repository
Connect to a third-party repository with AWS CodeStar Connections, or use an Amazon S3 Bucket.

Step 2: Create a customization
Amazon Q Developer will train a customization based on your codebase.

Step 3: Activate customization
Activate the customization and give access to your developers.

Customizations (0)   ⟳   Delete   **Create customization**
Choose an existing customization or create a new one.

| Q Find customizations by name or description | | ⟨ 1 ⟩ ⚙ |

| Name ▽ | Description ▽ | Status ▽ | Last updated ▼ |

No customization
No customization to display

Figure 10.2 – Amazon Q customizations – creation process

Let's look at these steps one by one.

Connecting to a repository

As soon as you click the **Create customization** button in the console, you will be presented with a page where you can provide the name and description for the customization along with the connection to the source repository, as shown in the following screenshot.

Amazon Q Developer > Customizations > **Create customization**

Create customization

Customization details

Customization name
The name that you give your customization should clearly describe the project, and should be easily identifiable by developers when they see it in the IDE.

> github-q-customization-connection

The customization name can have up to 100 characters. Valid characters: A-Z, a-z, 0-9, and - (hyphen)

Description - *optional*

The customization description can have up to 256 characters.

Connection to source provider   Info
Connect to a source provider to create a Amazon Q Developer customization with your data.

Source provider

● **AWS CodeStar Connections** Connect to Github, GitLab, or Bitbucket.	○ Amazon S3 Connect to S3 Bucket or folder (requires an S3 URI).

Select a connection
Choose an existing connection, or create a new connection.

> github-q-customization-for-book   ▼   ⟳   or   **Create new connection**

> ⓘ In order for code written in a particular language to be used to create a customization, there must be at least 10 files containing code in that language in your data source. The total size of all code files in your data source must be at least 20 MB. Valid file extensions are: .java, .py, .jsx .tsx, .js, .ts

Figure 10.3 – Amazon Q customizations – creating a connection

Always try to give a meaningful name and clear description here, as this information will be visible to the developers from the IDE.

The important aspect in the preceding screenshot is the source connection. There are two ways to make your source code available to Amazon Q for creating the customization. You can take all the source code in the organization, upload it into an Amazon S3 bucket, and provide the S3 URI in the source connection. Many organizations that store large amounts of enterprise code in code repositories as part of the DevOps process are less likely to use this option. The other option is indeed to connect to the code repositories where the code is hosted.

If your data source resides on GitHub, GitLab, or Bitbucket, you need to establish a connection to it using AWS CodeConnections. To demonstrate the customization feature in this chapter, I will connect to my GitHub repository, where I will be leveraging some public repositories that I will fork for this to work. For detailed steps on how to create a connection to your repository, I have provided the link in the *References* section at the end of this chapter so that you can go through it step by step to create a connection and put the connection in the Amazon Q customization creation page.

The following screenshot shows the bottom part of the same customization creation page, which allows you to emit the logs generated during this process to one of the log delivery options. In this case, I have picked AWS CloudWatch, and in the subsequent steps, I will show you the importance of logs during this process.

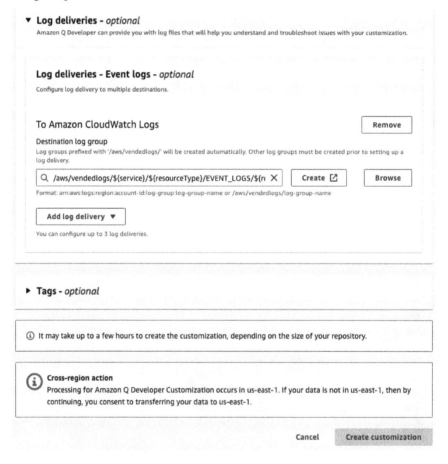

Figure 10.4 – Amazon Q customizations – log delivery option

Once you hit the **Create customization** button, it begins the training process. Depending on the total size of the code base, it may take anywhere from a few minutes to a few hours or even more.

Preparing customizations

In this step, Amazon Q will connect to your code repository and train a custom model based on your organization's coding practices so that it can utilize your content to provide suggestions tailored to the preferences of your organization's developers.

> **Security note**
>
> AWS will neither store nor utilize your content in any context that doesn't directly benefit your enterprise. Also, AWS won't leverage your content to offer code suggestions to other customers, and, of course, Amazon Q will not refer to security scans conducted for other customers.

During this process, Amazon Q retrieves the source code from the code repository, and after conducting sanity and quality checks, such as duplicate file checks and unsupported file formats, it attempts to establish whether it has enough context to create a customization that would benefit all the developers in the organization.

After running for a bit, the customization creation page failed and provided me with an error stating that the code size was not large enough for it to train an effective model. This issue is highlighted in the following screenshot.

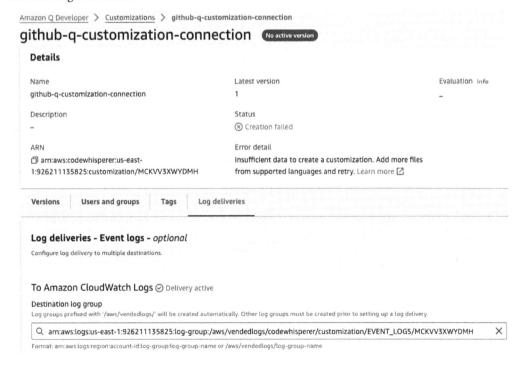

Figure 10.5 – Amazon Q customizations – creating a customization failure

The error doesn't exactly tell me what was insufficient and how it fell below the minimum threshold in our code base during this process. This is where the CloudWatch Logs will come in handy. For many possible troubleshooting error messages, you can follow the link in the *References* section, but I will cover one such error that I encountered during this process.

If you navigate to the CloudWatch log group link provided during the customization setup, as shown in the **Log deliveries** section in *Figure 10.4*, you will be able to see the exact cause of this error. For instance, in my case, the error log, as seen in the following screenshot, suggests that after the quality scrub process, Amazon Q could only gather about 1 MB of code. This limit is below the minimum threshold of 2 MB required by the engine to proceed toward the customization training process.

Figure 10.6 – Amazon Q customizations – failure details in AWS CloudWatch

The reason for this insufficient data error is that I had forked a popular code repository from the public code samples provided by AWS on GitHub, located at `https://github.com/aws-samples`.

In reality, as part of the organization, your administrator will connect the Amazon Q customization to your private enterprise repositories to allow Amazon Q to train customizations based on vast amounts of code used within your organization. To demonstrate the customization in this book, I cannot use any private code, so the best I can do is show this feature by connecting to a public repository with a license that allows anyone to use the code in any way they want.

Additionally, the public code found in the code repository may very well have been used to train Amazon Q Developer anyway, so technically, I will not be getting real customizations. However, to show you the steps of how it works, this should serve its purpose.

After receiving that failure message, I understood that I needed to provide a lot more code samples and possibly more variety for the training to work in the first place. So, I went ahead and forked a few more repositories in my GitHub account so that Amazon Q could make it past that error step. I reran the customization creation process and after a while, I looked at the CloudWatch logs again to see whether adding more code repositories helped it get past the minimum threshold.

The following screenshot shows that it barely made it past the 2 MB minimum threshold required.

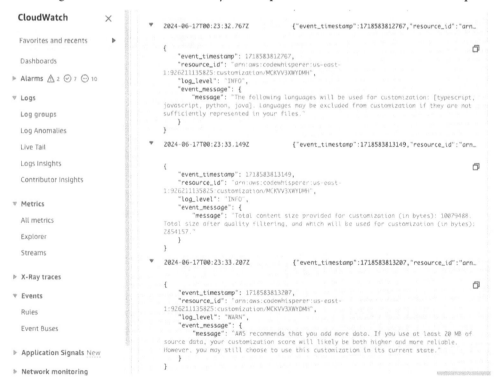

Figure 10.7 – Amazon Q customizations – AWS CloudWatch logs after modified re-run

So, even after adding 5 to 6 repositories, the code size was still just above 2 MB. The recommendation is to have at least 2 MB and this can go all the way up to 20 GB of code base. So, it's pretty evident that the customization process will complete, but I am not expecting a great evaluation from this.

The following screenshot shows the customization was completed successfully; however, it gave me an abysmal evaluation score of 1. In other words, it's telling me I better not roll out this customization feature to all the developers in my organization as it will yield poor results.

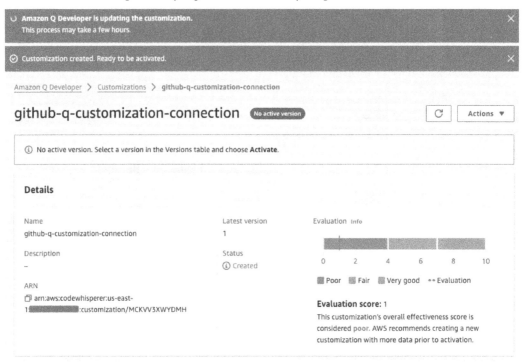

Figure 10.8 – Amazon Q customizations – customization creation is complete

We will discuss score evaluation and how to improve it in our next section, but let's complete the third step in this creation process, which is to activate the created customization.

Activating and deactivating customizations

Once the customization is created, it needs to be activated. Since this was our first version, we will activate it even though the evaluation is not great. However, in reality, you can keep iterating through the customization creation process until you get a good evaluation score, and then activate the version you want to use as the final one.

The following screenshot shows the **Activate** option once the customization creation is complete. Once activated, the status will show as activated. You can also deactivate any customization version by selecting the **Deactivate** option from the **Action** dropdown. This is useful for keeping only the version with the best score active for use.

Figure 10.9 – Amazon Q customizations – activate a version of the customization

Now, let's learn about the score evaluation process and how you can improve it.

Evaluating and optimizing the customization

When the customization process completed, it gave an evaluation score, and on the side, it also provided a detailed range of scores and a description of what they mean. Let's discuss them in detail.

Score evaluation

Based on your evaluation score, you should now decide whether to activate your customization. Consider the following factors:

- **Very good (8-10)**: Amazon Q recommends activating this customization.
- **Fair (5-7)**: Amazon Q recommends activating this customization. If you do not see a significant improvement, consider the following optimization suggestions. If those are not effective, consider switching to a different code source.
- **Poor (1-4)**: This customization is unlikely to be useful. Consider the optimization suggestions from the next section around optimizing the customization. If those are not effective, consider switching to a different code source.

This evaluation matrix is suggested in the AWS documentation, a link to which is also provided in the *References* section at the end of this chapter.

The following screenshot highlights this evaluation score for our version of the customization and on the side highlights what each range of score means.

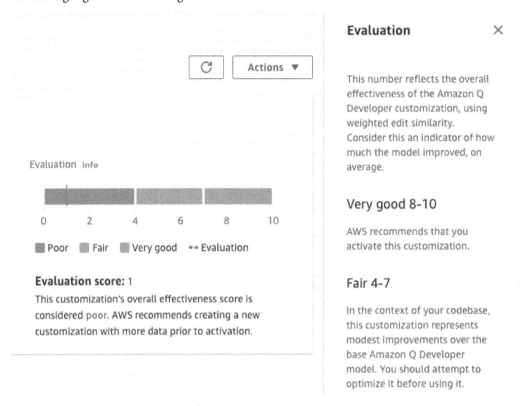

Figure 10.10 – Amazon Q customizations – score evaluation criteria

So, how do we go about bumping this score up? Let's take a look.

Optimizing the customization

Let's highlight some suggestions that may improve your evaluation score.

To tailor optimizations for the specific languages used in your organization, consider broadening your data source to encompass more code repositories. If your data set is limited to a few programming languages, try incorporating a wider variety of languages. Additionally, exclude auto-generated files and repositories or those created from templates, as training customizations for such files typically adds little value and introduces noise.

Assess whether your code base frequently employs internal libraries. If not, the core Amazon Q model may already be performing at its best. To fine-tune for particular languages, ensure you include at least 20 data files for each language, with a total size of at least 10 MB. Essentially, by increasing and improving the variety, quality, and quantity of code used for training customizations, the evaluation score may improve. A higher evaluation score will help generate better-customized code suggestions. Once you figure out how you are going to optimize the customization, you can create multiple versions of it.

Creating multiple versions of customization

As organizations create new code repositories and establish new coding standards, there may be a need to retain the Amazon Q Developer customizations to incorporate additional customizations and improve the evaluation score of an existing customization. This is where **versioning** of customizations comes in handy, allowing you to keep different versions of customizations based on training.

Creating multiple versions of customization is easy in Amazon Q Developer. From the created customization page where you see the current version, you can select **Create new version** from the **Actions** dropdown and start building a new version, as shown in the following screenshot.

Amazon Q Developer 〉 Customizations 〉 github-q-customization-connection 〉 Update customization

Create new version Info

Connection to source provider Info
Connect to a source provider to create a Amazon Q Developer customization with your data.

Source provider

⦿ AWS CodeStar Connections
Connect to Github, GitLab, or Bitbucket.

◯ Amazon S3
Connect to S3 Bucket or folder (requires an S3 URI).

Select a connection
Choose an existing connection, or create a new connection.

| github-q-customization-for-book ▼ | C | or | Create new connection |

ⓘ In order for code written in a particular language to be used to create a customization, there must be at least 10 files containing code in that language in your data source. The total size of all code files in your data source must be at least 20 MB. Valid file extensions are: .java, .py, .jsx. .tsx, .js, .ts

Cancel   **Create**

Figure 10.11 – Amazon Q customizations – creating a new version of customization

Amazon Q administrators can access up to three versions for each customization: the latest version, the currently active version in use, and the most recently active version that is no longer in use.

Once the customization version has an acceptable evaluation score, the admin can make it available to users or groups.

Adding users and groups to the customization

Adding users or groups is straightforward. This step will allow developers access to the customizations when they start coding in either the VS Code or JetBrains IDE, both of which would already have the Amazon Q extension installed.

The following screenshot shows the **Add Users/Groups** tab in the Amazon Q console.

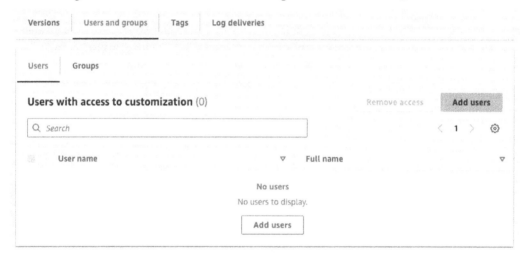

Figure 10.12 – Amazon Q customizations – add users/groups

In this case, as seen from the following screenshot, I'm adding myself so that I can start using the customizations in the VS Code IDE. Keep in mind this is the AWS IAM Identity Center user we configured using the steps listed in *Chapter 2*.

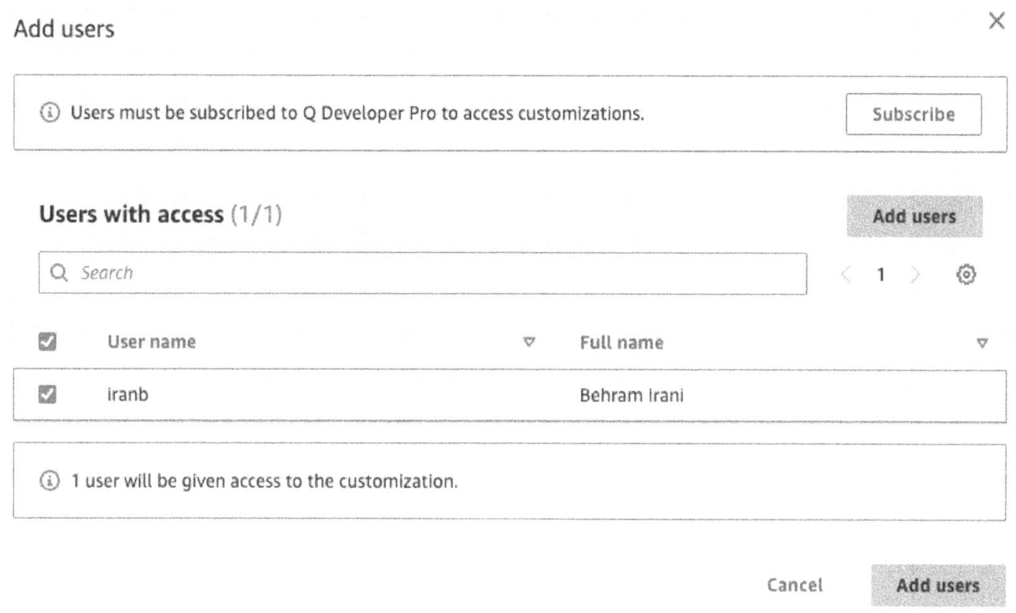

Add users   ✕

ⓘ Users must be subscribed to Q Developer Pro to access customizations. | Subscribe |

Users with access (1/1) | Add users |

🔍 Search   ‹ 1 › ⚙

☑	User name	▽	Full name	▽
☑	iranb		Behram Irani	

ⓘ 1 user will be given access to the customization.

Cancel | Add users |

Figure 10.13 – Amazon Q customizations – added a user

Now, we are all set to start using the customization from the IDE.

Using customization in IDEs

Once the admin gives the thumbs-up to start using a particular customization, it's straightforward to start using it from the IDE.

Just a reminder that customizations are only available in the Pro tier, which means that you have to log into the IDE using your IAM Identity Center credentials. The following screenshot highlights that as soon as I log in to the VS Code IDE using my IAM Identity Center credentials, it gives me a notification that I have access to a new Amazon Q customization.

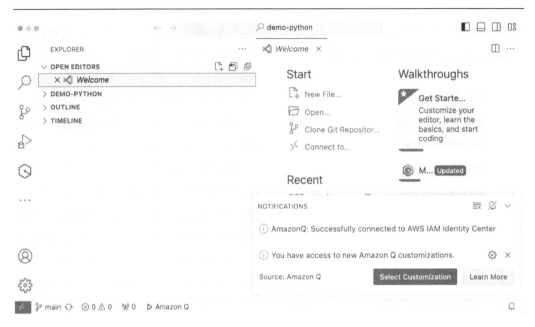

Figure 10.14 – Amazon Q customizations – new customization notification in VS Code IDE

As soon as I select it, I can see the name of the customization that was created earlier. The following screenshot highlights this aspect in VS Code IDE.

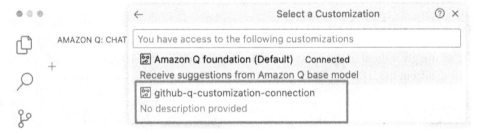

Figure 10.15 – Amazon Q customizations – selecting a customization in VS Code IDE

As soon as you select that, voilà, you are notified that all new Amazon Q code suggestions will be coming from the selected customization. This is highlighted in the following screenshot.

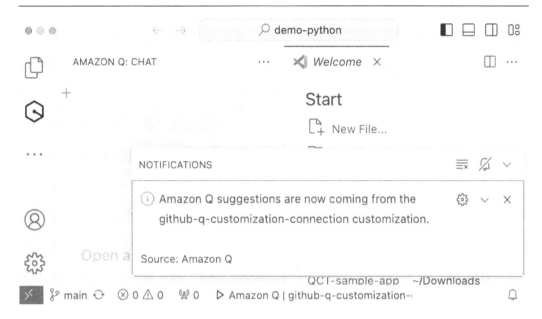

Figure 10.16 – Amazon Q customizations – customization selection confirmation in VS Code IDE

Now, keep in mind that Amazon Q will attempt to provide you with custom suggestions wherever it finds code logic that it would create from the custom training data. In case it's not able to relate to the customized model, Amazon Q will provide suggestions from the default model.

You can get custom code suggestions either through an inline prompt or by using the chat feature of Amazon Q Developer. For example, imagine you are working at a large e-commerce company with many customer-facing shopping applications, and as a developer for a new application, you have been tasked with building a feature that recommends items to customers. The logic for this feature may already exist in your organization. In your IDE, you can simply insert an inline prompt or use an appropriate function name such as `shoppingRecommendationEngine(customerId: String)`, and Amazon Q will try to provide the logic based on examples it has learned during the customization training process from your organization's private repository. All the complex logic in the function, which may have been established as a gold standard in your organization, is immediately utilized without you having to understand and create all the code by hand.

Amazon Q Developer can also answer your questions about your organization's custom code using the chat feature. You can simply ask the chat to generate or explain code based on the use case, and Q will try to infer from what it has learned from the organization's code base to provide the closest matching output. For example, you can ask the Amazon Q chat to generate a shopping recommendation engine implementation using the k-means clustering algorithm, and it will try to suggest code based on other similar examples it has learned from your organization's code base during the customization training process.

This brings us to the end of this chapter. Feel free to experiment with it if you have subscribed to the Pro tier or if your organization needs to set this up.

Summary

In this chapter, we covered what code customization is in Amazon Q Developer. We started by laying out the prerequisites for Amazon Q customizations. Then, we looked at how to create customization in Amazon Q. After the creation process, the scores need to be evaluated and optimization techniques need to be put in place so that it bumps up the score in subsequent runs of the customization creation process.

We also looked at how multiple versions of customizations are created and maintained. Once the customization is created, it's assigned to users or groups by the admin. Finally, the users can log into VS Code or JetBrains IDE using their IAM Identity Center credentials to select and start using the customizations.

In the next chapter, we will look at understanding the references of the code suggestion so that appropriate actions can be taken.

References

- Connection setup steps to code repositories: `https://docs.aws.amazon.com/dtconsole/latest/userguide/welcome-connections.html`

  Troubleshooting customization errors: `https://docs.aws.amazon.com/amazonq/latest/qdeveloper-ug/customizations-log-use-understand.html`

- Customization evaluations: `https://docs.aws.amazon.com/amazonq/latest/qdeveloper-ug/customizations-admin-activate.html`

11

Understanding
Code References

In this chapter, we will look at the following key topics:

- What are code references?
- Enabling, disabling, and opting out of code references
- Code reference example

This chapter will be a short one, but the topic deserves its own chapter so that it's not skipped over. So, let's get straight into it.

What are code references?

Development introduces innovative solutions; however, it also brings the responsibility of adhering to licensing requirements and ensuring proper attribution. Failing to comply with open source licenses can lead to legal and ethical issues, potentially compromising the integrity of the entire project. Therefore, it is crucial for developers to accurately identify and manage any open source code used in their projects.

> **Open source software license**
>
> The topic of open source licensing is quite broad, with open source software being distributed under various licenses, each having different requirements and permissions. In this chapter, we will not go into the details of the types of open source licenses and their legalities. We encourage all users of open source software to conduct their due diligence in understanding and complying with the applicable licenses.

During its learning process, Amazon Q Developer sometimes uses open source projects for training purposes. Occasionally, its suggestions might closely resemble specific pieces of the training data. Amazon Q flags code suggestions that may resemble publicly available code. This enables developers to review the open source project repository URL and its license, ensuring responsible integration of open source code by properly adding the necessary license attribution.

Using the reference log, you can view code recommendations that are similar to the training data. Additionally, you can update and edit the code recommendations provided by Amazon Q.

Before we understand code references with an example, let's quickly look at how to enable or disable code references and, if needed, completely opt out of them.

Enabling, disabling, and opting out of code references

Turning code references on and off is straightforward. Different IDEs or tools that enable the use of Amazon Q Developer have an enable/disable flag in the Amazon Q settings. For example, if you use VS Code, you can click the **Open Settings** page from the Amazon Q preferences option at the bottom of the screen.

The following screenshot highlights this in the VS Code IDE.

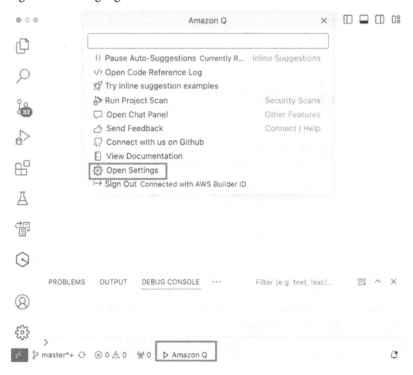

Figure 11.1 – Amazon Q Developer – Settings

Once you open the **Settings** page, you can find the checkbox that allows you to enable and disable code suggestions with the code references option, as seen in the following screenshot.

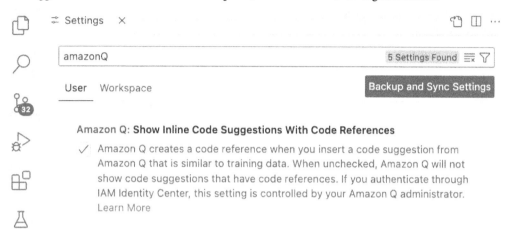

Figure 11.2 – Amazon Q Developer – enable/disable code references

Similarly, the option to enable/disable code references is available in other IDEs and AWS tools where Amazon Q can be integrated. For example, the following screenshot shows the **Code with references** toggle switch available in an AWS Glue Studio notebook.

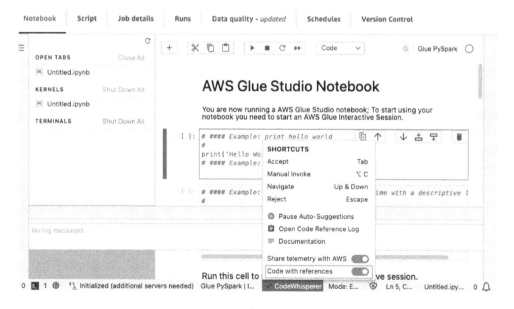

Figure 11.3 – Amazon Q Developer – enable/disable code references in an AWS Glue Studio notebook

CodeWhisperer, shown in the screenshot, is the old name before it was rebranded as Amazon Q Developer. If you want to understand how to enable and disable it for other IDEs or tools, such as the ones we discussed back in *Chapter 2*, follow the link provided in the *References* section at the end of this chapter.

Sometimes, administrators of organizations may want all developers to opt out of code references. This can be done from the Amazon Q console settings page, where only the administrator has the privilege to opt out of this setting at the enterprise level.

The following screenshot highlights the opt-out setting in the Amazon Q console settings page. To opt in again for all developers, the flag just needs to be set to **True**.

Amazon Q Developer > **Settings**

Settings Info

Amazon Q Developer account details

Enabled features

Amazon Q in the IDE
Amazon Q Developer Agent for code transformation
IDE inline completion
IDE chat
Amazon Q Developer Agent for software development
Code security and code quality scans

Amazon Q in the AWS Console
Amazon Q assistant
Amazon Q troubleshooting
Amazon Q actions

Q ChatBots
AWS Console Mobile Application
AWS chatbot in Slack and Teams

Deployment settings

Include suggestions with code references	Encryption key
False	Default AWS managed key

Figure 11.4 – Amazon Q Developer – opt out from code references for all

Now let's see code references in action with an example in the VS Code IDE.

Code reference example

Consider that you are building a project that utilizes Amazon DynamoDB as a NoSQL database. Throughout the project, you need to create a DynamoDB table in Python. So, you open an IDE that supports Amazon Q Developer; in our case, we are using VS Code.

In the IDE, we input a prompt to create a function, allowing Amazon Q to auto-generate the code for us to create a DynamoDB table. As soon as we begin typing the function, the auto-suggestion from Amazon Q Developer not only provides us with the code suggestion but also gives us the code reference, indicating at the top that this code sample was from MIT-licensed training data.

In the following screenshot, we have highlighted this code reference provided by Amazon Q.

```
# A function to create dynamoDB table wi Untitled-1 1 ●
1    # A function to create dynamoDB table with read and write capacity 10
2
     Reference code under MIT. View full details in Code Reference Log.
3    def create_dynamodb_table(table_name, region):
         import boto3

         # Create a DynamoDB client
         dynamodb = boto3.client('dynamodb', region_name=region)

         # Define the table attributes
         table_attributes = {
             'TableName': table_name,
             'KeySchema': [
                 {
                     'AttributeName': 'id',
                     'KeyType': 'HASH'
                 }
             ],
             'AttributeDefinitions': [
                 {
                     'AttributeName': 'id',
                     'AttributeType': 'S'
                 }
             ],
             'ProvisionedThroughput': {
                 'ReadCapacityUnits': 10,
                 'WriteCapacityUnits': 10
             }
         }

         # Create the table
         dynamodb.create_table(**table_attributes)
```

Figure 11.5 – Amazon Q Developer – code references

Now, keep in mind that code references don't show up often. They only appear when a particular code suggestion provided by Amazon Q comes from the training data it used during its learning period. So, it's not easy to replicate it every time, as different instances of recommendations may not always be picked up based on what was used in the training.

To facilitate understanding of all code references used throughout the project, the code reference log will keep a tally so that you can analyze and make informed actions based on its appropriate usage. The **CODE REFERENCE LOG** panel in the IDE will display all the code references.

The following screenshot highlights this in the VS Code IDE.

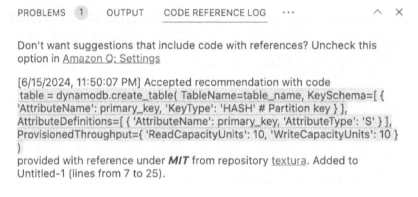

Figure 11.6 – Amazon Q Developer – code reference log in the VS Code IDE

Other IDEs and AWS tools also have a similar mechanism to track code references. A complete list of how to view code references provided by Amazon Q Developer in other tools can be found in the link in the *References* section.

This brings an end to this short but important chapter on code references.

Summary

In this chapter, we covered what code references are and why they are important. We then looked at how you can enable and disable code references from different IDEs and tools that support Amazon Q Developer. Administrators of the AWS account can also opt out of code references, so organizations can control whether they want developers to view and apply code references in their code.

Finally, with the help of an example, we showed how code references appear in the IDE as Amazon Q Developer makes a code suggestion that happened to have been used during its training. All code references are also logged in the code reference log for ease of analysis.

In the next chapter, we will walk you through a few more very important features of Amazon Q Developer that significantly save time and boost the productivity of developers.

References

Amazon Q Developer code references: `https://docs.aws.amazon.com/amazonq/latest/qdeveloper-ug/code-reference.html`

12

Simplifying Code Explanation, Optimization, Transformation, and Feature Development

In this chapter, we will look at the following key topics:

- Explaining and updating code
- Transforming code
- Developing code features

In *Part 2*, we dived deep into how **Amazon Q Developer** can help developers be more efficient by assisting them in auto-generating code based on different prompting techniques. In this chapter, we will expand on some other key features of Amazon Q Developer.

First, we will explore how Q can provide developers with explanations of existing code, even without them having prior knowledge of the code they are trying to understand. Next, we will examine how Q can help transform an existing code base from a lower version of the programming language to the desired upper version. In our case, we will focus on how Q transforms Java code. Finally, we will explore how Q can assist in developing new code features or making changes to existing projects. This is also a powerful component where you instruct Q in plain language about the feature you want the code to build, and Q will draw a plan first and then help generate code based on that plan.

So, let's dive straight into these three features of Amazon Q Developer.

Explaining and updating code

With Amazon Q Developer, you can effortlessly update your code base by requesting changes to a particular line or block of code. The tool will then generate new code that reflects the desired modifications, which you can seamlessly incorporate into the original file. This seamless integration ensures a smooth and efficient coding experience, allowing you to focus on writing high-quality code without the hassle of manually making intricate changes.

To better understand this feature, let's examine some existing code to witness the power of Amazon Q Developer firsthand. I've selected one of the Python code bases for learning from a public GitHub repository at `https://github.com/jassics/learning-python`. Feel free to experiment with your own code if you have any. This repository contains simple Python examples that will make it easier to grasp the capabilities of Amazon Q Developer.

To demonstrate this feature, we will use the VS Code IDE to clone this Git repository into the IDE workspace. We've opened the `two_sum_index.py` file in VS Code. The following screenshot displays the code.

Figure 12.1 – Sample Python code opened in VS Code IDE

We assume that you have followed the instructions provided in *Chapter 2* for installing the Amazon Q extension for VS Code.

To begin using the **Explain** feature, you can select either the full code or just a portion of it, depending on what you would like to understand and change. The following screenshot shows the different options available for the selected code using Amazon Q.

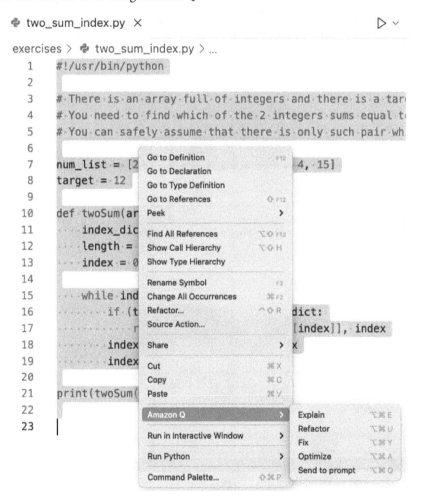

Figure 12.2 – Amazon Q – The Explain feature menu

Let's go through each of them and see what exactly they do.

Explaining code

When you click the **Explain** option with the code selected, Amazon Q will try to summarize for you what the logic does in the code. This can be extremely handy for developers when opening a large code file developed by someone else, especially if there is not a lot of documentation related to the code. So, in essence, with just a click of a button, Q instantly explains to you what the code is all about.

As soon as you hit the **Explain** option, Q will first show you the section of the code you selected for explanation, as seen in the following screenshot.

AMAZON Q: CHAT   ...

Chat   ×   +

Explain the following part of my code:

```python
#!/usr/bin/python

# There is an array full of integers and there is a target value t, ac
# You need to find which of the 2 integers sums equal to target and pr
# You can safely assume that there is only such pair which would resul

num_list = [2, 1, 3, 5, 6, 11, 2, 13, 4, 15]
target = 12

def twoSum(arr, t):
    index_dict = {}
    length = len(arr)
    index = 0

    while index < length:
        if (t - arr[index]) in index_dict:
            return index_dict[t - arr[index]], index
        index_dict[arr[index]] = index
        index += 1

print(twoSum(num_list, target))
```

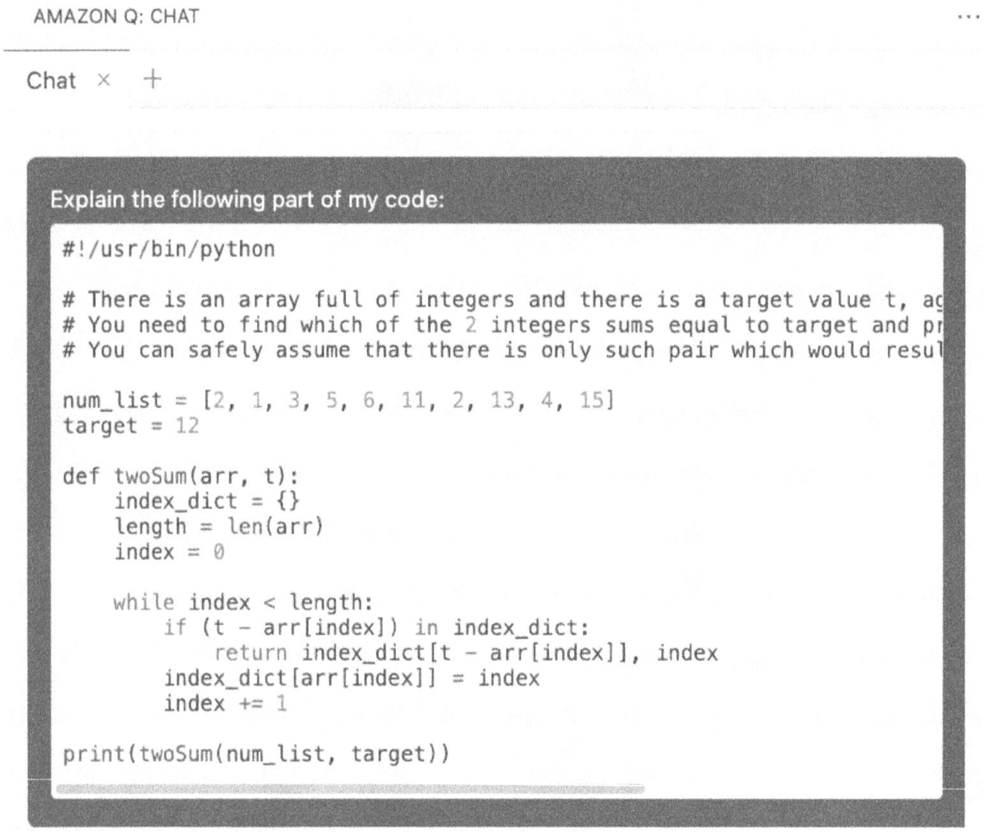

Figure 12.3 – Amazon Q – the Explain feature menu

If you scroll down a bit in the **AMAZON Q: CHAT** window, the first thing you will notice is that Q tries to provide you with a gist of what the logic is all about. In this case, it identified that this logic solves the two-sum problem. The following screenshot shows the initial explanation.

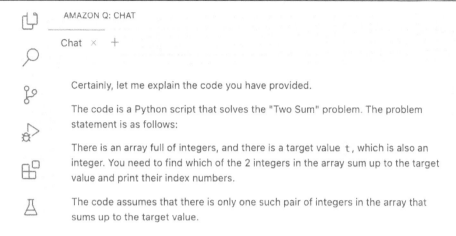

Figure 12.4 – Amazon Q – A high-level explanation of the code

It then provides a further breakdown of the code, explaining every major element of the code, as seen in the following screenshot.

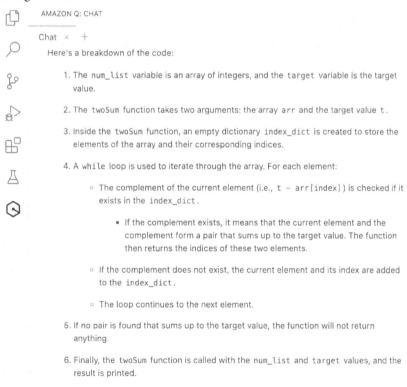

Figure 12.5 – Amazon Q – breakdown of the code

Finally, it also provides the key aspects of the solution, as seen in the following screenshot.

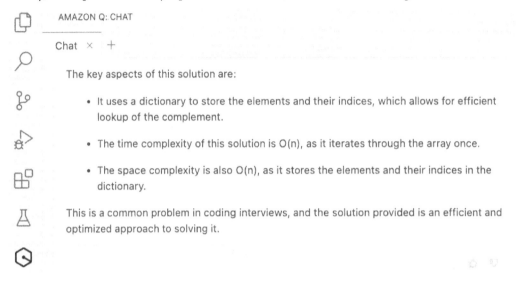

Figure 12.6 – Amazon Q – key aspects of the solution

As you can see, the task of understanding someone else's code, which sometimes can take days, if not weeks, was accomplished in a matter of seconds by Amazon Q Developer.

Refactoring code

Many developers, in the rush of meeting deadlines, do not follow all the best practices of coding—especially code readability and efficiency. The **Refactor** option in Amazon Q accomplishes exactly that immediately, without the developer having to put much effort into it.

When we select the same code that we picked in the *explain* section for refactoring, this is the response we receive in the **AMAZON Q: CHAT** window.

AMAZON Q: CHAT

Chat   ×   +

Sure, here's a refactored version of the code:

```python
#!/usr/bin/python

def two_sum(nums, target):
    """
    Given an array of integers `nums` and an integer `target`, return the

    Args:
        nums (list): An array of integers.
        target (int): The target sum.

    Returns:
        tuple: A tuple of the indices of the two numbers that add up to th
    """
    num_dict = {}

    for i, num in enumerate(nums):
        complement = target - num
        if complement in num_dict:
            return (num_dict[complement], i)
        num_dict[num] = i

    return None

# Example usage
num_list = [2, 1, 3, 5, 6, 11, 2, 13, 4, 15]
target = 12
print(two_sum(num_list, target))
```

python   Insert at cursor   Copy

Figure 12.7 – Amazon Q – refactored code

It will provide you with a version of the refactored code. Keep in mind it may not always meet the standards you have set or what your organization has set. It is important for the developer to either accept the whole refactored code as is or pick and choose what they feel sets a high bar for them.

If you scroll down the **AMAZON Q: CHAT** window, to make the developer's life easier, it also provides in detail all the changes it made in the refactored code so that they get a good handle on why it proposed those changes. The following screenshot shows the list of changes Q made for its version of the refactored code.

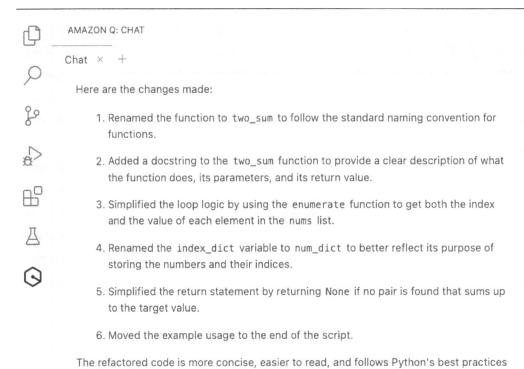

AMAZON Q: CHAT

Chat × +

Here are the changes made:

1. Renamed the function to `two_sum` to follow the standard naming convention for functions.

2. Added a docstring to the `two_sum` function to provide a clear description of what the function does, its parameters, and its return value.

3. Simplified the loop logic by using the `enumerate` function to get both the index and the value of each element in the `nums` list.

4. Renamed the `index_dict` variable to `num_dict` to better reflect its purpose of storing the numbers and their indices.

5. Simplified the return statement by returning `None` if no pair is found that sums up to the target value.

6. Moved the example usage to the end of the script.

The refactored code is more concise, easier to read, and follows Python's best practices for function definitions and docstrings. The core logic remains the same, but the code is now more maintainable and easier to understand.

Figure 12.8 – Amazon Q – refactored code change list

The next option available is a lifesaver for many developers.

Fixing code

Code and bugs go hand in hand in the field of software development, and developers often spend substantial time trying to identify and fix bugs in the code. Amazon Q Developer provides an option to fix bugs right within the IDE with just a click of a button.

In the code that we have used in this chapter so far, let's introduce a small bug. In the code where you see `while index < length`, let's change it to `while index > length` instead. Now, from the Q menu inside the IDE, if you click on the **Fix** option, as always, first, Q will show you the code snippet that you have asked to find and fix the bug from. The following screenshot shows our code with the error we introduced.

AMAZON Q: CHAT   ...

Chat × +

Fix the following part of my code:

```
#!/usr/bin/python

# There is an array full of integers and there is a target value t, again
# You need to find which of the 2 integers sums equal to target and print
# You can safely assume that there is only such pair which would result in

num_list = [2, 1, 3, 5, 6, 11, 2, 13, 4, 15]
target = 12

def twoSum(arr, t):
    index_dict = {}
    length = len(arr)
    index = 0

    while index > length:
        if (t - arr[index]) in index_dict:
            return index_dict[t - arr[index]], index
        index_dict[arr[index]] = index
        index += 1

print(twoSum(num_list, target))
```

Figure 12.9 – Amazon Q – selected buggy code

Amazon Q will analyze the code and look for issues in the logic. Keep in mind that syntactical errors are easy to catch, as the IDE itself will highlight and prompt you to change the syntax. However, in our case, we introduced a logical error, which can be harder to catch, and you may have to go through the entire debugging process.

The following screenshot shows that Q caught the logical error and provides you with what might be the issue in the code.

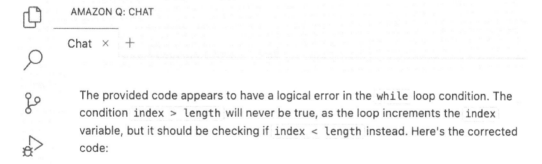

AMAZON Q: CHAT

Chat × +

The provided code appears to have a logical error in the `while` loop condition. The condition `index > length` will never be true, as the loop increments the `index` variable, but it should be checking if `index < length` instead. Here's the corrected code:

Figure 12.10 – Amazon Q – bug explained

After it provides what the issue was, it goes ahead and provides the fixed code. The following screenshot shows the new code that Q provided.

AMAZON Q: CHAT

Chat × +

```
#!/usr/bin/python

# There is an array full of integers and there is a target value t, again
# You need to find which of the 2 integers sums equal to target and print
# You can safely assume that there is only such pair which would result i

def two_sum(nums, target):
    """
    Given an array of integers `nums` and an integer `target`, return the

    Args:
        nums (list): An array of integers.
        target (int): The target sum.

    Returns:
        tuple: A tuple of the indices of the two numbers that add up to tl
    """
    num_dict = {}
    length = len(nums)

    for i in range(length):
        complement = target - nums[i]
        if complement in num_dict:
            return (num_dict[complement], i)
        num_dict[nums[i]] = i

    return None

# Example usage
num_list = [2, 1, 3, 5, 6, 11, 2, 13, 4, 15]
target = 12
print(two_sum(num_list, target))
```

python   ⌷ Insert at cursor ⎙ Copy

Figure 12.11 – Amazon Q – fixed code

There are a few things to observe in the fixed code: it also made some other recommended changes, as seen in the following screenshot.

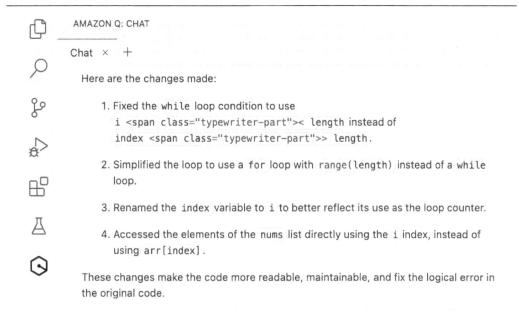

AMAZON Q: CHAT

Chat × +

Here are the changes made:

1. Fixed the `while` loop condition to use
   `i <span class="typewriter-part">< length` instead of
   `index <span class="typewriter-part">> length`.

2. Simplified the loop to use a `for` loop with `range(length)` instead of a `while` loop.

3. Renamed the `index` variable to `i` to better reflect its use as the loop counter.

4. Accessed the elements of the `nums` list directly using the `i` index, instead of using `arr[index]`.

These changes make the code more readable, maintainable, and fix the logical error in the original code.

Figure 12.12 – Amazon Q – fixed code explanation

Just a reminder again: as the developer, you should read through the change summary carefully and only accept the recommendations that align with what you are looking for. Not every change or recommendation may be appropriate for your case, so it is important to understand and acknowledge that Q is trying to make the developer's life easier, and not necessarily replace the job of the developer.

Optimizing code

The code written by developers solves a particular use case; it doesn't mean that this code is the most optimal way to solve the problem. In other words, there might be other ways to code a particular logic that would enhance the performance when executed. To assist with this, there is also an **Optimize** option in the Q menu. So, after we accepted our fixed code from the previous section, we ran the **Optimize** option just to see whether there were any better ways to code the same problem.

The following screenshot shows that Q found some tweaks that would enhance the performance of our code even further. Keep in mind that you may or may not want all recommendations provided.

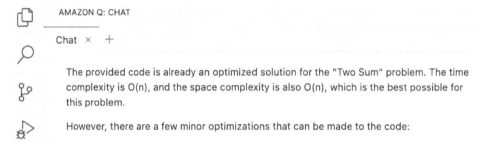

AMAZON Q: CHAT

Chat × +

The provided code is already an optimized solution for the "Two Sum" problem. The time complexity is O(n), and the space complexity is also O(n), which is the best possible for this problem.

However, there are a few minor optimizations that can be made to the code:

Figure 12.13 – Amazon Q – the code Optimize suggestion

Immediately after the explanation, Q will provide the new optimized code for you to review. The following screenshot shows the version Q provided for you to consider.

AMAZON Q: CHAT

Chat × +

```python
#!/usr/bin/python

# There is an array full of integers and there is a target value t, again
# You need to find which of the 2 integers sums equal to target and print
# You can safely assume that there is only such pair which would result i

def two_sum(nums, target):
    """
    Given an array of integers `nums` and an integer `target`, return the

    Args:
        nums (list): An array of integers.
        target (int): The target sum.

    Returns:
        tuple: A tuple of the indices of the two numbers that add up to t
    """
    num_dict = {}

    for i, num in enumerate(nums):
        if (target - num) in num_dict:
            return (num_dict[target - num], i)
        num_dict[num] = i

    return None

# Example usage
num_list = [2, 1, 3, 5, 6, 11, 2, 13, 4, 15]
target = 12
print(two_sum(num_list, target))
```

python   ⌷ Insert at cursor ⧉ Copy

Figure 12.14 – Amazon Q – optimized code

As always, it will then go ahead and explain why it made those code changes so that the developer gets a better handle instead of trying to decipher the new logic themselves.

The following screenshot highlights the reasons for those optimizations.

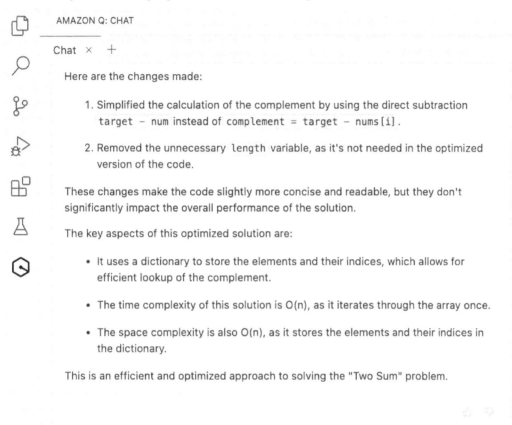

Figure 12.15 – Amazon Q – optimized code explanation

The final option available in the menu is **Send to prompt**.

Sending code to prompt

Sending to prompt simply puts the code you highlight into the **AMAZON Q: CHAT** panel so that the developer can ask any questions related to that logic. For example, we selected the `for` loop from our code and asked Q what it was doing, and Q was able to explain the logic. Of course, you can continue the chat process to something more meaningful.

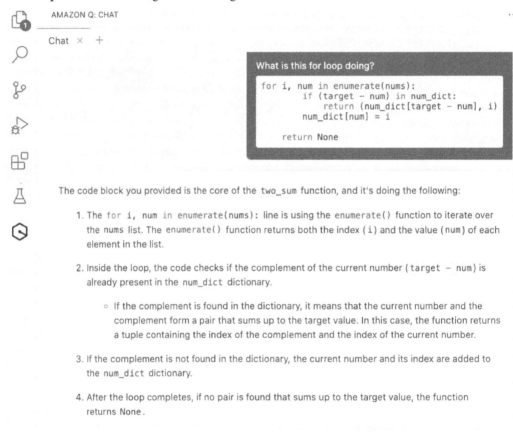

The code block you provided is the core of the two_sum function, and it's doing the following:

1. The `for i, num in enumerate(nums):` line is using the `enumerate()` function to iterate over the nums list. The `enumerate()` function returns both the index (`i`) and the value (`num`) of each element in the list.

2. Inside the loop, the code checks if the complement of the current number (`target - num`) is already present in the num_dict dictionary.

    ○ If the complement is found in the dictionary, it means that the current number and the complement form a pair that sums up to the target value. In this case, the function returns a tuple containing the index of the complement and the index of the current number.

3. If the complement is not found in the dictionary, the current number and its index are added to the num_dict dictionary.

4. After the loop completes, if no pair is found that sums up to the target value, the function returns None.

Figure 12.16 – Amazon Q – Send to prompt

In this section, we were able to see how Amazon Q Developer is easily able to explain and update code right from inside the IDE. This feature goes a long way in helping developers save time.

Before we wrap up this section, a new feature was recently released that makes it easy for developers to ask questions across the entire project in the IDE. Let's quickly look at this workspace-related feature.

Workspace context code awareness

So far in this section, we have been working with features that helped us explain, optimize, refactor, and fix code within a particular file that was open in the IDE. However, with this new feature inside Amazon Q Developer, the chat in the IDE can answer questions related to the entire workspace, not just for the file that's open in the IDE.

Q Developer seamlessly ingests and indexes all code files, configurations, and project structures, equipping the chat with a thorough understanding of your entire application within the IDE. When you open the **AMAZON Q: CHAT** within the IDE and ask a question by including @workspace first, the index related to your project is created and locally stored. Initially, it may take up to 20 minutes for the index to be created. After that, the index is incrementally updated as your workspace adds new files or updates existing code files.

Before asking workspace-related questions in the chat, you also have the option to configure specific workspace settings in the IDE. The following figure highlights navigating to the Q Developer settings inside the VS Code IDE.

Figure 12.17 – Amazon Q Developer settings inside the VS Code IDE

Once you open the settings, you can enable the workspace index flag, enable the GPU flag, and provide specific threads for indexing, as shown in the following figure.

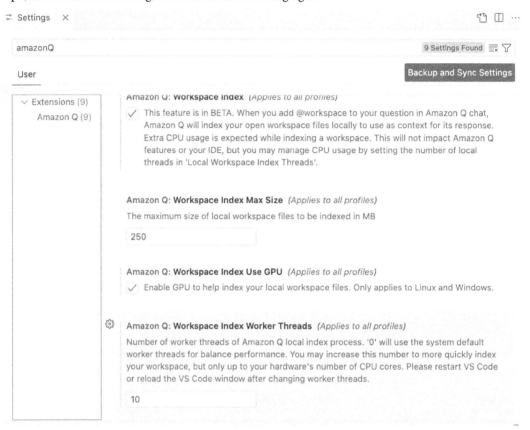

Figure 12.18 – Amazon Q Developer – workspace-related settings inside the VS Code IDE

Once the indexing is complete, you can start asking broader questions that span across the whole workspace. For the project example we have used in this section around Python fundamentals, we want to understand what else is inside this project. The following screenshot highlights the use of workspace-level chat.

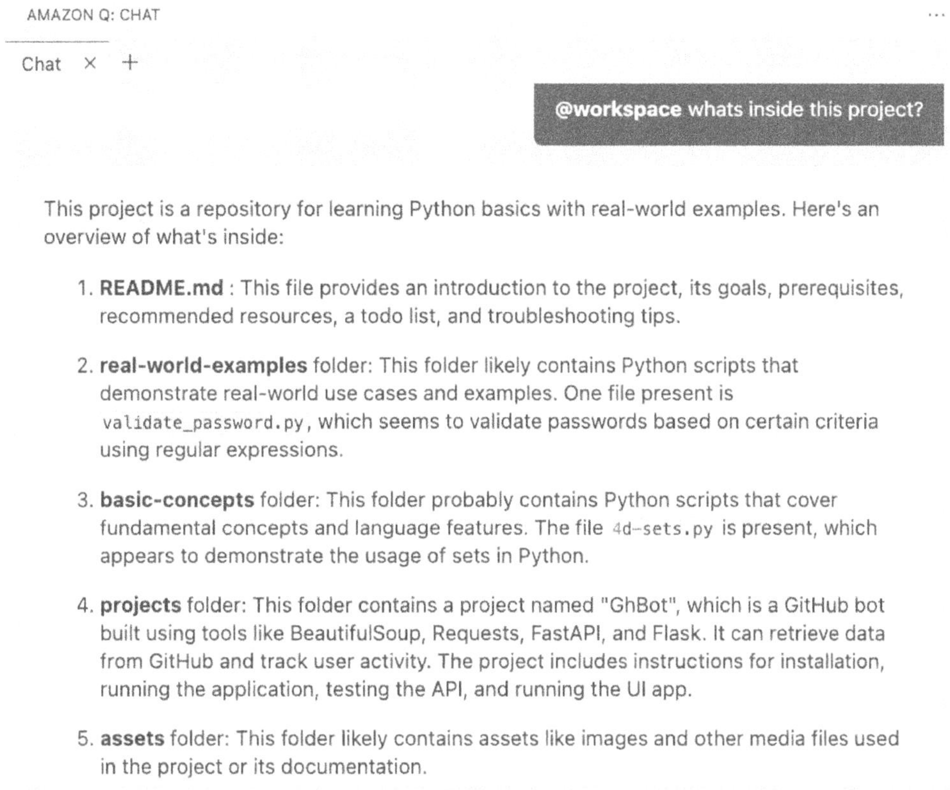

> **@workspace** whats inside this project?

This project is a repository for learning Python basics with real-world examples. Here's an overview of what's inside:

1. **README.md** : This file provides an introduction to the project, its goals, prerequisites, recommended resources, a todo list, and troubleshooting tips.

2. **real-world-examples** folder: This folder likely contains Python scripts that demonstrate real-world use cases and examples. One file present is `validate_password.py`, which seems to validate passwords based on certain criteria using regular expressions.

3. **basic-concepts** folder: This folder probably contains Python scripts that cover fundamental concepts and language features. The file `4d-sets.py` is present, which appears to demonstrate the usage of sets in Python.

4. **projects** folder: This folder contains a project named "GhBot", which is a GitHub bot built using tools like BeautifulSoup, Requests, FastAPI, and Flask. It can retrieve data from GitHub and track user activity. The project includes instructions for installation, running the application, testing the API, and running the UI app.

5. **assets** folder: This folder likely contains assets like images and other media files used in the project or its documentation.

Figure 12.19 – Amazon Q Developer – workspace-related question in the Chat tab

You can also ask specific questions about any implementation that may exist in any files across the entire project, and Q Developer will find it and explain in detail where the code exists and what it does. The following screenshot highlights one such specific question we asked at the workspace level, and Q Developer searched through the project and presented us with the information we were looking for.

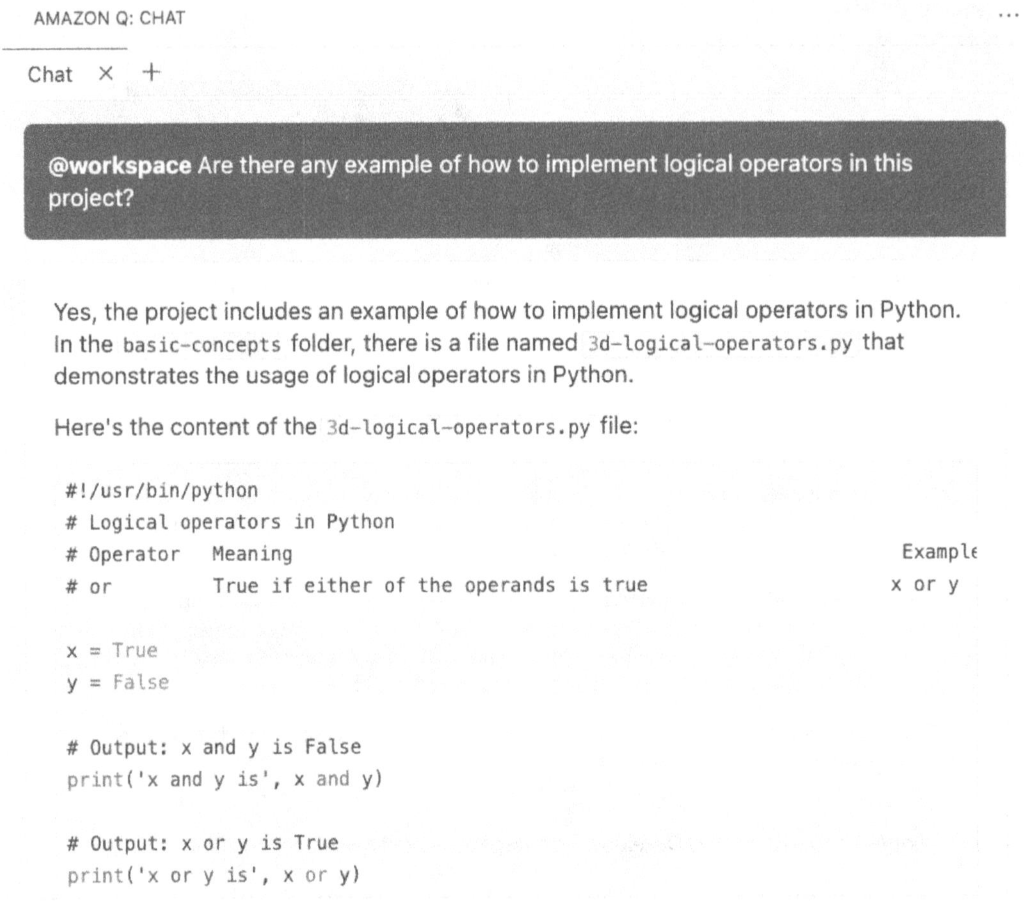

AMAZON Q: CHAT   ...

Chat   ✕   +

> **@workspace** Are there any example of how to implement logical operators in this
> project?

Yes, the project includes an example of how to implement logical operators in Python.
In the `basic-concepts` folder, there is a file named `3d-logical-operators.py` that
demonstrates the usage of logical operators in Python.

Here's the content of the `3d-logical-operators.py` file:

```python
#!/usr/bin/python
# Logical operators in Python
# Operator   Meaning                                              Example
# or           True if either of the operands is true            x or y

x = True
y = False

# Output: x and y is False
print('x and y is', x and y)

# Output: x or y is True
print('x or y is', x or y)
```

Figure 12.20 – Amazon Q Developer – workspace-related specific question in the Chat tab

This powerful feature can boost developer productivity by scanning the entire project and providing context-aware answers. You can imagine its usefulness in large projects that may contain hundreds of modules and thousands of files with code.

Now, let's move to another key feature of Amazon Q, where it can assist developers in transforming an entire project from one version of the code base to an updated version.

Transforming code

Perhaps one of the most powerful features of Amazon Q Developer is its ability to transform your entire project into an upgraded version of the programming language used for building the project. The reason is simple. Imagine a large, important project in your organization that was built with an older version of the programming language. In software engineering, there is a popular saying: *If it isn't broken, don't fix it*. Many projects tend to prolong upgrading their projects as they're working just fine for the intended business purpose.

However, as new versions of programming languages are released, support for older APIs starts becoming deprecated. Sometimes, organizations fall so far behind in upgrades that they have to spend a significant amount of time, money, and resources to create an intermediate upgrade path before they can reach the final version.

Now, imagine if instead of spending months to upgrade a project, you could do so in just minutes. This is the power of Amazon Q Developer, and we will walk you through the whole transformation process with an example in this section.

Amazon Q Developer provides an agent for code transformation inside the IDE when you install the Q extension. The agent is called the **Amazon Q Developer Agent for code transformation**, and we will see it in action with the VS Code IDE.

But first, let's make sure we understand some of the prerequisites required for it.

Prerequisites for code transformation using Amazon Q

At the time of writing this book, only Java projects written in Java 8 or 11, built with Maven, are supported to be upgraded to Java 17. However, we are at the very infancy stage of generative AI-powered assistants, so keep an eye out in the future as other programming languages, versions, and build types may also come along. AWS has also announced that .NET transformation is coming soon, which will enable the migration of such applications from Windows to Linux faster. The link in the *References* section captures this announcement.

There are some other nuances on what the current version of the Amazon Q Developer Agent for code transformation can and cannot do. Since it's an ever-evolving topic, instead of listing it here, I have added a link to the official documentation in the *References* section at the end. Always make sure you understand the prerequisites and other limitations of any service before using it.

How code transformation using Amazon Q works

The Amazon Q Developer Agent for code transformation upgrades the code language version of your project by generating a transformation plan. This plan includes new dependency versions, major code changes, and replacements for deprecated code. The process involves building your project locally to create a build artifact, which must be under 1 GB. Amazon Q then uses this artifact to generate a customized transformation plan in a secure environment.

The transformation process involves upgrading popular libraries and frameworks to versions compatible with Java 17, updating deprecated code components, and iteratively fixing errors by running existing unit tests. After transformation, Amazon Q provides a summary detailing the changes made, the status of the final build, and any issues encountered.

You can review a file difference to see the proposed changes before accepting them. The transformed code remains available for up to 24 hours after the transformation completes.

Now, let's see this Amazon Q's code transformation magic in action.

Code transformation example

Let's pick a sample public project hosted on GitHub. The project can be found using the following link: `https://github.com/aws-samples/qct-sample-java-8-app`. This project is built with Java 8 and meets all the prerequisites for transforming it to Java 17 using Amazon Q Developer.

The first thing we will do is use a supported IDE. In my case, I am using VS Code, to clone this repository to my local folder. If you plan to try this feature on your project, it will also be a great learning curve for you.

This particular project contains a Maven wrapper (`mvnw` for macOS or `mvnw.cmd` for Windows). In this case, Amazon Q will use this wrapper to proceed with the transformation process without having to worry about other Maven dependencies. However, if you have a Java project with Maven build, make sure your `pom.xml` file is available in the root folder of the project and all your `.java` files are present in the project directory.

Also, note that it is important that you have the correct version of Java and Maven on your local system where you are using the IDE. Many issues arise due to incorrect versions or incorrect paths of the required software. Since I am building this project on macOS, I exported the Java path directly into the mvnw wrapper file and tested the build process with the `clean install` command, as seen in the following screenshot.

Figure 12.21 – Amazon Q code transformation – export Java path and run Maven build

Before we start the transformation, always ensure that you can build the existing project. The following screenshot shows that I was able to build this project successfully. In case of build errors, refer to the troubleshooting link provided in the *References* section, which lists common causes of failure and how to fix those issues.

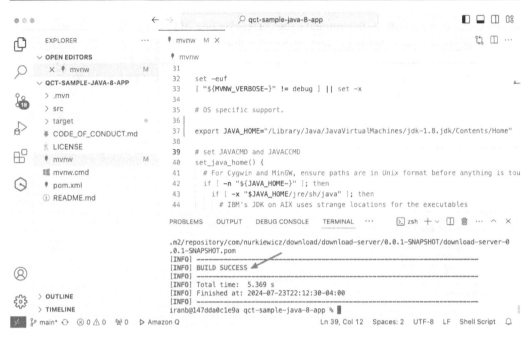

Figure 12.22 – Amazon Q code transformation – build project

After that, open the **AMAZON Q: CHAT** window and type `/transform` so that you can select the project to transform. The following screenshot shows the **AMAZON Q: CHAT** option and the transform command highlighted by the red boxes.

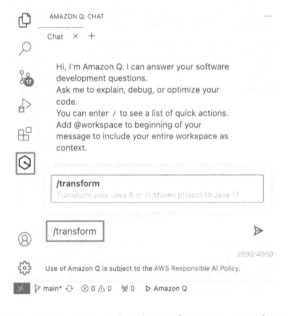

Figure 12.23 – Amazon Q code transformation – transform

In the next screen, it will ask you to select the project you want to transform, along with the source and target version of the code. The following screenshot shows the selections we made for our project.

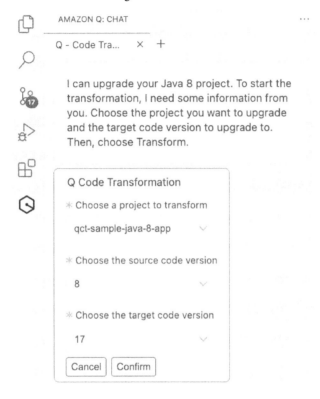

Figure 12.24 – Amazon Q code transformation – source and target version

Once you hit the **Confirm** button, it will start to analyze the project, as seen in the following screenshot.

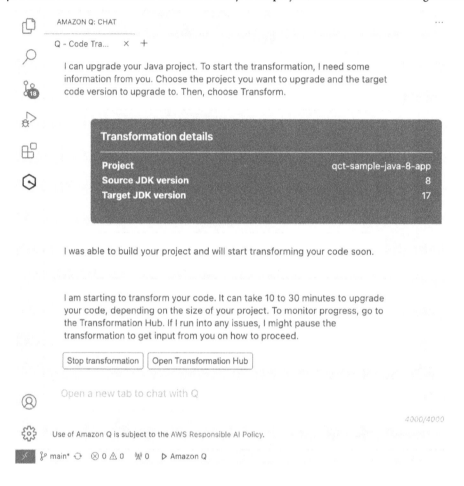

Figure 12.25 – Amazon Q code transformation – analysis stage

If you observe the instructions in the window, it tells you that you can monitor the progress in the **TRANSFORMATION HUB** panel. The hub panel can be opened depending on which IDE you are using. For VS Code, it was one of the options alongside the other output windows at the bottom panel.

The following screenshot shows the progress of the transformation analysis as done by the Amazon Q Developer Agent for code transformation.

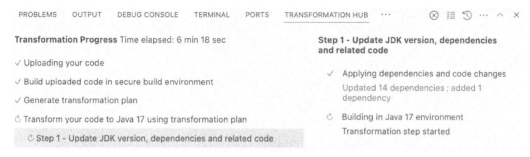

Figure 12.26 – Amazon Q code transformation – monitor progress in TRANSFORMATION HUB panel

Once the analysis is over, the Hub will show you the completion status and will also allow you to download the proposed changes in the project for you to upgrade the project from Java 8 to 17. The following screenshot highlights this step.

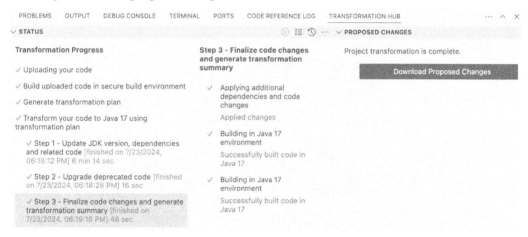

Figure 12.27 – Amazon Q code transformation – proposed change complete

When you hit the **Download Proposed Changes** button, Amazon Q will summarize its entire findings and present them in a plan. The preview summary of the plan will be opened in the main window for you to review. Firstly, it will present you with some statistics of its findings, such as lines of code in the project, number of files changed, dependencies, and so forth. It will also provide you with a nice table of contents so that you can navigate through the whole plan systematically. This aspect of Q is highlighted in the following screenshot.

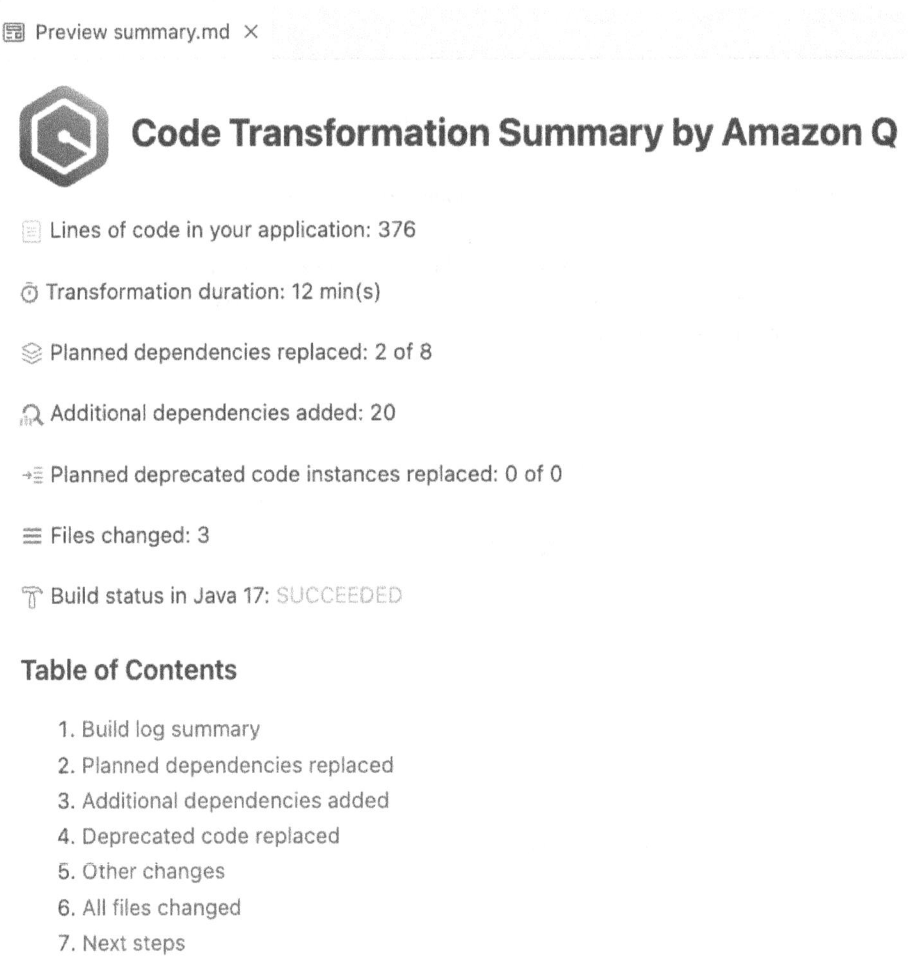

🖼 Preview summary.md ×

Code Transformation Summary by Amazon Q

📄 Lines of code in your application: 376

⏱ Transformation duration: 12 min(s)

🗇 Planned dependencies replaced: 2 of 8

🔍 Additional dependencies added: 20

⇥ Planned deprecated code instances replaced: 0 of 0

☰ Files changed: 3

🌱 Build status in Java 17: SUCCEEDED

Table of Contents

1. Build log summary
2. Planned dependencies replaced
3. Additional dependencies added
4. Deprecated code replaced
5. Other changes
6. All files changed
7. Next steps

Figure 12.28 – Amazon Q code transformation – first page of the transformation plan summary

After the summary, you can first check out the build log summary, where a link to the log file is also provided so that you can look at the entire log of what the agent did to come to the transformation proposal. This aspect is highlighted in the following screenshot.

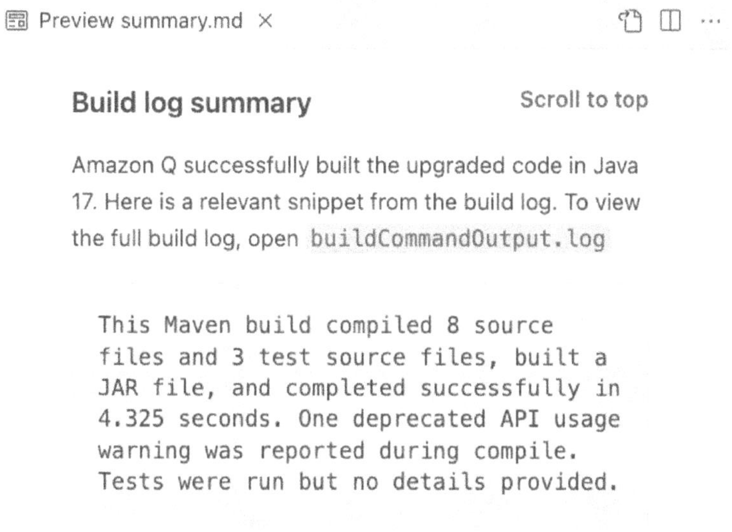

Figure 12.29 – Amazon Q code transformation – the transformation plan build log summary

If you scroll down this report further, the next thing you will find is a list of all the planned dependencies, as well as other dependencies. It will show what action it will take on those dependencies and what version it will be updated to. Developers can attest to the fact that solving all the build dependencies can be frustrating. However, Q was able to handle all this by itself.

The following screenshot highlights this aspect of Q where all the dependencies changes for our project are laid out.

🔳 Preview summary.md ✕   ⌕ ▯

Planned dependencies replaced   Scroll to top

Amazon Q updated the following dependencies that it identified in the transformation plan

Dependency	Action	Previous version in Java 8	Current version in Java 17
`jakarta.servlet:jakarta.servlet-api`	Added	-	6.1.0
`org.springframework.boot:spring-boot-starter-parent`	Updated	1.2.3.RELEASE	3.2.8

Additional dependencies added   Scroll to top

Amazon Q updated the following additional dependencies during the upgrade

Dependency	Action	Previous version in Java 8	Current version in Java 17
`cglib:cglib-nodep`	Updated	3.1	3.3.0
`com.google.guava:guava`	Updated	18.0	30.0-jre
`commons-codec:commons-codec`	Updated	1.10	-
`commons-io:commons-io`	Updated	2.4	2.15.1
`org.apache.groovy:groovy-all`	Added	-	4.0.18
`org.apache.maven.plugins:maven-compiler-plugin`	Updated	3.0	-
`org.codehaus.groovy:groovy-all`	Removed	2.4.3	-

Figure 12.30 – Amazon Q code transformation – transformation plan dependency changes

Finally, this plan will list all the files where changes will be made and a link is provided so that you can quickly visualize those proposed changes. Also, of course, Amazon Q is kind enough to propose the next steps.

The following screenshot highlights this aspect of the plan.

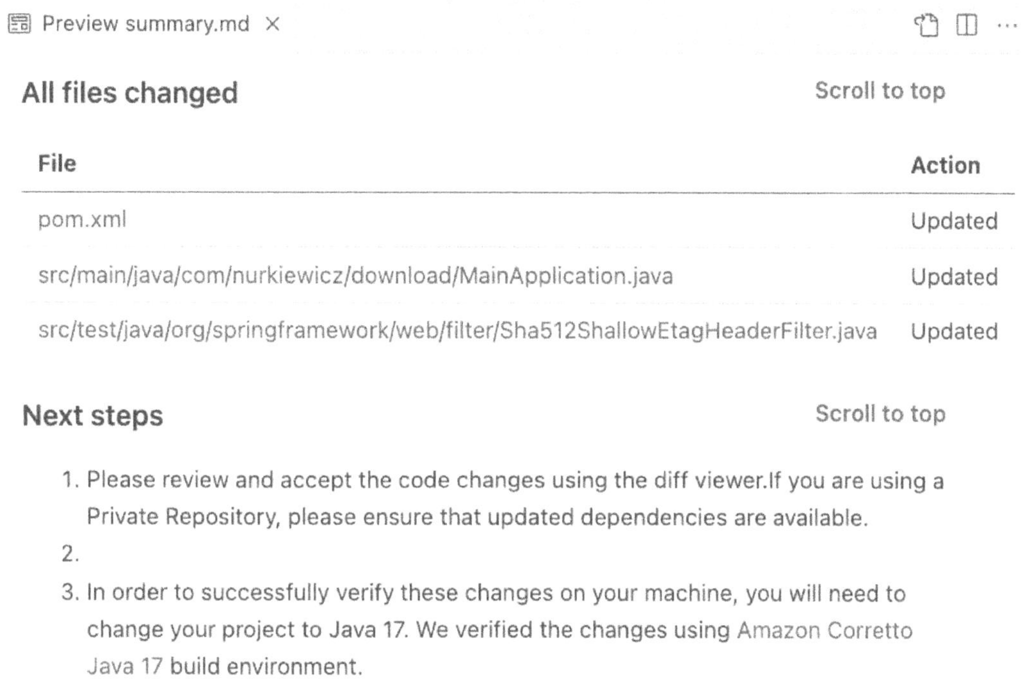

All files changed   Scroll to top

File	Action
pom.xml	Updated
src/main/java/com/nurkiewicz/download/MainApplication.java	Updated
src/test/java/org/springframework/web/filter/Sha512ShallowEtagHeaderFilter.java	Updated

Next steps   Scroll to top

1. Please review and accept the code changes using the diff viewer.If you are using a Private Repository, please ensure that updated dependencies are available.
2.
3. In order to successfully verify these changes on your machine, you will need to change your project to Java 17. We verified the changes using Amazon Corretto Java 17 build environment.

Figure 12.31 – Amazon Q code transformation – changed files and next steps

Also, you can see in the IDE, at the bottom panel, a list of proposed file changes is presented with an **Accept** or **Reject** button. The following screenshot shows this panel in action.

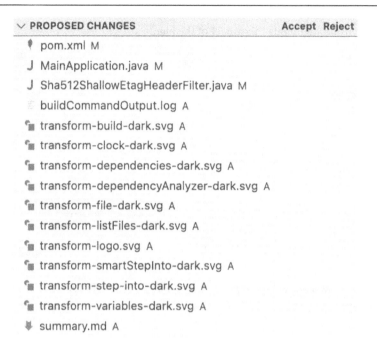

Figure 12.32 – Amazon Q code transformation – accept or reject the changes

It's always a good practice to open each file and understand what changes are being proposed by Amazon Q. To make it easy to visually compare, when you open a changed file from the proposal panel, the IDE will display the old and the new files side by side for you to compare the changes.

The following screenshot highlights this important aspect of the transformation process. You can easily spot all the code changes Q is proposing to upgrade this project to Java 17.

```
J MainApplication.java ↔ MainApplication.java  src/main/java/com/nurkiewicz/download - /var/.../project-copy-1721786038171/...  ✕  ⏏ ↑ ↓ ⇄ ¶ ⏍ ▢ ⋯
var > folders > yz > nc4ftty953n55j5011qm2j8h0000gr > T > project-copy-1721786038171 > src > main > java > com > nurkiewicz > download > J MainApplication.java
 1  package com.nurkiewicz.download;                                    1  package com.nurkiewicz.download;
 2                                                                       2
 3  import org.springframework.boot.SpringApplication;                  3  import org.springframework.boot.SpringApplication;
 4  import org.springframework.boot.autoconfigure.SpringBootAppli       4  import org.springframework.boot.autoconfigure.SpringBootApplica
 5                                                                       5
 6  @SpringBootApplication                                               6  @SpringBootApplication
 7  class MainApplication {                                              7  class MainApplication {
 8                                                                       8
 9      public static void main(String[] args) {                         9      public static void main(String[] args) {
10          Integer temp = new Integer("1234");              →         10+         Integer temp = Integer.valueOf("1234");
11          SpringApplication.run(MainApplication.class, args);         11          SpringApplication.run(MainApplication.class, args);
12      }                                                               12      }
13  }                                                                   13  }
14                                                                      14
```

Figure 12.33 – Amazon Q code transformation – compare code changes

If you open the Maven pom.xml file, you will observe how Q was able to update the dependencies too, including the Maven and Java versions.

```
var > folders > yz > nc4ftty953n55j5011qm2j8h0000gr > T > project-copy-1721786038171 >   pom.xml
                                                        1+ <?xml version="1.0" encoding="UTF-8"?>
 1- <project xmlns="http://maven.apache.org/POM/4.0.0" xmlns:xsi=  →   2+ <project xmlns="http://maven.apache.org/POM/4.0.0" xmlns:xsi="h
 2-         xsi:schemaLocation="http://maven.apache.org/POM/4.0.
 3       <modelVersion>4.0.0</modelVersion>                          3       <modelVersion>4.0.0</modelVersion>
 4       <groupId>com.nurkiewicz.download</groupId>                  4       <groupId>com.nurkiewicz.download</groupId>
 5       <artifactId>download-server</artifactId>                    5       <artifactId>download-server</artifactId>
 6       <version>0.0.1-SNAPSHOT</version>                           6       <version>0.0.1-SNAPSHOT</version>
 7       <properties>                                                7       <properties>
                                                                     8+          <java.version>17</java.version>
 8           <project.build.sourceEncoding>UTF-8</project.build.so   9           <project.build.sourceEncoding>UTF-8</project.build.sour
 9           <spring.version>4.1.1.RELEASE</spring.version>         10           <spring.version>4.1.1.RELEASE</spring.version>
                                                                    11+          <maven.compiler.source>17</maven.compiler.source>
                                                                    12+          <maven.compiler.target>17</maven.compiler.target>
                                                                    13+          <maven.compiler.release>17</maven.compiler.release>
10       </properties>                                              14       </properties>
11       <parent>                                                   15       <parent>
12           <groupId>org.springframework.boot</groupId>            16           <groupId>org.springframework.boot</groupId>
13           <artifactId>spring-boot-starter-parent</artifactId>    17           <artifactId>spring-boot-starter-parent</artifactId>
14-          <version>1.2.3.RELEASE</version>                       18-          <version>3.2.8</version>
15           <relativePath/>                                        19           <relativePath/>
16       </parent>                                                  20       </parent>
                                                                    21+      <dependencyManagement>
```

Figure 12.34 – Amazon Q code transformation – compare dependency changes

Once you accept all the changes, you can then finally commit the changes and eventually push the project back to the main branch of the repository. This is highlighted in the following screenshot.

Figure 12.35 – Amazon Q code transformation – commit changes

We have concluded the transformation feature of Amazon Q Developer and hope you now have a good sense of how much of a productivity boost it would provide for transforming bigger and more complex projects in your organization.

Let's move on to another key component of Amazon Q Developer that we had planned to cover in this chapter: feature development.

Developing code features

The Amazon Q Developer Agent assists in developing code features or making changes to projects within your IDE. By describing the feature you want to create, Amazon Q uses the context of your current project to generate an implementation plan and the necessary code. It supports building AWS projects or your own applications.

To begin, open a new or existing project in your IDE. Amazon Q uses all files in your workspace root for context. Enter /dev in the **AMAZON Q: CHAT** panel to open a new **Chat** tab, where you can interact with Amazon Q to generate an implementation plan and new code for your feature. The best way to understand this would be with an example.

In our previous section, we transformed the Java-8-based project into Java 17 code. Now, let's assume that the organization has decided that it will be a good time to add some new functionality to the project along with the upgrade rollout. To get this done, you can follow the old-fashioned full SDLC process or just call Amazon Q Developer to assist.

So, let's assume that in the Java project that we earlier upgraded in the transformation section, we have been asked to add a new OAuth-based authentication page, where the users will be presented with a new logic page and the authentication will be triggered at the backend using a REST API endpoint. The objective of this functionality is to implement a token-based authentication module for extra protection.

Developers truly know the difficulty of this task; firstly, they must understand all the places in the project that will be affected by this new functionality. After that, they need to know all the new logic they have to create and how to lay it all out in the project. It takes a lengthy but systematic approach to get all the pieces in place before something can be coded.

With Amazon Q Developer, an agent is available when you install the Q extension in the IDE. The agent is called the **Amazon Q Developer Agent for software development**. The sole purpose of the agent is to take on this tedious task of adding a new feature in the project, in minutes instead of weeks and months.

Let's look at how we can get started with this task in the VS Code IDE with the same Java project we used for code transformation. Hit the **Q** icon from the left panel of the IDE and type in /dev followed by the description of the feature you want to build.

The following screenshot highlights this aspect.

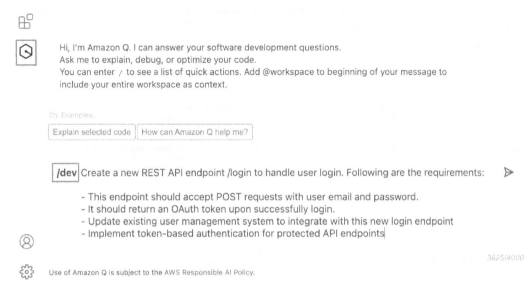

Hi, I'm Amazon Q. I can answer your software development questions.
Ask me to explain, debug, or optimize your code.
You can enter / to see a list of quick actions. Add @workspace to beginning of your message to include your entire workspace as context.

Try Examples.

| Explain selected code | How can Amazon Q help me? |

/dev Create a new REST API endpoint /login to handle user login. Following are the requirements:

- This endpoint should accept POST requests with user email and password.
- It should return an OAuth token upon successfully login.
- Update existing user management system to integrate with this new login endpoint
- Implement token-based authentication for protected API endpoints

3625/4000

Use of Amazon Q is subject to the AWS Responsible AI Policy.

main*   ⊗ 0 ⚠ 0   0   ▷ Amazon Q

Figure 12.36 – Amazon Q feature development – ask Q what you want to build

To maximize the benefits of Amazon Q feature development capabilities, some best practices need to be followed. Firstly, provide a detailed description of the new feature or code changes you want to implement, including specific objectives and outcomes. This helps Amazon Q generate a comprehensive implementation plan and the necessary code changes. Also, limit your feature to updates affecting no more than five files at a time. Requesting Amazon Q to handle larger changes might compromise the quality and manageability of the feature's implementation. If the file difference shows modifications to numerous files, consider narrowing the scope of your feature description.

Once you request what feature you want to develop, Q will create a very detailed plan that lays down all the steps, including files where the logic will be created, what the logic will do, what functions it will add, different imports it will perform, whether the class file will be new or a modification to an existing one, and will also provide test case generation.

The next few series of screenshots will show you the plan the Amazon Q Developer Agent for software development came up with. The following screenshot highlights the creation of a new class file for implementing the user model along with the authentication service.

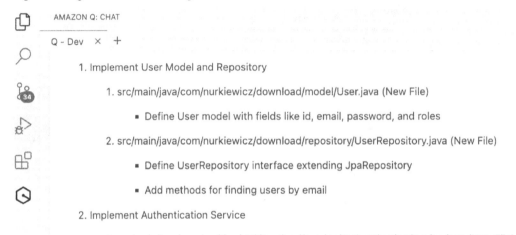

AMAZON Q: CHAT

Q - Dev   ×   +

1. Implement User Model and Repository

    1. src/main/java/com/nurkiewicz/download/model/User.java (New File)

       ▪ Define User model with fields like id, email, password, and roles

    2. src/main/java/com/nurkiewicz/download/repository/UserRepository.java (New File)

       ▪ Define UserRepository interface extending JpaRepository

       ▪ Add methods for finding users by email

2. Implement Authentication Service

    1. src/main/java/com/nurkiewicz/download/service/AuthenticationService.java (New File)

       ▪ Define AuthenticationService interface

       ▪ Add methods for authenticating users and generating tokens

    2. src/main/java/com/nurkiewicz/download/service/AuthenticationServiceImpl.java (New File)

       ▪ Implement AuthenticationService interface

       ▪ Authenticate users by checking email and password with UserRepository

       ▪ Generate JWT tokens using a library like java-jwt

Figure 12.37 – Amazon Q feature development – plan Steps 1 and 2

The next screenshot shows the plan for adding a new authentication controller file. Additionally, a new file is added to update the security configuration.

AMAZON Q: CHAT

Q - Dev   ×   +

3. Implement Authentication Controller

  1. src/main/java/com/nurkiewicz/download/controller/AuthenticationController.java (New File)

    ▪ Define AuthenticationController class

    ▪ Inject AuthenticationService

    ▪ Add @PostMapping("/login") method

      ▪ Accept request body with email and password

      ▪ Call AuthenticationService to authenticate and generate token

      ▪ Return token in response

4. Update Security Configuration

  1. src/main/java/com/nurkiewicz/download/config/SecurityConfig.java (New File)

    ▪ Configure JWT authentication filter

    ▪ Configure paths requiring authentication

    ▪ Import required classes (e.g., JwtAuthenticationFilter, AuthenticationManager)

Figure 12.38 – Amazon Q feature development – plan Steps 3 and 4

The next screenshot shows that the existing download controller file is updated to enable the authentication mechanism. It also indicates that the agent will create unit test cases for the new logic.

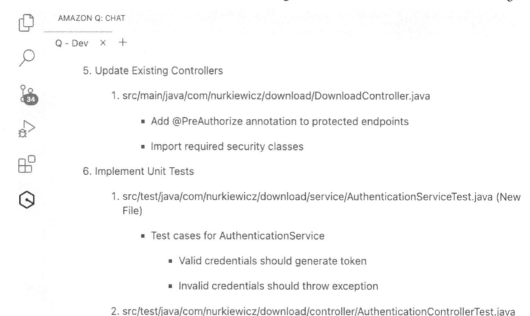

5. Update Existing Controllers

    1. src/main/java/com/nurkiewicz/download/DownloadController.java

        ■ Add @PreAuthorize annotation to protected endpoints

        ■ Import required security classes

6. Implement Unit Tests

    1. src/test/java/com/nurkiewicz/download/service/AuthenticationServiceTest.java (New File)

        ■ Test cases for AuthenticationService

            ■ Valid credentials should generate token

            ■ Invalid credentials should throw exception

    2. src/test/java/com/nurkiewicz/download/controller/AuthenticationControllerTest.java (New File)

        ■ Test cases for AuthenticationController

            ■ Valid request should return token

            ■ Invalid request should return error

Figure 12.39 – Amazon Q feature development – plan Steps 5 and 6

Finally, the following screenshot shows the last step of the plan, where it even proposes to add integration test logic. As you have seen from the plan, Q does a thorough job of following the best practices of building software. Instead of developers spending weeks to come up with this plan, Q did it in minutes.

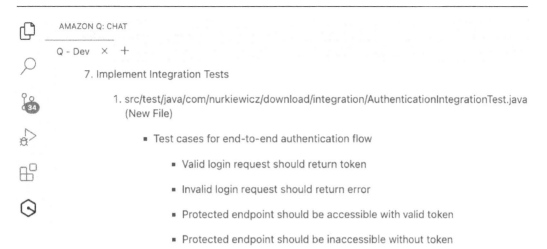

7. Implement Integration Tests

    1. src/test/java/com/nurkiewicz/download/integration/AuthenticationIntegrationTest.java (New File)

        ■ Test cases for end-to-end authentication flow

          ■ Valid login request should return token

          ■ Invalid login request should return error

          ■ Protected endpoint should be accessible with valid token

          ■ Protected endpoint should be inaccessible without token

8. Update Documentation and README

    ○ Update project documentation and README with details about authentication and authorization

You have 9 out of 10 plan iterations remaining.

Note: Q will use its largest, most capable model for code generation. This model may continue to optimize this plan as it generates your diff.

Figure 12.40 – Amazon Q feature development – plan Steps 7 and 8

The creation of the plan is cool by itself, but what's even cooler is the fact that the Q agent can auto-generate all this logic by itself and place it in the project for you to review and accept. The following screenshot shows the **Generate code** button offered at the end of the plan in the **AMAZON Q: CHAT** window.

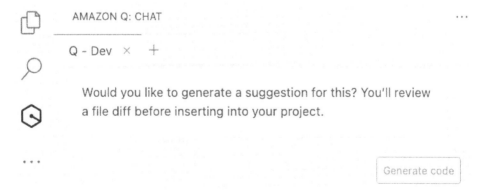

Figure 12.41 – Amazon Q feature development – option to generate code

Once this button is clicked, Q will provide a summary of all the code changes in the respective files. The following screenshot highlights this aspect.

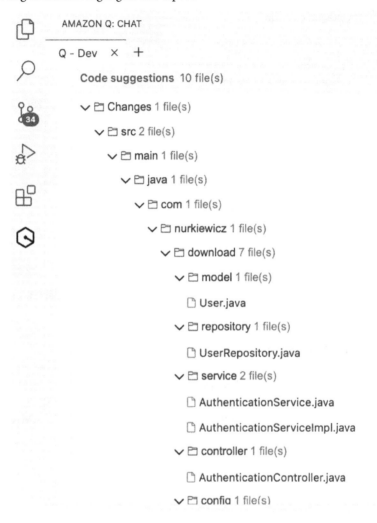

Figure 12.42 – Amazon Q feature development – summarize code changes

At the end of this summary, you will be presented with an **Insert code** button, which will instruct Q to place all this code back in the project. The following screenshot shows the **Insert code** option in the **AMAZON Q: CHAT** window.

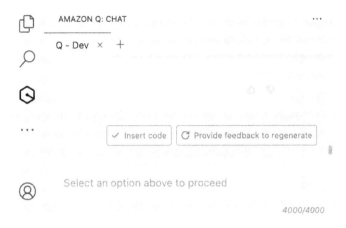

Figure 12.43 – Amazon Q feature development – insert code changes

Once the code is inserted into the project, you can see a list of all the changes ready for review and commit, as shown in the following screenshot.

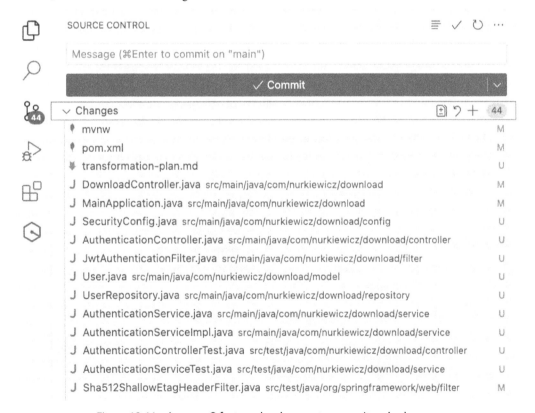

Figure 12.44 – Amazon Q feature development – commit code changes

However, as always, it's good practice to review the changes before committing. When you open the code files, the IDE will present the before and after changes side by side, allowing you to compare them. Some files were added as brand new in the process, so the agent generated the entire class file based solely on the feature development request.

The following screenshot shows a side-by-side comparison of the code before and after the change. This gives you a chance to review it before committing and pushing it to the code repository.

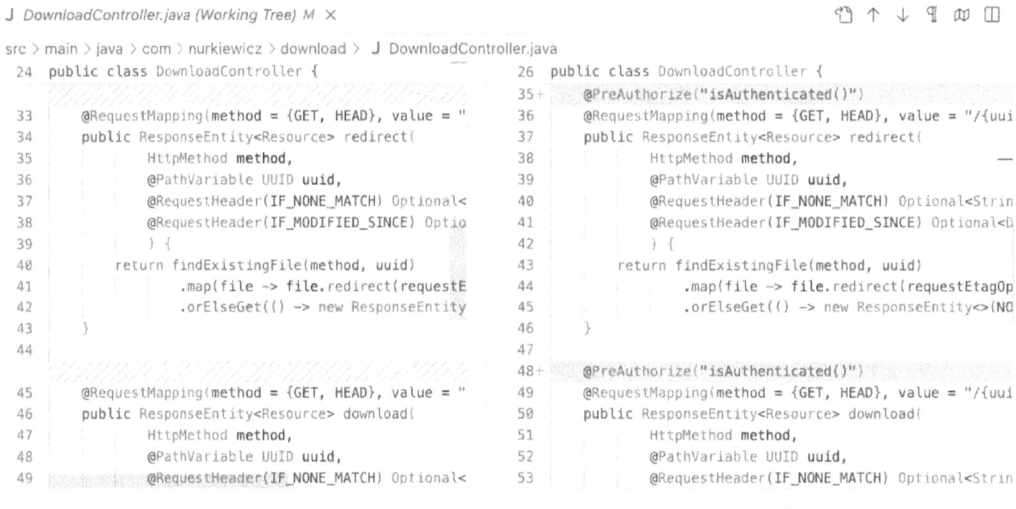

Figure 12.45 – Amazon Q feature development – compare code changes

This was just one example of how the feature development component of Amazon Q Developer can speed up developer productivity. Give it a try to generate many more features for your real-world projects in your organization.

With this, we have come to the end of this important chapter.

Summary

In this chapter, we covered three important components of Amazon Q Developer: code explanation and update, code transformation, and feature development.

The code explanation feature allows developers to understand the logic in the code by providing explanations. This makes it easier for them to comprehend the code and make necessary updates. Additionally, this feature enables developers to refactor, optimize, and fix code directly within the IDE. We also looked at how Amazon Q Developer can answer questions for the whole project with the workspace context-aware feature.

Next, we explored the code transformation feature, which enables Q to upgrade Java projects from version 8 or 11 to version 17 in a matter of minutes. The agent responsible for transformation analysis generates a detailed transformation plan and implements the necessary changes for developers to accept and commit in the code repository.

Finally, we examined the feature development component of Q, where developers can request Q to build new features in the project by describing the use case in the chat. The software development agent in Q analyzes and devises a detailed plan for updating the existing project to incorporate the new feature.

All these features of Amazon Q Developer tremendously boost the productivity of developers. In the next chapter, we will explore another crucial aspect of software development: code scanning for identifying and fixing security vulnerabilities in the code.

References

- Amazon Q Developer Agent for code transformation: `https://aws.amazon.com/q/developer/code-transformation/`

- Upgrading language versions with the Amazon Q Developer Agent for code transformation: `https://docs.aws.amazon.com/amazonq/latest/qdeveloper-ug/code-transformation.html`

- Troubleshooting issues with the Amazon Q Developer Agent for code transformation: `https://docs.aws.amazon.com/amazonq/latest/qdeveloper-ug/troubleshooting-code-transformation.html`

13

Simplifying Scanning and Fixing Security Vulnerabilities in Code

In this chapter, we will go through the following key topics:

- The importance of code-security scanning
- Types of code scans using Amazon Q
- Initiating security scans
- Addressing vulnerability findings

This chapter is short; however, one could argue that anything related to software security is never brief and takes top priority over everything else. Before we explore the features related to code-security scanning using Amazon Q Developer, let's reiterate the importance of this subject.

The importance of code-security scanning

Code-security scanning is a critical practice in software development, aimed at identifying and mitigating potential security vulnerabilities in the code base before they can be exploited. Here are several reasons why code-security scanning is essential:

- **Preventing security breaches**: Regular code-security scanning helps in identifying vulnerabilities that could be exploited by attackers. By catching these issues early in the development process, organizations can prevent data breaches, financial losses, and damage to their reputation.

- **Compliance with regulations**: Many industries are subject to strict regulatory requirements regarding data protection and security, such as the **General Data Protection Regulation (GDPR)**, the **Health Insurance Portability and Accountability Act (HIPAA)**, and the **Payment Card Industry Data Security Standard (PCI-DSS)**. Code-security scanning ensures that the software complies with these regulations, thereby avoiding legal penalties and ensuring customer trust.

- **Protecting sensitive information**: Applications often handle sensitive information such as personal data, financial records, and proprietary business information. Security vulnerabilities in the code can lead to unauthorized access to this data. Regular scanning helps to safeguard this sensitive information by identifying and addressing security flaws.

- **Reducing costs**: Fixing security vulnerabilities early in the development cycle is significantly less costly than addressing them after deployment. Post-deployment fixes can involve not only code changes but also compensations, legal fees, and damage control measures.

- **Maintaining software integrity**: Code-security scanning helps maintain the integrity of the software by ensuring that no malicious code or backdoors are present. This is particularly important for software that is distributed to end users, as it helps maintain trust and reliability in the product.

- **Facilitating continuous improvement**: By integrating code-security scanning into the development process, organizations can continuously improve their security posture. Scanning tools can provide feedback and recommendations, allowing developers to learn from past mistakes and avoid introducing similar vulnerabilities in the future.

- **Supporting secure development practices**: Regular security scanning reinforces the importance of secure coding practices among developers. It encourages a security-first mindset, making developers more aware of common security pitfalls and best practices to avoid them.

- **Enhancing customer confidence**: Customers and clients are increasingly aware of cybersecurity threats. Demonstrating a commitment to security through regular code scanning can enhance customer confidence and trust in the organization's products and services.

Code-security scanning for vulnerabilities is a proactive measure that plays a crucial role in the overall security strategy of any software development organization. By identifying and addressing security issues early, companies can protect sensitive data, comply with regulations, reduce costs, and maintain the integrity and trustworthiness of their software products.

Now let's shift our focus to what Amazon Q Developer can assist with when it comes to code scanning and fixing vulnerabilities in the code.

Types of code scans using Amazon Q

Amazon Q scans utilize security detectors built on years of Amazon standards and security best practices. As security policies evolve and new detectors are introduced, scans automatically integrate these updates to ensure your code remains compliant with the latest policies.

The security detectors are powered by Amazon CodeGuru, a developer tool that analyzes code and offers smart recommendations to enhance code security and quality. A link to all the different detector libraries for different programming languages is provided in the *References* section at the end of this chapter. Note that the detectors cover hundreds of code recommendations for each of the supported programming languages. For the purpose of understanding the security scan feature of Amazon Q Developer, we will just pick a couple of vulnerabilities detected in our sample code in the next section.

Amazon Q ensures your code's security by identifying policy violations and vulnerabilities through **static application security testing** (**SAST**), secrets detection, and **infrastructure as code** (**IaC**) scanning. The AWS documentation also highlights this list, the link to which is provided in the *References* section. This list may expand to other types of scans in the future, so keep a close eye on the official documentation for updates:

- **SAST scanning**: This type of scan is done before the code complication stage and is used by the application security team to catch vulnerabilities in the source code itself. It is also referred to as white-box testing. Amazon Q uncovers issues in the source code, such as resource leaks, SQL injection, and cross-site scripting.

- **Secrets scanning**: Safeguard sensitive information from exposure in your code base. Amazon Q checks your code and text files for secrets such as hardcoded passwords, database connection strings, and usernames, providing details about the vulnerabilities and recommendations for securing them.

- **IaC scanning**: Assess the security of your infrastructure files. Amazon Q reviews your IaC files to identify misconfigurations, compliance issues, and security vulnerabilities.

Now let's jump into how you can start using the security scan features of Amazon Q.

Initiating security scans

Amazon Q Developer provides two variations of the code-security scans.

Scanning while coding

Some developers prefer the IDE to warn them of any security vulnerabilities in their code as they type. Amazon Q's auto-scan feature continuously monitors the file you're actively working on, generating findings immediately as they are detected in your code.

> **Note**
>
> This feature is only available for use in the Pro tier and is enabled by default when using Amazon Q Developer. If you have subscribed to the Pro tier of Amazon Q Developer, then the auto-scan feature will appear in the Amazon Q menu option, where you can even pause it if you prefer to run the entire scan at once at the project level.

The following screenshot illustrates this feature in the VS Code IDE.

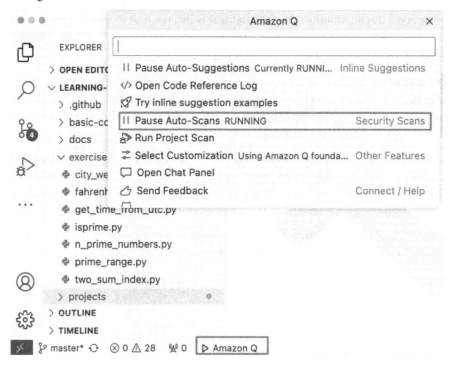

Figure 13.1 – Amazon Q security scan – auto-scan settings

If the auto-scan feature is running, as soon as you code something that would be considered a security vulnerability, Q will immediately flag it and let you know the issue, providing an explanation for it.

The following screenshot shows an obvious security violation we introduced in our code, and Q immediately flagged it and provided us with the reason on the spot.

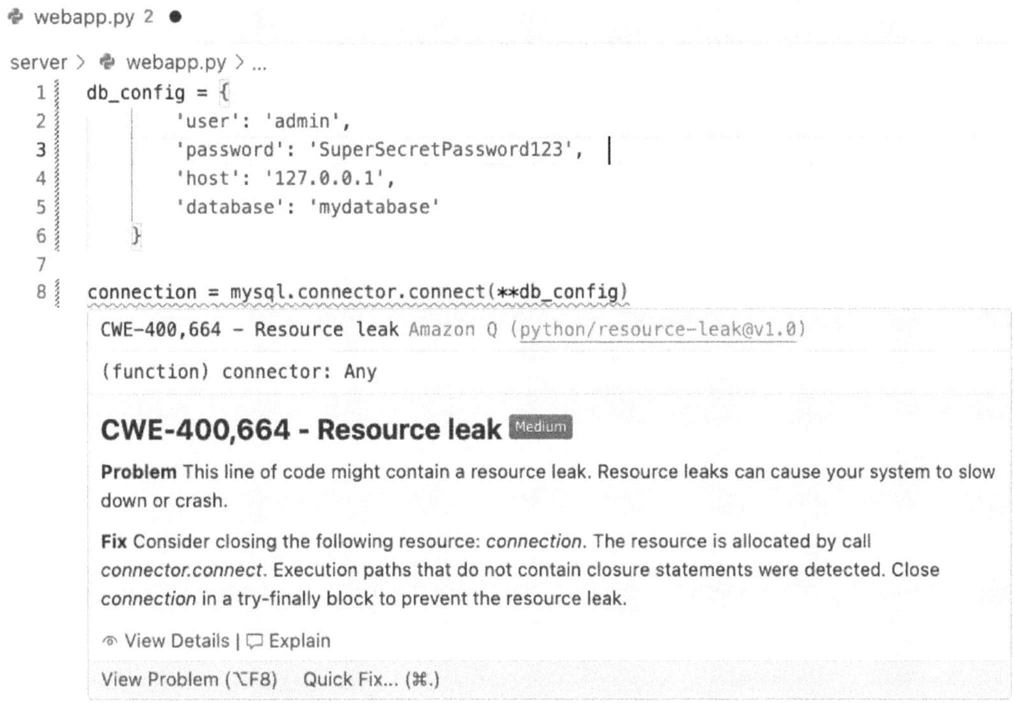

```
webapp.py 2 ●

server > webapp.py > ...
  1  db_config = {
  2          'user': 'admin',
  3          'password': 'SuperSecretPassword123',  |
  4          'host': '127.0.0.1',
  5          'database': 'mydatabase'
  6      }
  7
  8  connection = mysql.connector.connect(**db_config)
```

CWE-400,664 – Resource leak Amazon Q (python/resource-leak@v1.0)

(function) connector: Any

CWE-400,664 - Resource leak `Medium`

Problem This line of code might contain a resource leak. Resource leaks can cause your system to slow down or crash.

Fix Consider closing the following resource: *connection*. The resource is allocated by call *connector.connect*. Execution paths that do not contain closure statements were detected. Close *connection* in a try-finally block to prevent the resource leak.

👁 View Details | 💬 Explain

View Problem (⌥F8)   Quick Fix... (⌘.)

Figure 13.2 – Amazon Q security scan – auto-scan detected an issue

Now let's look at the open scan option Amazon Q provides.

Scanning the whole project

If the developer prefers to scan the project after it's complete, then they can use this option provided by Amazon Q, which scans the entire project and provides all the security vulnerabilities at once. This option is available in both tiers of Amazon Q Developer.

To initiate this, open the project and simply run the **Run Project Scan** option for Amazon Q, as shown in the following screenshot.

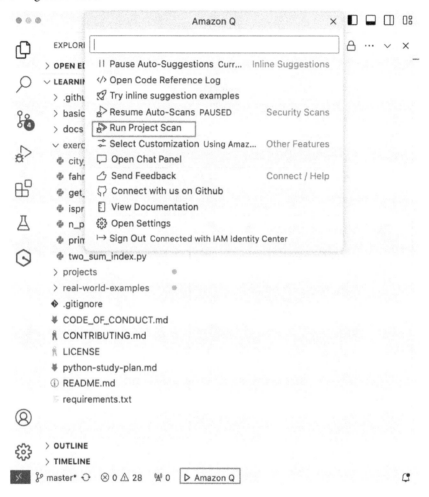

Figure 13.3 – Amazon Q security scan – scan whole project

Once the scan is complete, Amazon Q will provide a summary of all the security problems for the developer to review and take action to address them. The following screenshot shows all the issues we found in the Python project we used in *Chapter 12*. In case you skipped that chapter, here is the GitHub link to the project for you to try it out again in this chapter: `https://github.com/jassics/learning-python`.

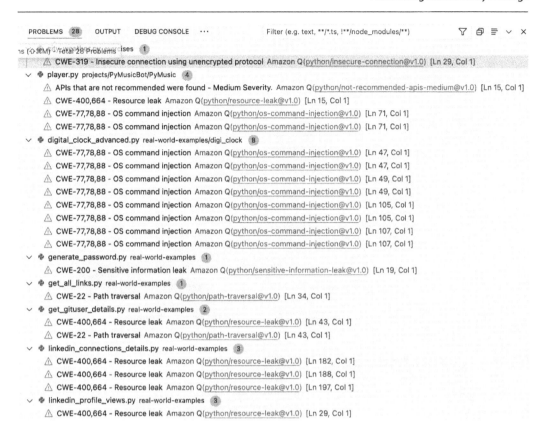

Figure 13.4 – Amazon Q security scan – security scan findings

Getting a report of issues in the code is one thing, but taking appropriate action to address them is the main reason we discovered them in the first place. Amazon Q Developer can also help address the vulnerability findings, so let's take a look at the options we have.

Addressing vulnerability findings

Amazon Q Developer provides a few options once it identifies the issue. It can either provide you with further details about the issue or you can use the Q chat feature to explain the finding along with recommendations to fix it. Or sometimes, you will also see an automatic fix option when Q is able to resolve it on its own.

The following screenshot shows the actions we can take for the issues we discussed in the project from the previous section. Of course, in this case, the printing of the password was done on purpose to showcase the password generation feature, but Q's scan was able to detect the issue and present actions you can take on it.

```
🐍 generate_password.py 1 ✕

real-world-examples > 🐍 generate_password.py > ...
  9    alphabet = letters + digits + special_chars
 10
 11    # define password length
 12    password_len = 12
 13
 14    password = ''
 15
 16    for i in range(password_len):
 17        password += ''.join(secrets.choice(alphabet))
 18    💡
 19    print(f"Password is: {password}")
 20
 21        ┌─────────────────────────────────────────────────────────────┐
           │ Quick Fix                                                     │
           │                                                               │
           │ 💡 Amazon Q: View details for "CWE-200 - Sensitive information leak" │
           │ 💡 Amazon Q: Explain "CWE-200 - Sensitive information leak"   │
           └─────────────────────────────────────────────────────────────┘
```

Figure 13.5 – Amazon Q security scan – options to address findings

Viewing security findings in detail

Looking at *Figure 13.5*, if you choose the **View details** option, Amazon Q will describe in detail what the issue is and how you can approach coming up with a resolution. It basically lets you make an informed decision to improve your code. The following screenshot shows this option.

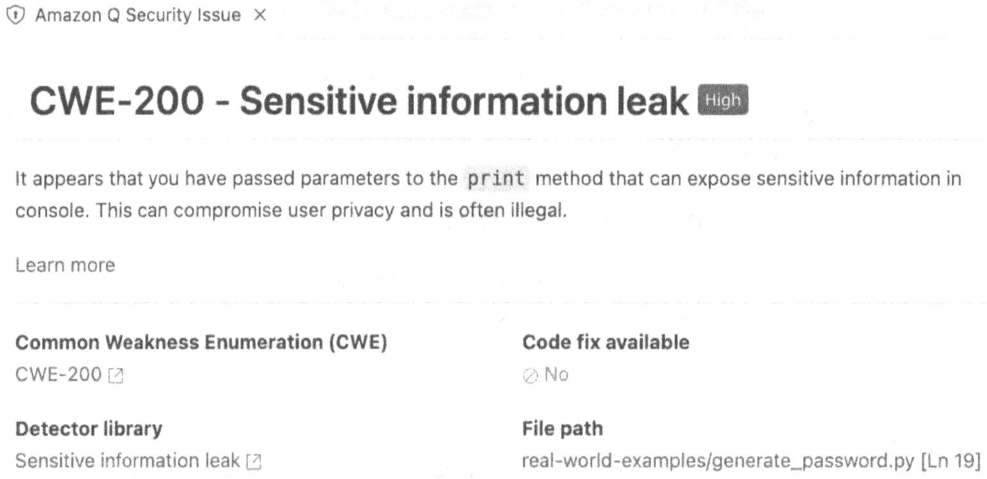

ⓘ Amazon Q Security Issue   ✕

CWE-200 - Sensitive information leak High

It appears that you have passed parameters to the `print` method that can expose sensitive information in console. This can compromise user privacy and is often illegal.

Learn more

Common Weakness Enumeration (CWE)	**Code fix available**
CWE-200 ↗	⊘ No
Detector library	**File path**
Sensitive information leak ↗	real-world-examples/generate_password.py [Ln 19]

Figure 13.6 – Amazon Q security scan – detailed view of the finding

Let's now look at the Q explanation option you can choose.

Seeking an explanation of the issue

Looking again at *Figure 13.5*, if you choose the **Explain** option, Amazon Q will let you seek an interactive explanation for the issue, where Q will invoke its chat feature and provide you with the explanation along with recommended fixes that you can insert back into your code. The following screenshot shows the explanation for the same issue we just looked at in the previous section.

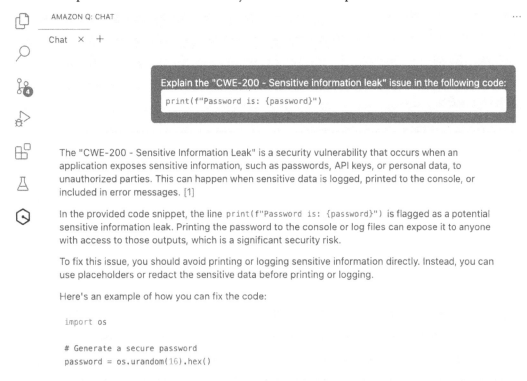

Figure 13.7 – Amazon Q security scan – explain finding

This brings us to the end of this important chapter. Feel free to try this feature on your own projects to see what you discover. The mechanism for detecting and viewing issues using Amazon Q Developer, along with seeking explanations and fixing them, remains the same across all types of security issues, hence we will keep this chapter short.

Summary

In this chapter, we covered why code security scans are vital for every project. No software project can be put into production without a clear strategy for finding security issues in the code and addressing them.

We then looked at the different types of scans that Amazon Q Developer provides. We also explored how to initiate security scans in Amazon Q, either with the auto-scan option or by scanning at the project level. Finally, we examined how Amazon Q can help address all the findings.

In the next chapter, we will pivot toward using Amazon Q Developer to create solutions in AWS environments.

References

- Amazon CodeGuru Detector Library: `https://docs.aws.amazon.com/codeguru/detector-library/`
- Types of security scans by Amazon Q Developer: `https://docs.aws.amazon.com/amazonq/latest/qdeveloper-ug/security-scans.html`

Part 4:
Accelerate Build on AWS

In this part, we will look at how Amazon Q Developer can assist in building applications faster on AWS. Amazon Q integrates with various AWS services and tools to provide a holistic assistant experience. Amazon Q not only boosts the productivity of AWS builders but also provides an integrated and seamless working experience.

This part contains the following chapters:

14

Accelerate Data Engineering on AWS

In this chapter, we will look at the following key topics:

- Code assistance options with AWS services
- Code assistance integration with AWS Glue
- Code assistance integration with Amazon EMR
- Code assistance integration with AWS Lambda
- Code assistance integration with Amazon SageMaker
- Code assistance integration with Amazon Redshift

In the previous part of the book, we explored auto-code generation techniques and the integration of a code companion with **integrated development environments** (**IDEs**) and provided examples using JetBrains PyCharm IDE with Amazon Q Developer for different languages that developers use very often. In this chapter, we will specifically focus on how Amazon is expanding in the area of assisting code developers by integrating with core AWS services.

Code assistance options with AWS services

AWS users select diverse services, considering factors such as the unique requirements of their projects, use cases, developers' technical needs, developer preferences, and the characteristics of AWS services. To cater to various developer personas, such as data engineers, data scientists, application developers, and so on, AWS has integrated code assistance with many of its code services. If you are an application builder, software developer, data engineer, or data scientist working with AWS services, you would frequently use builder-friendly tools such as Amazon SageMaker as a platform for building AI / **machine learning** (**ML**) projects, Amazon EMR as a platform for building big data processing projects,

AWS Glue for building **extract, transform, and load** (ETL) pipelines, AWS Lambda as a serverless compute service for application development. All these services provide tools that help builders and developers write code.

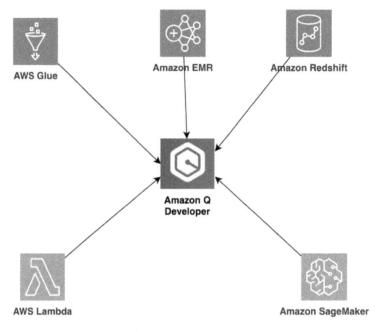

Figure 14.1 – Code assistance options with AWS services

As of the writing of this book, AWS has integrated Amazon Q Developer with AWS Glue, Amazon EMR, AWS Lambda, Amazon SageMaker, and Amazon Redshift. However, we anticipate that the list of services benefiting from code assistance, such as Amazon Q Developer, will continue to expand in the future.

In the following sections, we will dive deep into each of these services, examining their integration with Amazon Q in detail. We will provide examples that will be helpful for data engineers to accelerate development on AWS.

> **Note**
>
> **Large language models** (LLMs), by nature, are non-deterministic, so you may not get the same code blocks shown in the code snapshots. However, logically, the generated code should meet the requirements.
>
> **CodeWhisperer** is a legacy name from a service that merged with Amazon Q Developer. As of the time of writing this book, some of the integrations are still referred to as CodeWhisperer in the AWS console, which may change in the future.

Code assistance integration with AWS Glue

Before we start diving deep into code assistance support for AWS Glue service, let's quickly go through an overview of AWS Glue. **AWS Glue** is a serverless data integration service designed to simplify the process of discovering, preparing, moving, and integrating data from diverse sources, catering to analytics, ML, and application development needs. At the very high level, AWS Glue has the following major components, and each of them has multiple features to support data engineers:

- **Glue Data Catalog**: It's a centralized technical metadata repository. It stores metadata about data sources, transformations, and targets, providing a unified view of the data.

- **Glue Studio**: AWS Glue Studio offers a graphical interface that facilitates the seamless creation, execution, and monitoring of data integration jobs within AWS Glue. Additionally, it provides Jupyter notebooks for advanced developers.

AWS Glue Studio is seamlessly integrated with Amazon Q Developer. Let's explore the further functionality by considering a very common use case of data enrichment.

Use case for AWS Glue

Features and functionalities of any service or tool are best understood when we have a use case to solve. So, let's start with one of the easy and widely used use cases of data enrichment using lookups.

Data enrichment using lookup: In a typical scenario, business analysts often require data enrichment by incorporating details associated with codes/IDs found in a column through a lookup table. The desired result is a comprehensive and denormalized record containing both the code and corresponding details in the same row. To address this specific use case, data engineers develop ETL jobs to join the tables, creating the final structure with a denormalized dataset.

To illustrate this use case, we will use yellow taxi trip records that encompass details such as the date and time of pick-up and drop-off, the locations for pick-up and drop-off, the trip distance, comprehensive fare breakdowns, various rate types, utilized payment methods, and passenger counts reported by the driver. Additionally, trip information incorporates passenger location codes for both pick-up and drop-off.

The business objective is to enhance the dataset with zone information based on the pick-up location code.

To meet this requirement, data engineers must develop a PySpark ETL script. This script should perform a lookup for zone information corresponding to the pick-up location code. Subsequently, the engineers create denormalized/enriched data by amalgamating yellow taxi trip data with detailed pick-up zone information and save the result as a file.

As a code developer / data engineer, you will need to convert the preceding business objectives into technical requirements.

Solution blueprint

1. Write a PySpark code to handle technical requirements.

2. Read the `yellow_tripdata_2023-01.parquet` file from the S3 location in a DataFrame and display a sample of 10 records.

3. Read the `taxi+_zone_lookup.csv` file from the S3 location in a DataFrame and display a sample of 10 records.

4. Perform a left outer join on `yellow_tripdata_2023-01.parquet` and `taxi+_zone_lookup.csv` on `PULocationID = LocationID` to gather pick-up zone information.

5. Save the preceding dataset as a CSV file in the preceding Amazon S3 bucket in a new `glue_notebook_yellow_pick_up_zone_output` folder.

6. For verification, download and check the files from the `glue_notebook_yellow_pick_up_zone_output` folder.

Now that we have a use case defined, let's go through the step-by-step solution for it.

Data preparation

The first step will be to prepare the data. To illustrate its functionality, in the following sections, we will utilize the publicly available NY Taxi dataset from TLC Trip Record Data. `https://www.nyc.gov/site/tlc/about/tlc-trip-record-data.page`.

Firstly, we will download the required files on a local machine and then upload them in one of Amazon's S3 buckets:

1. Download the Yellow Taxi Trip Records data for the Jan 2023 Parquet file (`yellow_tripdata_2023-01.parquet`) on a local machine from `https://d37ci6vzurychx.cloudfront.net/trip-data/yellow_tripdata_2023-01.parquet`.

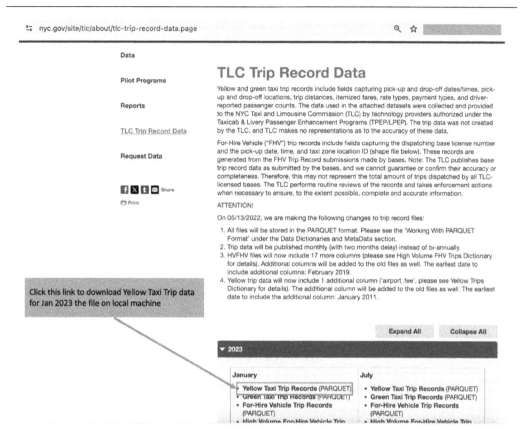

Figure 14.2 – The Yellow Taxi Trip Records data for Jan 2023 Parquet file

2. Download the Taxi Zone Lookup Table CSV file (taxi+_zone_lookup.csv) on a local machine from https://d37ci6vzurychx.cloudfront.net/misc/taxi+_zone_lookup.csv.

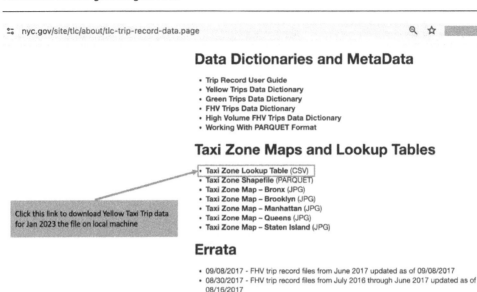

Figure 14.3 – The Zone Lookup Table CSV file

3. Create the two `yellow_taxi_trip_records` and `zone_lookup` folders in Amazon S3, which we can reference in our Glue notebook job.

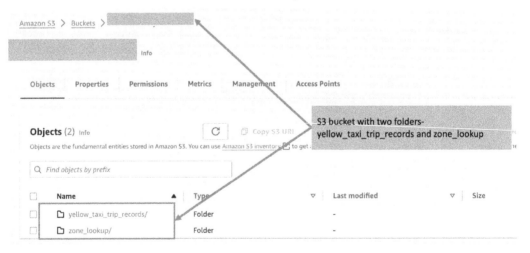

Figure 14.4 – S3 folders structure

4. Upload the `yellow_tripdata_2023-01.parquet` file to the `yellow_taxi_trip_records` folder.

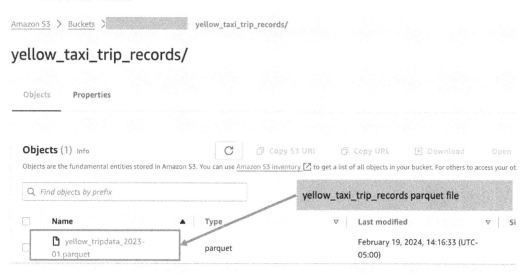

Figure 14.5 – The yellow_taxi_tripdata_record file

5. Upload the `taxi+_zone_lookup.csv` file to the `zone_lookup` folder.

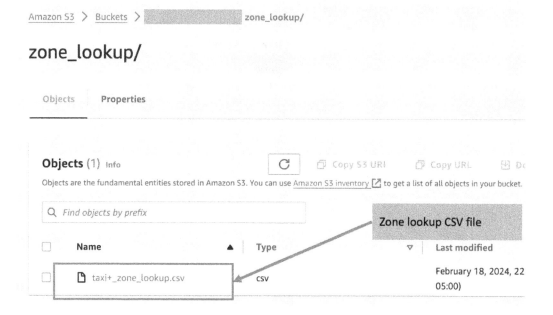

Figure 14.6 – The zone_lookup file

> **Note**
>
> We will use the same dataset and use case to discover solutions using AWS Glue and Amazon EMR. For illustrative purposes, we have prepared the data manually. However, in a production environment, file transfers can be automated by leveraging various AWS services and/or third-party software.

Now, let's dive deep into a detailed exploration of the solution using the integration of Amazon Q Developer with an AWS Glue Studio notebook for the preceding use case.

Solution – Amazon Q Developer with an AWS Glue Studio notebook

Let's first enable Amazon Q Developer with an AWS Glue Studio notebook.

Prerequisites to enable Amazon Q Developer with an AWS Glue Studio notebook

The developer is required to modify the **identity and access management** (**IAM**) policy associated with the IAM user or role to grant permissions for Amazon Q Developer to initiate recommendations in a Glue Studio notebook. Reference *Chapter 2* for the details to enable Amazon Q Developer with an AWS Glue Studio notebook.

To fulfill the previously mentioned solution blueprint, we will use various auto-code generation techniques that were discussed in *Chapter 3*. Mainly, we will focus on single-line prompts, multi-line prompts, and chain-of-thought prompts for auto-code generation.

Let's use Amazon Q Developer to auto-generate an end-to-end script in an AWS Glue Studio notebook. Here is the step-by-step solution walk-through for the previously defined solution blueprint.

Requirement 1

First, you need to write some PySpark code.

While creating a Glue Studio notebook, select the **Spark (Python)** engine and the role that has the Amazon Q Developer policy attached.

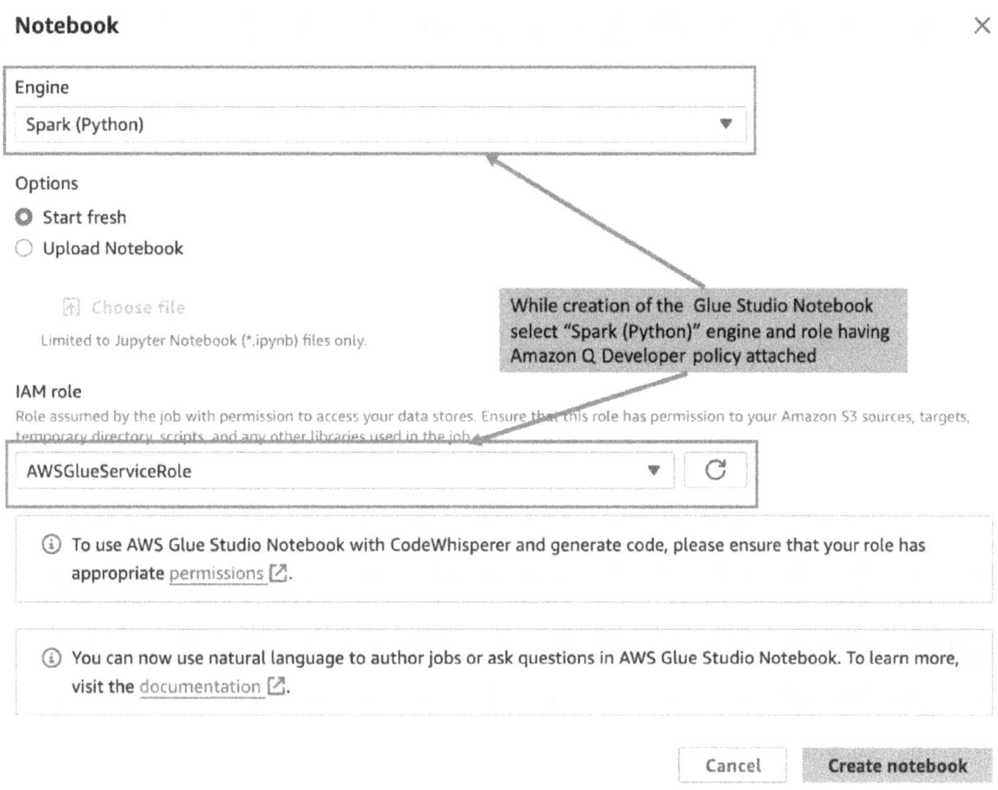

Notebook   ✕

Engine

```
Spark (Python)                                                            ▼
```

Options

◉ Start fresh
○ Upload Notebook

⬆ Choose file

Limited to Jupyter Notebook (*.ipynb) files only.

> While creation of the Glue Studio Notebook
> select "Spark (Python)" engine and role having
> Amazon Q Developer policy attached

IAM role

Role assumed by the job with permission to access your data stores. Ensure that this role has permission to your Amazon S3 sources, targets, temporary directory, scripts, and any other libraries used in the job.

```
AWSGlueServiceRole                                        ▼        ↻
```

ⓘ To use AWS Glue Studio Notebook with CodeWhisperer and generate code, please ensure that your role has appropriate permissions ⬈.

ⓘ You can now use natural language to author jobs or ask questions in AWS Glue Studio Notebook. To learn more, visit the documentation ⬈.

Cancel   **Create notebook**

Figure 14.7 – Create a Glue Studio notebook with PySpark

Once you create the notebook, observe the kernel named Glue PySpark.

Figure 14.8 – A Glue Studio notebook with the Glue PySpark kernel

Requirement 2

Read the yellow_tripdata_2023-01.parquet file from the S3 location in a DataFrame and display a sample of 10 records.

Let's use a chain-of-thought prompt technique with multiple single-line prompts in different cells to achieve the preceding requirement:

```
Prompt # 1:
# Read s3://<your-bucket-name-here>/yellow_taxi_trip_records/yellow_
```

```
tripdata_2023-01.parquet file in a dataframe


Prompt # 2:
# display a sample of 10 records from dataframe
```

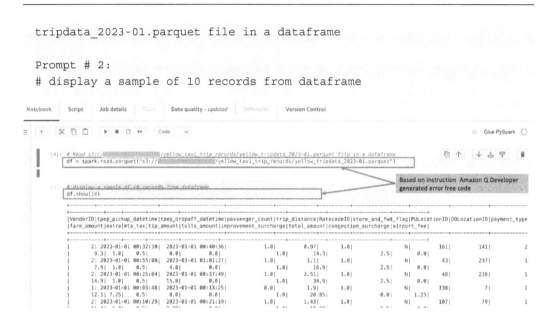

Figure 14.9 – PySpark code to read the Yellow Taxi Trip Records data using single-line prompts

Observe that upon entering the Amazon Q Developer-enabled Glue Studio notebook prompt, it initiates code recommendations. Q Developer recognizes the file format as Parquet and suggests using the spark.read.parquet method. You can directly execute each cell/code from the notebook. Furthermore, as you move to the next cell, Q Developer utilizes "line-by-line recommendations" to suggest displaying the schema.

Figure 14.10 – Line-by-line recommendations to display schema

Requirement 3

Read the `taxi+_zone_lookup.csv` file from the S3 location in a DataFrame and display a sample of 10 records.

We already explored the chain-of-thought prompt technique with multiple single-line prompts for *Requirement 2*. Now, let's try with a multi-line prompt to achieve the preceding requirement and we will try to customize the code for the DataFrame name:

```
Prompt:
"""
Read s3://<your-bucket-name-here>/zone_lookup/taxi+_zone_lookup.csv in
a dataframe name zone_df.
Show sample 10 records from zone_df.
"""
```

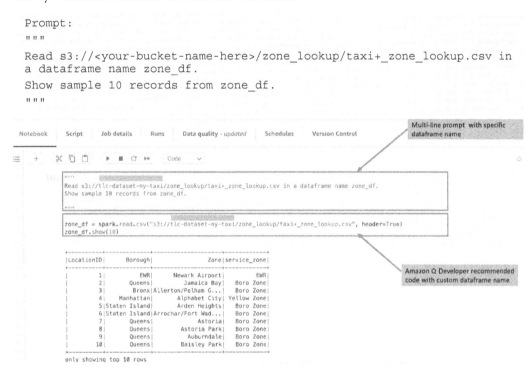

Figure 14.11 – PySpark code to read the Zone Lookup file using a multi-line prompt

Observe that Amazon Q Developer understood the context behind the multi-line prompt and also the specific DataFrame name instructed in the prompt. It auto-generated multiple lines of code with the DataFrame name as `zone_df` and file format as CSV, suggesting the use of the `spark.read.csv` method to read CSV files. You can directly execute each cell/code from the notebook.

Requirement 4

Perform a left outer join on `yellow_tripdata_2023-01.parquet` and `taxi+_zone_lookup.csv` on `pulocationid = LocationID` to gather pick-up zone information.

We will continue using multi-line prompts and some code customization to achieve the preceding requirement:

```
Prompt:
"""

Perform a left outer join on dataframe df and dataframe zone_df on
PULocationID = LocationID to save in dataframe name yellow_pu_zone_df.
Show sample 10 records from yellow_pu_zone_df and show schema.
"""
```

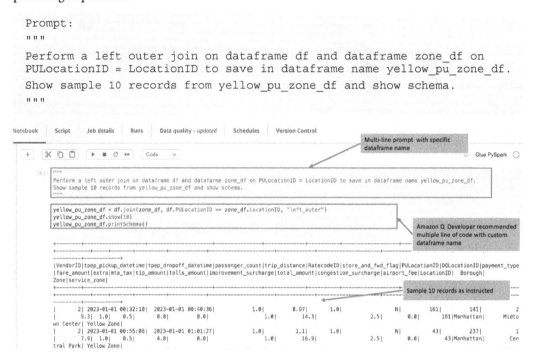

Figure 14.12 – Left outer join df and dataframe zone_df – multi-line prompt

Now, let's review the schema of the DataFrame returns by the code execution.

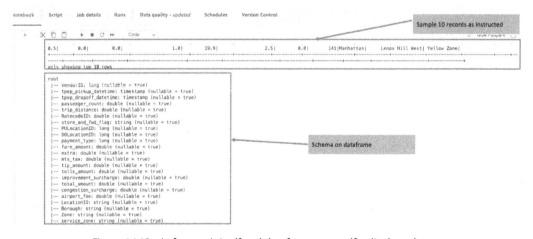

Figure 14.13 – Left outer join df and dataframe zone_df – display schema

Observe that, as instructed in the multi-line prompt, Amazon Q Developer understood the context and auto-generated error-free code with the exact specifications we provided related to the DataFrame name of `yellow_pu_zone_df`. You can directly execute each cell/code from the notebook.

Requirement 5

Save the preceding dataset as a CSV file in the preceding Amazon S3 bucket in a new folder called `glue_notebook_yellow_pick_up_zone_output`.

Since the preceding requirement is straightforward and can be encapsulated in a single sentence, we will use a single-line prompt to generate the code, and we will also include a header to facilitate easy verification:

```
Prompt:
# Save dataframe yellow_pu_zone_df as CSV file at location s3://<your-
bucket-name-here>/tlc-dataset-ny-taxi/glue_notebook_yellow_pick_up_
zone_output/ with header information
```

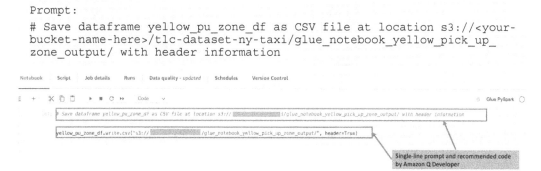

Figure 14.14 – Save the CSV file with enrichment pick-up location data

Requirement 6

For verification, download and check files from the `glue_notebook_yellow_pick_up_zone_output` folder.

Let's go to the Amazon S3 console to verify the files. Select one of the files and click **Download**.

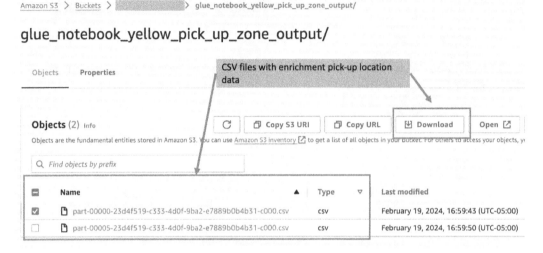

Figure 14.15 – Save a CSV file with enrichment pick-up location data

After downloading the file, you can use any text editor to review the file contents.

```
VendorID,tpep_pickup_datetime,tpep_dropoff_datetime,passenger_count,trip_distance,RatecodeID,store_and_fwd_flag,PULocationID,DOLocationID,payment_type,fare_amount,extra,m
ta_tax,tip_amount,tolls_amount,improvement_surcharge,total_amount,congestion_surcharge,airport_fee,LocationID,Borough,Zone,service_zone
2,2023-01-01T00:32:10.000Z,2023-01-01T00:40:36.000Z,1.0,0.97,1.0,N,161,141,2,9.3,1.0,0.5,0.0,0.0,1.0,14.3,2.5,0.0,161,Manhattan,Midtown Center,Yellow Zone
2,2023-01-01T00:55:08.000Z,2023-01-01T01:01:27.000Z,1.0,1.1,1.0,N,43,237,1,7.9,1.0,0.5,4.0,0.0,1.0,16.9,2.5,0.0,43,Manhattan,Central Park,Yellow Zone
2,2023-01-01T00:25:04.000Z,2023-01-01T00:37:49.000Z,1.0,2.51,1.0,N,48,238,1,14.9,1.0,0.5,15.0,0.0,1.0,34.9,2.5,0.0,48,Manhattan,Clinton East,Yellow Zone
1,2023-01-01T00:03:48.000Z,2023-01-01T00:13:25.000Z,0.0,1.9,1.0,N,138,7,1,12.1,7.25,0.5,0.0,0.0,1.0,20.85,0.0,1.25,138,Queens,LaGuardia Airport,Airports
2,2023-01-01T00:10:29.000Z,2023-01-01T00:21:19.000Z,1.0,1.43,1.0,N,107,79,1,11.4,1.0,0.5,3.28,0.0,1.0,19.68,2.5,0.0,107,Manhattan,Gramercy,Yellow Zone
2,2023-01-01T00:50:34.000Z,2023-01-01T01:02:52.000Z,1.0,1.84,1.0,N,161,137,1,12.8,1.0,0.5,10.0,0.0,1.0,27.8,2.5,0.0,161,Manhattan,Midtown Center,Yellow Zone
2,2023-01-01T00:09:22.000Z,2023-01-01T00:19:49.000Z,1.0,1.66,1.0,N,239,143,1,12.1,1.0,0.5,3.42,0.0,1.0,20.52,2.5,0.0,239,Manhattan,Upper West Side South,Yellow Zone
2,2023-01-01T00:27:12.000Z,2023-01-01T00:49:56.000Z,1.0,11.7,1.0,N,142,200,1,45.7,1.0,0.5,10.74,3.0,1.0,64.44,2.5,0.0,142,Manhattan,Lincoln Square East,Yellow Zone
2,2023-01-01T00:21:44.000Z,2023-01-01T00:36:40.000Z,1.0,2.95,1.0,N,164,236,1,17.7,1.0,0.5,5.68,0.0,1.0,28.38,2.5,0.0,164,Manhattan,Midtown South,Yellow Zone
2,2023-01-01T00:39:42.000Z,2023-01-01T00:50:36.000Z,1.0,3.01,1.0,N,141,107,2,14.9,1.0,0.5,0.0,0.0,1.0,19.9,2.5,0.0,141,Manhattan,Lenox Hill West,Yellow Zone
2,2023-01-01T00:53:01.000Z,2023-01-01T01:01:45.000Z,1.0,1.8,1.0,N,234,68,1,11.4,1.0,0.5,3.28,0.0,1.0,19.68,2.5,0.0,234,Manhattan,Union Sq,Yellow Zone
1,2023-01-01T00:43:37.000Z,2023-01-01T01:17:18.000Z,4.0,7.3,1.0,N,79,264,1,33.8,3.5,0.5,7.75,0.0,1.0,46.55,2.5,0.0,79,Manhattan,East Village,Yellow Zone
2,2023-01-01T00:34:44.000Z,2023-01-01T01:04:25.000Z,1.0,3.23,1.0,N,164,143,1,26.1,1.0,0.5,6.22,0.0,1.0,37.32,2.5,0.0,164,Manhattan,Midtown South,Yellow Zone
2,2023-01-01T00:09:29.000Z,2023-01-01T00:29:23.000Z,2.0,11.43,1.0,N,138,33,1,44.3,6.0,0.5,13.26,0.0,1.0,66.31,0.0,1.25,138,Queens,LaGuardia Airport,Airports
2,2023-01-01T00:33:53.000Z,2023-01-01T00:49:15.000Z,1.0,2.95,1.0,N,33,61,1,17.7,1.0,0.5,4.84,0.0,1.0,24.24,0.0,0.0,33,Brooklyn,Brooklyn Heights,Boro Zone
2,2023-01-01T00:13:04.000Z,2023-01-01T00:22:10.000Z,1.0,1.52,1.0,N,79,186,1,10.0,1.0,0.5,1.25,0.0,1.0,16.25,2.5,0.0,79,Manhattan,East Village,Yellow Zone
2,2023-01-01T00:45:11.000Z,2023-01-01T01:07:39.000Z,1.0,2.23,1.0,N,90,48,1,19.8,1.0,0.5,4.96,0.0,1.0,29.76,2.5,0.0,90,Manhattan,Flatiron,Yellow Zone
1,2023-01-01T00:04:33.000Z,2023-01-01T00:19:22.000Z,1.0,4.5,1.0,N,113,255,1,20.5,3.5,0.5,4.0,0.0,1.0,29.5,2.5,0.0,113,Manhattan,Greenwich Village North,Yellow Zone
1,2023-01-01T00:03:36.000Z,2023-01-01T00:09:36.000Z,3.0,1.2,1.0,N,237,239,2,8.6,3.5,0.5,0.0,0.0,1.0,13.6,2.5,0.0,237,Manhattan,Upper East Side South,Yellow Zone
1,2023-01-01T00:15:23.000Z,2023-01-01T00:29:41.000Z,2.0,2.5,1.0,N,143,229,2,15.6,3.5,0.5,0.0,0.0,1.0,20.6,2.5,0.0,143,Manhattan,Lincoln Square West,Yellow Zone
```

Figure 14.16 – Verify the CSV file with enrichment pick-up location data

Observe that the CSV file has additional columns with zone information based on the pick-up location ID. In the next section, we will explore Amazon Q Developer integration with AWS Glue and use the chat assistant technique.

> **Think challenge**
>
> To fulfill *Requirement 6*, if you are interested, attempt to utilize the same Glue Studio notebook for reading a CSV file, displaying sample records, and adding a header.
>
> **Hint**: Use the multi-line prompt technique, similar to the one we used when reading the Zone Lookup file.

Solution – Amazon Q Developer with AWS Glue

Amazon Q Developer provides a chat-style interface in the AWS Glue console. Now, let's explore the integration between Amazon Q Developer and AWS Glue for the same use case and solution blueprint that we handled using Amazon Q Developer and an AWS Glue Studio notebook integration.

Let's now look at the prerequisites to enable Amazon Q with AWS Glue.

To enable Amazon Q Developer integration with AWS Glue, we will need to update the IAM policy. Please refer to *Chapter 2* for additional details on initiating interaction with Amazon Q in AWS Glue.

Now, let's dive deep into a detailed exploration of the integration of Amazon Q Developer with AWS Glue Studio for the preceding use case.

To fulfill the mentioned requirements, we will mainly use the chat companion that was discussed in *Chapter 3*.

Here is a step-by-step solution walk-through that we'll use as a prompt for all of the preceding requirements:

```
Instruction to Amazon Q:
Write a Glue ETL job.
Read the 's3://<your bucket name>/yellow_taxi_trip_records/yellow_
tripdata_2023-01.parquet' file in a dataframe and display a sample of
10 records.
Read the 's3://<your bucket name>/zone_lookup/taxi+_zone_lookup.csv'
file in a dataframe and display a sample of 10 records.
Perform a left outer join on 'yellow_tripdata_2023-01.parquet' and
'taxi+_zone_lookup.csv' on DOLocationID = LocationID to gather pick-up
zone information.
Save the above dataset as a CSV file in above Amazon S3 bucket in a
new folder 'glue_notebook_yellow_drop_off_zone_output'.
```

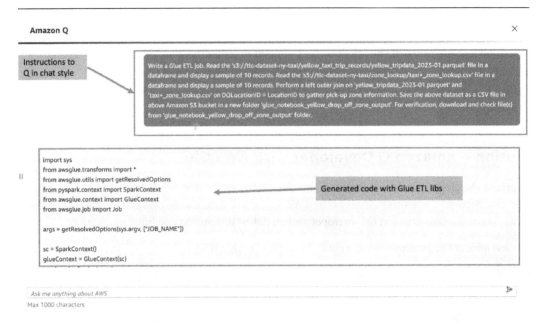

Figure 14.17 – The AWS Glue ETL code suggested by Amazon Q Developer – part 1

You can see that, based on the instruction provided to Amazon Q, it generated the skeleton on the ETL code. It generated code structure with Glue-PySpark libraries, a s3node with create dynamic dataframe to read parquet file, and a s3node with write dynamic dataframe to write CSV file.

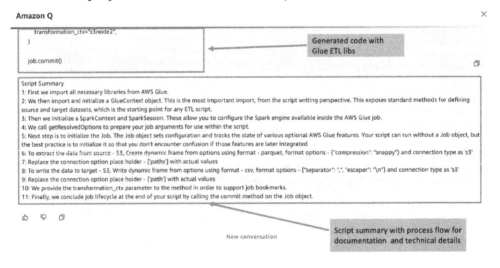

Figure 14.18 – AWS Glue ETL code suggested by Amazon Q Developer – part 2

Observe that Amazon Q also provided technical details to explain the script flow. This can also be used to meet the in-script documentation needs.

Script Summary

1: First we import all necessary libraries from AWS Glue.

2: We then import and initialize a GlueContext object. This is the most important import, from the script writing perspective. This exposes standard methods for defining source and target datasets, which is the starting point for any ETL script.

3: Then we initialize a SparkContext and SparkSession. These allow you to configure the Spark engine available inside the AWS Glue job.

4: We call getResolvedOptions to prepare your job arguments for use within the script.

5: Next step is to initialize the Job. The Job object sets configuration and tracks the state of various optional AWS Glue features. Your script can run without a Job object, but the best practice is to initialize it so that you don't encounter confusion if those features are later integrated.

6: To extract the data from source - S3, Create dynamic frame from options using format - parquet, format options - {"compression": "snappy"} and connection type as 's3'

7: Replace the connection option place holder - ['paths'] with actual values

8: To write the data to target - S3, Write dynamic frame from options using format - csv, format options - {"separator": ",", "escaper": "\n"} and connection type as 's3'

9: Replace the connection option place holder - ['path'] with actual values

10: We provide the transformation_ctx parameter to the method in order to support job bookmarks.

11: Finally, we conclude job lifecycle at the end of your script by calling the commit method on the Job object.

Figure 14.19 – AWS Glue ETL code suggested by Amazon Q Developer – script summary

Data engineers with coding experience can easily reference the script summary and script skeleton to write end-to-end scripts to meet the solution blueprint. LLMs, by nature, are non-deterministic, so you may not get the same code blocks shown in the code snapshots.

Based on the preceding use case illustration, AWS Glue integration with Amazon Q Developer with prompting techniques can be used by data engineers at a relatively lower experience level, while AWS Glue integration with Amazon Q Developer using the chat assistant can be utilized by ETL developers with relatively more experience.

Summary – Amazon Q Developer with an AWS Glue Studio notebook

As illustrated, we can automatically generate end-to-end, error-free, and executable code simply by providing prompts with specific requirements. Amazon Q Developer, integrated with an AWS Glue Studio notebook, comprehends the context and automatically generates PySpark code that can be run directly from the notebook without the need to provision any hardware upfront. This marks a significant advancement for many data engineers, relieving them from concerns about the technical intricacies associated with PySpark libraries, methods, and syntax.

Next, we will explore code assistance integration with Amazon EMR.

Code assistance integration with Amazon EMR

Before we dive deep into the details of code assistance support for Amazon EMR, let's quickly go through an overview of Amazon EMR. **Amazon EMR** is a cloud-based big data platform that simplifies the deployment, management, and scaling of various big data frameworks such as Apache Hadoop, Apache Spark, Apache Hive, and Apache HBase. At a high level, Amazon EMR comprises the following major components, each with multiple features to support data engineers and data scientists:

- **EMR on EC2/EKS**: The Amazon EMR service provides two options, EMR on EC2 and EMR on EKS, allowing customers to provision clusters. Amazon EMR streamlines the execution of batch jobs and interactive workloads for data analysts and engineers.

- **EMR Serverless**: Amazon EMR Serverless is a serverless alternative within Amazon EMR. With Amazon EMR Serverless, users can access the full suite of features and advantages offered by Amazon EMR, all without requiring specialized expertise for cluster planning and management.

- **EMR Studio**: EMR Studio supports data engineers and data scientists in developing, visualizing, and debugging applications within an IDE. It also provides a Jupyter Notebook environment for interactive coding.

Use case for Amazon EMR Studio

For simplicity and ease of following Amazon Q Developer integration with Amazon EMR, we will use the same use case and data that we used in this chapter under the *Code assistance integration with AWS Glue* section. Refer to the *Use case for AWS Glue* section, which covers details related to the solution blueprint and data preparation.

Solution – Amazon Q Developer with Amazon EMR Studio

Let's first enable Amazon Q Developer with Amazon EMR Studio. To enable Amazon Q Developer integration with Amazon EMR Studio, we will need to update the IAM policy.

Prerequisite to enable Amazon Q Developer with Amazon EMR Studio

The developer is required to modify the IAM policy associated with the role to grant permissions for Amazon Q Developer to initiate recommendations in EMR Studio. Please reference *Chapter 2* for additional details on initiating interaction with Amazon Q Developer in Amazon EMR Studio.

To fulfill the mentioned requirements, we will use various auto-code generation techniques that were discussed in *Chapter 3*. Mainly, we will focus on single-line prompts, multi-line prompts, and chain-of-thought prompts for auto-code generation techniques.

Let's use Amazon Q Developer to auto-generate end-to-end scripts, which can achieve the following requirements in Amazon EMR Studio. Here is a step-by-step solution walk-through of the solution.

> **Note**
> You can observe lots of similarities between a Glue Studio notebook and an EMR Studio notebook when it comes to code recommended by Amazon Q Developer.

Requirement 1

You will need to write a PySpark code to handle technical requirements.

Once you open Amazon EMR Studio, use **Launcher** to select **PySpark** from the **Notebook** section.

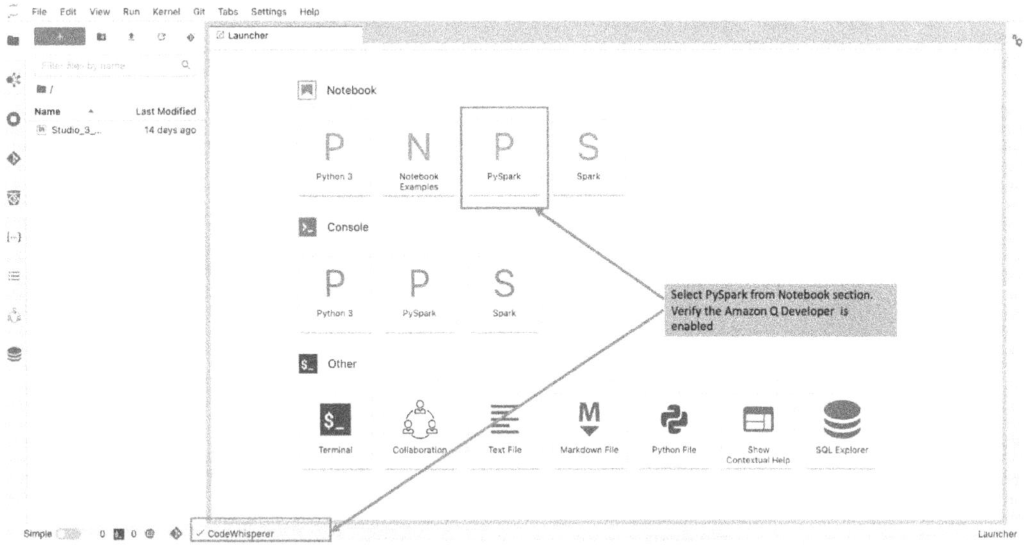

Figure 14.20 – Create an EMR Studio notebook with PySpark

Once you create the notebook, you can see a kernel named PySpark. The kernel is a standalone process that runs in the background and executes the code you write in your notebooks. For more information, refer to the *References* section at the end of the chapter.

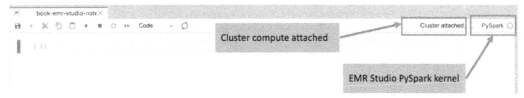

Figure 14.21 – The EMR Studio notebook with a PySpark kernel

I have already attached a cluster to my notebook, but you can explore different options to attach the compute to EMR studio in AWS documentation at https://docs.aws.amazon.com/emr/latest/ManagementGuide/emr-studio-create-use-clusters.html.

Requirement 2

Read the yellow_tripdata_2023-01.parquet file from the S3 location in a DataFrame and display a sample of 10 records.

Let's use a chain-of-thought prompts technique with multiple single-line prompts in different cells to achieve this requirement:

```
Prompt # 1:
# Read s3://<your-bucket-name-here>/yellow_taxi_trip_records/yellow_
tripdata_2023-01.parquet file in a dataframe


Prompt # 2:
# Display a sample of 10 records from dataframe
```

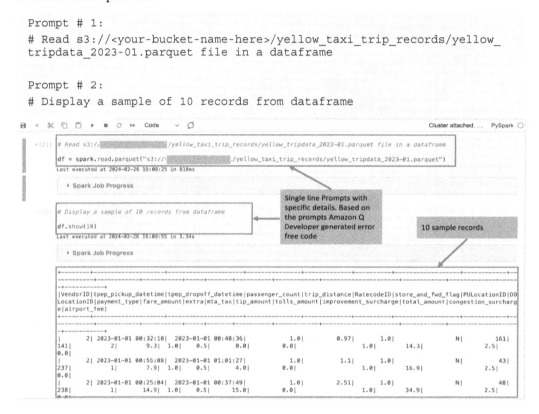

Figure 14.22 – PySpark code to read the Yellow Taxi Trip Records data using single-line prompts

Observe that upon entering the Amazon Q Developer-enabled EMR Studio notebook prompt, it initiates code recommendations. Amazon Q Developer recognizes the file format as Parquet and suggests using the spark.read.parquet method. You can directly execute each cell/code from the notebook. Furthermore, as you move to the next cell, Amazon Q Developer utilizes "line-by-line recommendations" to suggest displaying the schema.

Requirement 3

Read the taxi+_zone_lookup.csv file from the S3 location in a DataFrame and display a sample of 10 records.

We already explored the chain-of-thought prompts technique with multiple single-line prompts for *Requirement 2*. Now, let's try a multi-line prompt to achieve this requirement and we will try to customize the code for the DataFrame name:

```
Prompt:
" " "

Read s3://<your-bucket-name-here>/zone_lookup/taxi+_zone_lookup.csv in
a dataframe name zone_df. Show sample 10 records from zone_df.
" " "
```

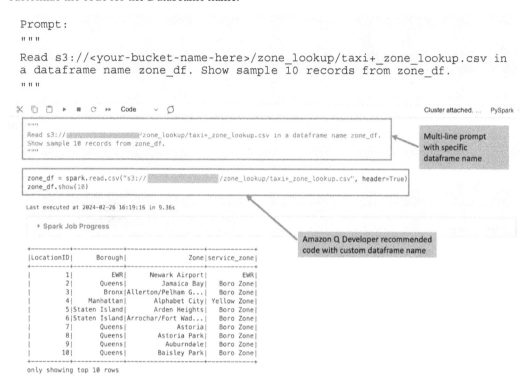

Figure 14.23 – PySpark code to read the Zone Lookup file using a multi-line prompt

Observe that Amazon Q Developer understood the context behind the multi-line prompt and also the specific DataFrame name instructed in the prompt. It auto-generated multiple lines of code with the DataFrame name of `zone_df` and file format as CSV, suggesting the use of the `spark.read.csv` method to read CSV files. You can directly execute each cell/code from the notebook.

Requirement 4

Perform a left outer join on `yellow_tripdata_2023-01.parquet` and `taxi+_zone_lookup.csv` on `pulocationid = LocationID` to gather pick-up zone information.

We will continue using multi-line prompts and some code customization to achieve the preceding requirement:

```
Prompt:
"""

Perform a left outer join on dataframe df and dataframe zone_df on
PULocationID = LocationID to save in dataframe name yellow_pu_zone_df.
Show sample 10 records from yellow_pu_zone_df and show schema.
"""
```

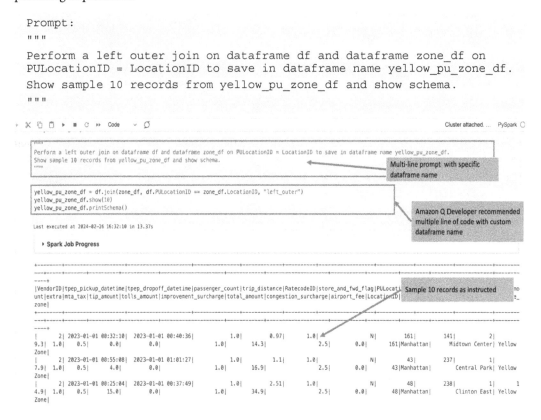

Figure 14.24 – Left outer join df and dataframe zone_df – multi-line prompt

Now, let's review the schema of the DataFrame printed by the code.

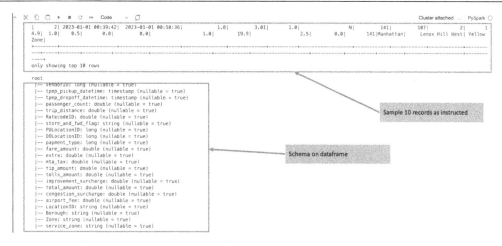

Figure 14.25 – Left outer join df and dataframe zone_df – display schema

Observe that, as instructed in the multi-line prompt, Amazon Q Developer understood the context and auto-generated error-free code with the exact specifications we provided related to the DataFrame named `yellow_pu_zone_df`. You can directly execute each cell/code from the notebook.

Requirement 5

Save the preceding dataset as a CSV file in the previous Amazon S3 bucket in a new folder called `glue_notebook_yellow_pick_up_zone_output`.

Since this requirement is straightforward and can be encapsulated in a single sentence, we will use a single-line prompt to generate the code, and we will also include a header to facilitate easy verification:

```
Prompt:
# Save dataframe yellow_pu_zone_df as CSV file at location s3://<your-
bucket-name-here>/tlc-dataset-ny-taxi/glue_notebook_yellow_pick_up_
zone_output/ with header information
```

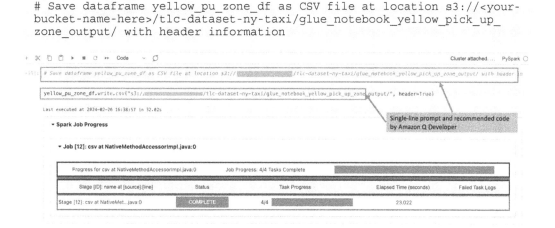

Figure 14.26 – Save the CSV file with enrichment pick-up location data

Requirement 6

For verification, download and check files from the `glue_notebook_yellow_pick_up_zone_output` folder.

Let's go to the Amazon S3 console to verify the files. Select one of the files and click **Download**.

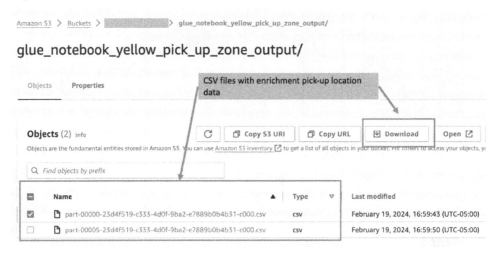

Figure 14.27 – Verify final result set – Amazon Q Developer with Amazon EMR Studio

After downloading, you can use a text editor to review the file contents.

```
VendorID,tpep_pickup_datetime,tpep_dropoff_datetime,passenger_count,trip_distance,RatecodeID,store_and_fwd_flag,PULocationID,DOLocationID,payment_type,fare_amount,extra,m
ta_tax,tip_amount,tolls_amount,improvement_surcharge,total_amount,congestion_surcharge,airport_fee,LocationID,Borough,Zone,service_zone
2,2023-01-01T00:32:10.000Z,2023-01-01T00:40:36.000Z,1.0,0.97,1.0,N,161,141,2,9.3,1.0,0.5,0.0,0.0,1.0,14.3,2.5,0.0,161,Manhattan,Midtown Center,Yellow Zone
2,2023-01-01T00:55:08.000Z,2023-01-01T01:01:27.000Z,1.0,1.1,1.0,N,43,237,1,7.9,1.0,0.5,4.0,0.0,1.0,16.9,2.5,0.0,43,Manhattan,Central Park,Yellow Zone
2,2023-01-01T00:25:04.000Z,2023-01-01T00:37:49.000Z,1.0,2.51,1.0,N,48,238,1,14.9,1.0,0.5,15.0,0.0,1.0,34.9,2.5,0.0,48,Manhattan,Clinton East,Yellow Zone
1,2023-01-01T00:03:48.000Z,2023-01-01T00:13:25.000Z,0.0,1.9,1.0,N,138,7,1,12.1,7.25,0.5,0.0,0.0,1.0,20.85,0.0,1.25,138,Queens,LaGuardia Airport,Airports
2,2023-01-01T00:10:29.000Z,2023-01-01T00:21:19.000Z,1.0,1.43,1.0,N,107,79,1,11.4,1.0,0.5,3.28,0.0,1.0,19.68,2.5,0.0,107,Manhattan,Gramercy,Yellow Zone
2,2023-01-01T00:50:34.000Z,2023-01-01T01:02:52.000Z,1.0,1.84,1.0,N,161,137,1,12.8,1.0,0.5,10.0,0.0,1.0,27.8,2.5,0.0,161,Manhattan,Midtown Center,Yellow Zone
2,2023-01-01T00:09:22.000Z,2023-01-01T00:19:49.000Z,1.0,1.66,1.0,N,239,143,1,12.1,1.0,0.5,3.42,0.0,1.0,20.52,2.5,0.0,239,Manhattan,Upper West Side South,Yellow Zone
2,2023-01-01T00:27:12.000Z,2023-01-01T00:49:56.000Z,1.0,11.7,1.0,N,142,200,1,45.7,1.0,0.5,10.74,3.0,1.0,64.44,2.5,0.0,142,Manhattan,Lincoln Square East,Yellow Zone
```

Figure 14.28 – Verify the CSV file contents of Amazon Q Developer with Amazon EMR Studio

Observe that the CSV file has additional columns with zone information based on the pick-up location ID.

Summary – Amazon Q Developer with Amazon EMR Studio

As illustrated, we can automatically generate end-to-end, error-free, and executable code simply by providing prompts with specific requirements. Amazon Q Developer, integrated with an Amazon EMR Studio notebook, comprehends the context and automatically generates PySpark code that can be run directly from the notebook. This marks a significant advancement for many data engineers, relieving them from concerns about the technical intricacies associated with PySpark libraries, methods, and syntax.

> **Think challenge**
>
> To fulfill *Requirement 6*, if you are interested, attempt to utilize the same EMR Studio notebook for reading a CSV file, displaying sample records, and adding a header.
>
> **Hint**: Use the multi-line prompt technique, similar to the one we used when reading the Zone Lookup file.

In the next section, we will consider an application developer persona to explore code assistance integration with AWS Lambda.

Code assistance integration with AWS Lambda

Before we start diving deep into code assistance support for the AWS Lambda service, let's quickly go through an overview of AWS Lambda. **AWS Lambda** is a serverless computing service that allows users to run code without provisioning or managing servers. With Lambda, you can upload your code or use the available editor from the Lambda console. During the runtime of the code, based on the provided configurations, the service automatically takes care of the compute resources needed for execution. It is designed to be highly scalable, cost effective, and suitable for event-driven applications.

AWS Lambda supports multiple programming languages, including Node.js, Python, Java, Go, and .NET Core, allowing you to choose the language that best fits your application. Lambda can be easily integrated with other AWS services, enabling you to build complex and scalable architectures. It works seamlessly with services such as Amazon S3, DynamoDB, and API Gateway.

The AWS Lambda console is integrated with Amazon Q Developer to make it easy for developers to get coding assistance/recommendations.

Use case for AWS Lambda

Let's start with one of the easy and widely used use cases of converting file format.

File format conversion: In a typical scenario, once a file is received from an external team and/or source, it may not be in the target location and have the required name expected by the application. In that case, AWS Lambda can be used to quickly copy the file from the source location to the target location and rename the file at the target location.

To illustrate this use case, let's copy the NY Taxi Zone lookup file from the source location (`s3://<your-bucket-name>/zone_lookup/`) to the target location (`s3://<your-bucket-name>/source_lookup_file/`). Also, remove the special character (+) from the filename to save it as `taxi_zone_lookup.csv`.

To meet this requirement, application developers must develop a Python script. This script should copy and rename the Zone Lookup file from the source to the target location.

As a code developer / data engineer, you will need to convert the preceding business objectives into the solution blueprint.

Solution blueprint

1. Write a Python script to handle technical requirements.
2. Copy the `taxi+_zone_lookup.csv` file from S3 to the `zone_lookup` folder to the `source_lookup_file` folder.
3. During copying, change `taxi+_zone_lookup.csv` to `taxi_zone_lookup.csv` in the target `source_lookup_file` folder.
4. For verification, check the contents of the `source_lookup_file/taxi_zone_lookup.csv` file.

Now that we have a use case defined, let's go through the step-by-step solution for it.

Data preparation

We are using the same lookup file that we provisioned in this chapter under the *Code assistance integration with AWS Glue* section. Please refer to the *Use case for AWS Glue* section, which covers details related to data preparation.

Solution – Amazon Q Developer with AWS Lambda

Let's first enable Amazon Q Developer with the AWS Lambda console. To enable Amazon Q Developer integration with AWS Lambda, we will need to update the IAM policy.

Prerequisite to enable Amazon Q Developer with AWS Lambda

The developer is required to modify the IAM policy associated with the IAM user or role to grant permissions for Amazon Q Developer to initiate recommendations in the AWS Lambda console. Please reference *Chapter 2* for additional details on initiating interaction with Amazon Q Developer in AWS Lambda.

To let Amazon Q Developer start code suggestions, make sure to choose **Tools | Amazon CodeWhisperer Code Suggestions**.

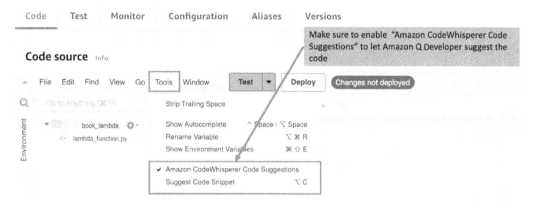

Figure 14.29 – AWS Lambda console with Amazon Q Developer for Python runtime

To fulfill the mentioned requirements, we will use auto-code generation techniques that were discussed in *Chapter 3*. Mainly, we will focus on the multi-line prompt for auto-code generation. Let's use Amazon Q Developer to auto-generate an end-to-end script that can achieve the following requirements in AWS Lambda Console and EMR Studio. Here is a step-by-step solution walk-through of the solution.

Requirement 1

You need to write a Python script to handle technical requirements.

Once you open the AWS Lambda console, use the launcher to select a Python runtime.

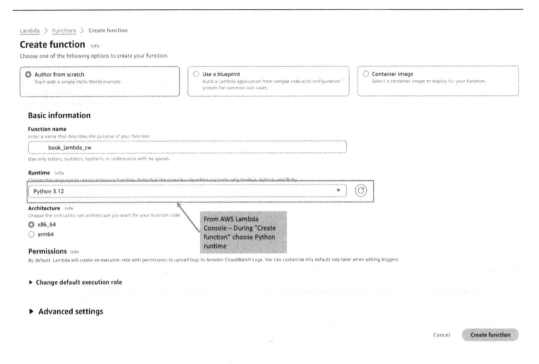

Figure 14.30 – Create a Python runtime from AWS Lambda

Once you successfully create a Lambda function, observe that AWS Lambda creates a `lambda_function.py` file with some sample code. We can safely delete the sample code for this exercise, as we will use Amazon Q Developer to generate end-to-end code.

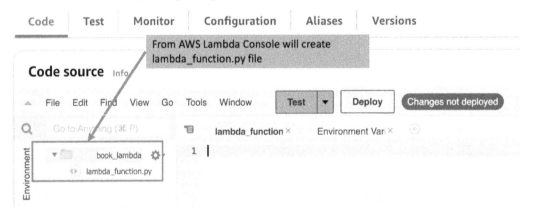

Figure 14.31 – The AWS Lambda console with Amazon Q Developer for Python runtime

Let's combine *Requirements 2* and *3*, as we are planning to use a multi-line prompt.

Requirements 2 and 3

Copy the `taxi+_zone_lookup.csv` file from S3 to the `zone_lookup` folder to the `source_lookup_file` folder.

During copying, change the file name from `taxi+_zone_lookup.csv` to `taxi_zone_lookup.csv` in the target `source_lookup_file` folder.

Let's use multi-line prompts to auto-generate the code:

```
Prompt:
"""
write a lambda function.
copy s3://<your-bucket-name>/zone_lookup/taxi+_zone_lookup.csv
as s3://<your-bucket-name>/source_lookup_file/taxi_zone_lookup.csv
"""
```

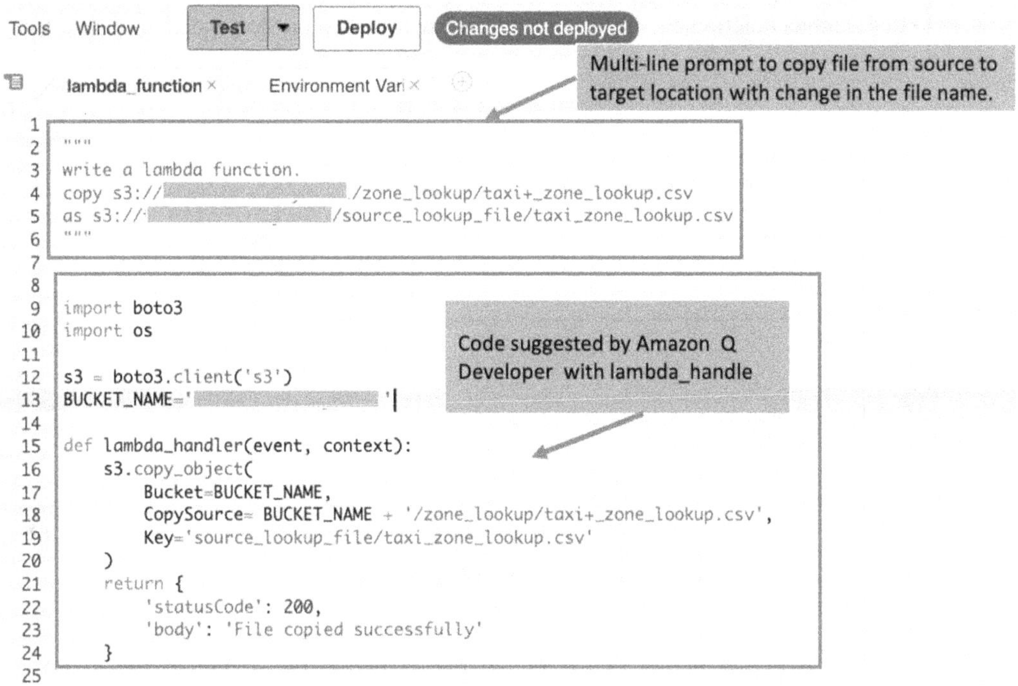

Figure 14.32 – Amazon Q Developer generated code for the AWS Lambda console

Observe that Amazon Q Developer created a `lambda_handler` function and added `return code of 200` with a success message.

Requirement 4

For verification, check the contents of the source_lookup_file/taxi_zone_lookup.csv file.

Let's deploy and use a test event to run Lambda code generated by Amazon Q Developer.

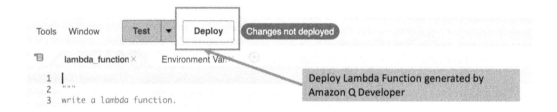

Figure 14.33 – Deploy AWS Lambda code

Now, let's test the code by going to the **Test** tab and clicking the **Test** button. Since we are not passing any values to this Lambda function, the JSON event values from the **Test** tab do not matter in our case.

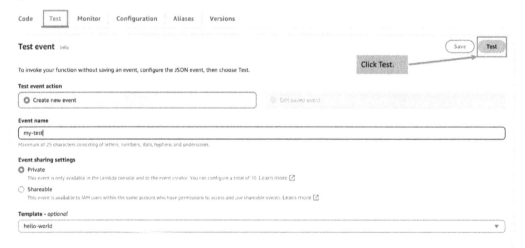

Figure 14.34 – Test AWS Lambda code

Once the Lambda code executes successfully, it will provide you with the details of the execution. Observe that the code is executed successfully and displays the returned code with a success message.

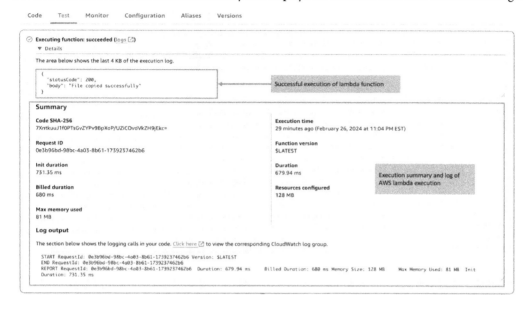

Figure 14.35 – AWS Lambda code execution

Let's use the Amazon S3 console to download and verify s3://<your-bucket-name>/ source_lookup_file/taxi_zone_lookup.csv.

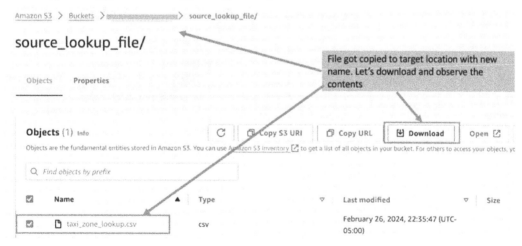

Figure 14.36 – Target lookup file from Amazon S3

```
  taxi_zone_lookup.csv    ✕
 1    "LocationID","Borough","Zone","service_zone"
 2    1,"EWR","Newark Airport","EWR"
 3    2,"Queens","Jamaica Bay","Boro Zone"
 4    3,"Bronx","Allerton/Pelham Gardens","Boro Zone"
 5    4,"Manhattan","Alphabet City","Yellow Zone"
 6    5,"Staten Island","Arden Heights","Boro Zone"
 7    6,"Staten Island","Arrochar/Fort Wadsworth","Boro Zone"
 8    7,"Queens","Astoria","Boro Zone"
 9    8,"Queens","Astoria Park","Boro Zone"
10    9,"Queens","Auburndale","Boro Zone"
11    10,"Queens","Baisley Park","Boro Zone"
```

Figure 14.37 – The Zone Lookup file

Summary – Amazon Q Developer with AWS Lambda

As illustrated, we can automatically generate end-to-end, error-free, and executable code simply by providing prompts with specific requirements. Amazon Q Developer, integrated with AWS Lambda, automatically generates the `lambda_handler ()` function with return code based on the Lambda runtime environment selected. This integration can assist application developers with relatively limited coding experience in automatically generating Lambda functions with minor to no code changes.

Continuing with the application developer persona, next, we will explore the data scientist persona to investigate code assistance integration with Amazon SageMaker.

Code assistance integration with Amazon SageMaker

Before we start diving deep into code assistance support for the Amazon SageMaker service, let's quickly go through an overview of Amazon SageMaker. **Amazon SageMaker** is a fully managed service that simplifies the process of building, training, and deploying ML models at scale. It is designed to make it easier for developers and data scientists to build, train, and deploy ML models without the need for extensive expertise in ML or deep learning. It has multiple features such as end-to-end workflow, built-in algorithms, custom model training, automatic model tuning, ground truth, edge manager, augmented AI, and managed notebooks, just to name a few. Amazon SageMaker integrates with other AWS services, such as Amazon S3 for data storage, AWS Lambda for serverless inference, and Amazon CloudWatch for monitoring.

Amazon SageMaker Studio hosts the managed notebooks, which are integrated with Amazon Q Developer.

Use case for Amazon SageMaker

Let's use a very common business use case related to churn prediction for which data scientists use the XGBoost algorithm.

Churn prediction in business involves utilizing data and algorithms to forecast which customers are at risk of discontinuing their usage of a product or service. The term "churn" commonly denotes customers ending subscriptions, discontinuing purchases, or ceasing service utilization. The primary objective of churn prediction is to identify these customers before they churn, enabling businesses to implement proactive measures for customer retention.

We will use publicly available direct marketing bank data to illustrate the support provided by Amazon Q Developer for milestone steps such as data collection, feature engineering, model training, and model deployment using Amazon SageMaker.

Typically, data scientists need to write a complex script to carry out all of the preceding milestone steps from an Amazon SageMaker Studio notebook.

Solution blueprint

1. Set up an environment with the required set of libraries.

2. **Data collection**: Download and unzip direct marketing bank data from `https://sagemaker-sample-data-us-west-2.s3-us-west-2.amazonaws.com/autopilot/direct_marketing/bank-additional.zip`.

3. **Feature engineering**: To demonstrate the functionality, we will carry out the following commonly used feature engineering steps:

    *   Manipulate column data using default values

    *   Drop extra columns

    *   Carry out one-hot encoding

4. **Model training**: Let's use the XGBoost algorithm:

    *   Rearrange data to create training, validation, and test datasets/files

    *   Use XGBoost algorithms to train the model using the training dataset

5. **Model deployment**: Deploy the model as an endpoint to allow inferences.

In the preceding solution blueprint, we illustrate the integration of Amazon Q Developer with Amazon SageMaker by handling commonly used milestone steps. However, based on the complexity of your data and enterprise needs, there might be additional steps required.

Data preparation

We will utilize an AWS dataset publicly hosted for direct marketing bank data. The complete dataset is available at `https://sagemaker-sample-data-us-west-2.s3-us-west-2.amazonaws.com/autopilot/direct_marketing/bank-additional.zip`. All the data preparation steps will be conducted in the SageMaker Studio notebook as part of the data collection requirement.

Solution – Amazon Q with Amazon SageMaker Studio

Let's first enable Amazon Q Developer with Amazon SageMaker Studio. The following prerequisites are needed to allow Amazon Q Developer to auto-generate code inside Amazon SageMaker studio.

Prerequisite to enable Amazon Q Developer with Amazon SageMaker Studio

The developer is required to modify the IAM policy associated with the IAM user or role to grant permissions for Amazon Q Developer to initiate recommendations in for Amazon SageMaker Studio notebook. Refer *Chapter 2* for the details to enable Amazon Q Developer with Amazon SageMaker Studio notebook.

Once the Amazon Q Developer is activated for Amazon SageMaker Studio notebook, select **Create notebook** from the **Launcher** to verify that Amazon Q Developer is enabled.

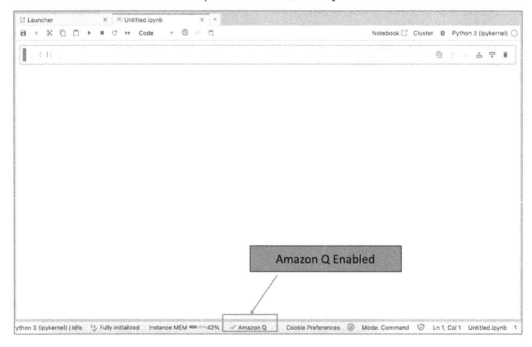

Figure 14.38 –An Amazon Q Developer-enabled notebook from SageMaker Studio

To fulfill the mentioned requirements, we will use auto-code generation techniques that were discussed in *Chapter 4*. Mainly, we will focus on single-line prompts, multi-line prompts, and chain-of-thought prompts for auto-code generation techniques.

Requirement 1

Set up an environment with the required set of libs.

Let's use single-line prompts:

```
Prompt 1:
# Fetch this data by importing the SageMaker library

Prompt 2:
# Defining global variables BUCKET and ROLE that point to the bucket
associated with the Domain and it's execution role
```

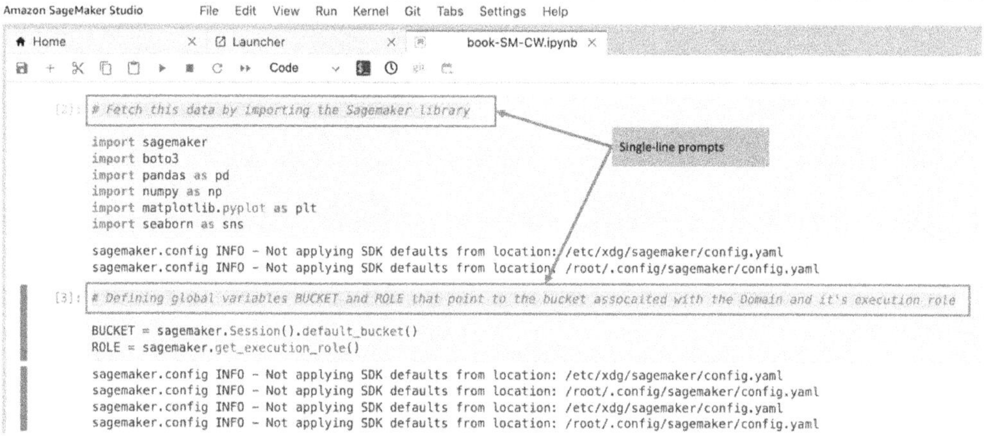

Figure 14.39 –Amazon Q Developer – SageMaker Studio setup environment

Observe that, based on our prompts, Amazon Q Developer generated code with a default set of libraries and variables. However, based on your needs, and account setup, you may need to update/add the code.

Requirement 2

For data collection, download and unzip direct marketing bank data from `https://sagemaker-sample-data-us-west-2.s3-us-west-2.amazonaws.com/autopilot/direct_marketing/bank-additional.zip`.

We will focus on multi-line prompts to achieve this requirement:

```
Prompt:
'''Using the requests library download the ZIP file from
the url "https://sagemaker-sample-data-us-west-2.s3-us-west-2.
amazonaws.com/autopilot/direct_marketing/bank-additional.zip"
and save it to current directory and unzip the archive
'''
```

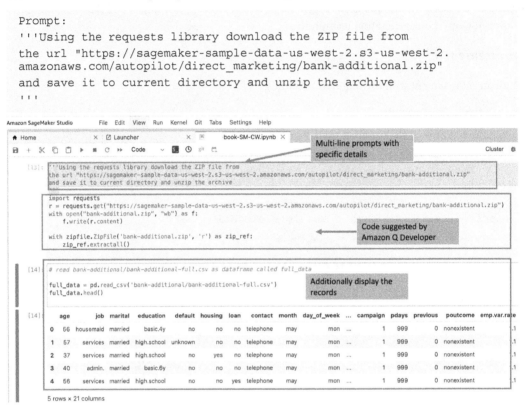

Figure 14.40 – Amazon Q Developer – SageMaker Studio data collection

Requirement 3

To demonstrate the functionality of feature engineering, we will carry out the following commonly used feature engineering steps, which will help us improve the model accuracy:

1. Manipulate column data using default values.

2. Drop extra columns.

3. Carry out one-hot encoding.

We will focus on multi-line prompts to achieve this requirement:

```
Prompt #1:
'''
Create a new dataframe with column no_previous_contact and populates
from existing dataframe column pdays using numpy when the condition
equals to 999, 1, 0 and show the table
'''


Prompt # 2:
# do one hot encoding for full_data


Prompt # 3:
'''
Drop the columns 'duration', emp.var.rate', 'cons.price.idx', 'cons.
conf.idx', 'euribor3m' and 'nr.employed'
from the dataframe and create a new dataframe with name model_data
'''
```

We get the following screen.

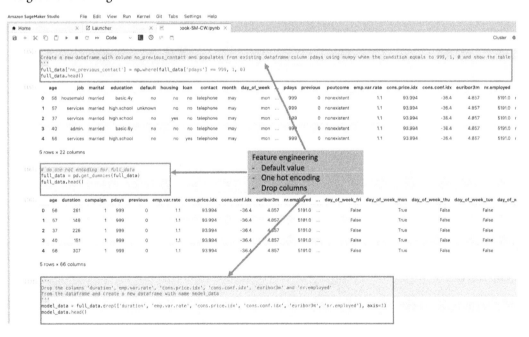

Figure 14.41 –Amazon Q Developer – SageMaker Studio feature engineering

Now, let's proceed with the generated code for the model training, testing, and validation:

```
Prompt # 4:
#split model_data for train, validation, and test

Prompt # 5:
'''
for train_data move y_yes as first column.
Drop y_no and y_yes columns from train_data.
save file as train.csv
'''


Prompt # 6:
'''
for validation_data move y_yes as first column.
Drop y_no and y_yes columns from validation_data.
save file as validation.csv
'''
```

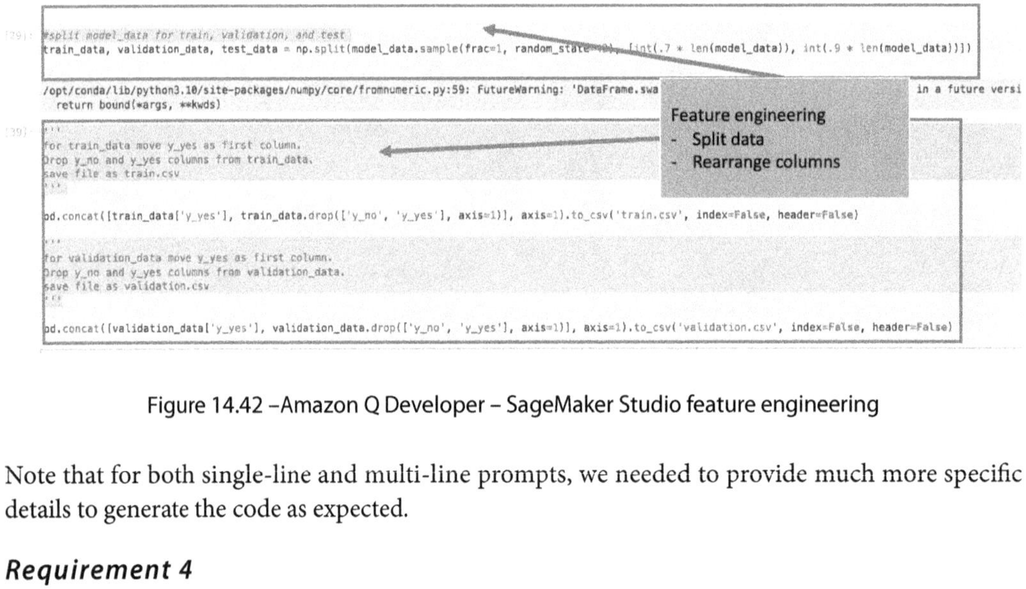

Figure 14.42 –Amazon Q Developer – SageMaker Studio feature engineering

Note that for both single-line and multi-line prompts, we needed to provide much more specific details to generate the code as expected.

Requirement 4

Model training: Let's use the XGBoost algorithm:

- Rearrange data to create training, validation, and test datasets/files
- Use XGBOOST algorithms to train the model using a training dataset

We will focus on multi-line prompts to achieve this requirement to start the model training activity:

```
Prompt #1:
''' upload train.csv to S3 Bucket train/train.csv prefix.
upload validation.csv to S3 Bucket validation/validation.csv prefix
'''
```

```
Prompt #2:
# pull latest xgboost model as a CONTAINER
```

```
Prompt #3:
# create TrainingInput from s3 train/train.csv and validation/
validation.csv
```

```
Prompt #3:
# create training job with hyper paramers max_depth=5,
eta=0.2, gamma=4, min_child_weight=6, subsample=0.8,
objective='binary:logistic', num_round=100
```

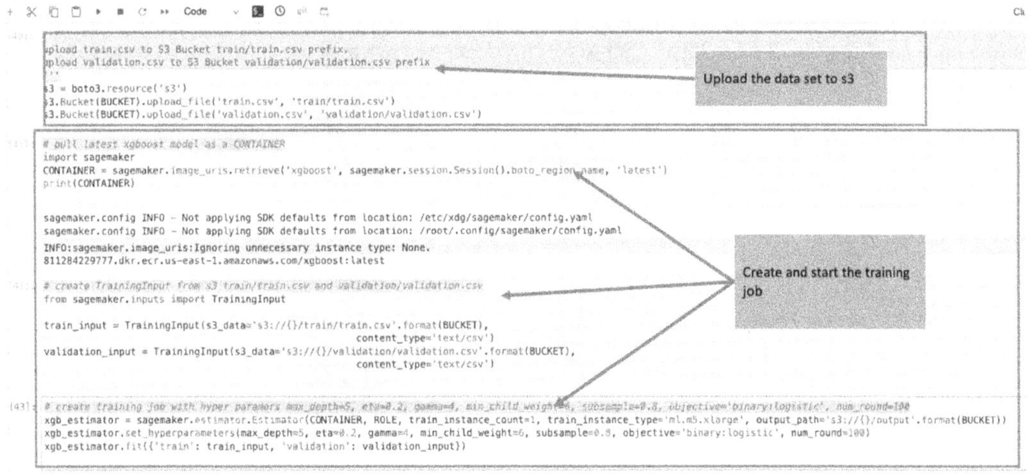

Figure 14.43 – Amazon Q Developer – SageMaker Studio model training

Requirement 5

Deploy the model as an endpoint to allow inferences.

We will focus on single-line prompts to achieve this requirement:

```
Prompt #1:
# Deploy a model that's hosted behind a real-time endpoint
```

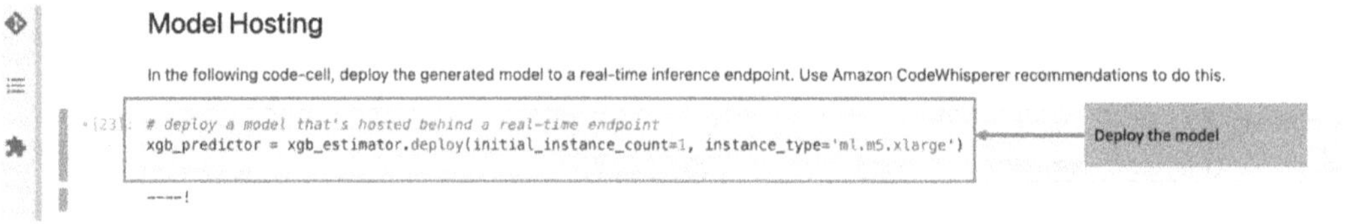

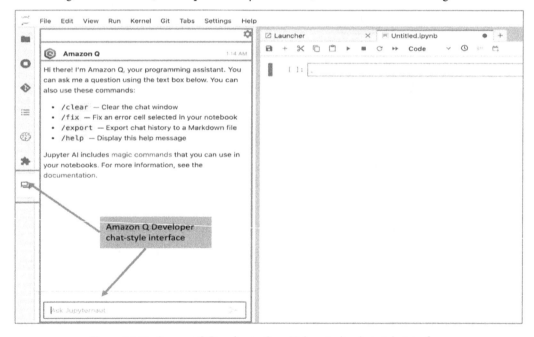

Figure 14.44 – Amazon Q Developer -SageMaker studio model training

Observe that Amazon Q Developer uses a default configuration for `instance_type` and `initial_instance_count`. You can check the hosted model from the Amazon SageMaker console by clicking the **Inference** dropdown and selecting the **Endpoints** option.

In the preceding examples, we extensively used inline prompts with single-line prompting, multi-line prompting, and chain of thought prompting techniques. If you wish to use a chat-style interface, you can leverage the Amazon Q Developer chat-style interface, as shown in the following screenshot.

Figure 14.45 –Amazon Q Developer -SageMaker studio chat style interface

Summary – Amazon Q Developer with Amazon SageMaker

As demonstrated, Amazon Q Developer seamlessly integrated with the Amazon SageMaker Studio notebook IDE, enables the automatic generation of end-to-end, error-free, and executable code. By supplying prompts with specific requirements, Q Developer can auto-generate code for essential milestone steps, including data collection, feature engineering, model training, and model deployment within the SageMaker Studio notebook.

While data scientists can utilize this integration to produce code blocks, customization may be necessary. Specific details must be provided in prompts to tailor the code. In some instances, adjustments may be required to align with enterprise standards, business requirements, and configurations. Users should possess expertise in prompt engineering, familiarity with scripting, and conduct thorough testing to ensure the scripts meet business requirements before deploying them into production.

Now, let's dive deep to see how data analysts can use code assistance while working with Amazon Redshift.

Code assistance integration with Amazon Redshift

Before we start diving deep into code assistance support for the Amazon Redshift service, let's quickly go through an overview of AWS Redshift. **Amazon Redshift** is an AI-powered, fully managed, cloud-based data warehouse service. It is designed for high-performance analysis and the processing of large datasets using standard SQL queries.

Amazon Redshift is optimized for data warehousing, providing a fast and scalable solution for processing and analyzing large volumes of structured data. It uses columnar storage and **massively parallel processing** (**MPP**) architecture, distributing data and queries across multiple nodes to deliver high performance for complex queries. This architecture allows it to easily scale from a few hundred gigabytes to petabytes of data, enabling organizations to grow their data warehouse as their needs evolve. It integrates with various data sources, allowing you to load data from multiple sources, including Amazon S3, Amazon DynamoDB, and Amazon EMR.

> **Note**
> To query the data, Amazon Redshift also provides a query editor. The Redshift query editor v2 has two modes to interact with databases: **Editor** and **Notebook**. Code assistance is integrated with the Notebook mode of the Redshift query editor v2.

Use case for Amazon Redshift

Let's start with one of the easy and widely used use cases of converting file format.

Identifying top performers: In a typical business use case, analysts are interested in identifying top performers based on certain criteria.

To illustrate this use case, we will be using the publicly available `tickit` database, which is readily available with Amazon Redshift. For more information about the `tickit` database, refer to the *References* section at the end of the chapter.

Analysts want to identify the top state where most of the venues are.

To meet this requirement, analyst developers must develop SQL queries to interact with different tables from the `tickit` database.

Solution blueprint

As we are considering the data analyst persona and using code assistance to generate the code, we do not need to further break down the business ask into the solution blueprint. This makes it easy for analysts to interact with databases without getting involved in table structures and relationship details:

- Write SQL to identify the top state where most of the venues are

Data preparation

We will be using the publicly available `tickit` database, which comes with Amazon Redshift. Let's import the data using Redshift query editor v2:

1. Connect to your Amazon Redshift cluster or Serverless endpoint from Redshift query editor 2.

2. Then, choose `sample_data_dev` and click on `tickit`.

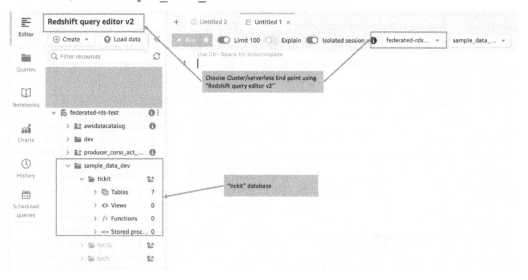

Figure 14.46 – Import the tickit database using Amazon Redshift

Solution – Amazon Q with Amazon Redshift

Let's first enable Amazon Q with Amazon Redshift. To allow Amazon Q to generate SQL inside Amazon Redshift, the admin needs to enable the **Generative SQL** option inside **Notebook** of Redshift query editor v2. Please reference *Chapter 3* for additional details on initiating interaction with Amazon Q in Amazon Redshift.

Prerequisite to enable Amazon Q with Amazon Redshift

Let's walk through the steps needed to enable the **Generative SQL** option inside **Notebook** of the Redshift query editor v2.

1. Log in with admin privileges to connect to your Amazon Redshift cluster or Serverless endpoint.

2. Choose **Notebook**.

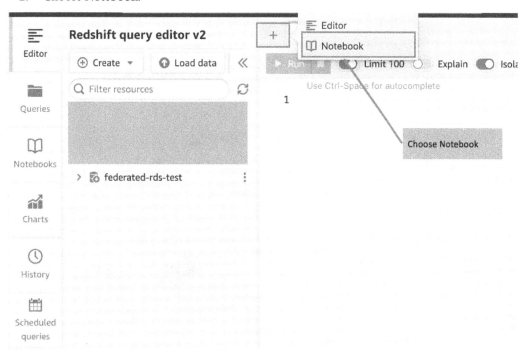

Figure 14.47 – Notebook using Redshift query editor v2

3.   Choose **Generative SQL**, then check the **Generative SQL** box, and click **Save**.

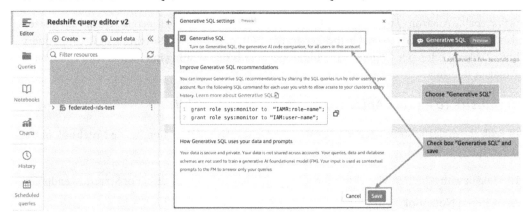

Figure 14.48 – Enable Generative SQL using Redshift query editor v2

To fulfill the mentioned requirements, we will use auto-code generation techniques that were discussed in *Chapter 4*. Mainly, we will focus on the chat companion for auto-code generation.

Requirement 1

Write SQL to identify the top state where most of the venues are.

Use Amazon Q's interactive session to ask the following question:

```
Q:Which state has most venues?
```

Observe that we did not provide database details to the Amazon Q Developer, but it was still able to identify the required table, `tickit.venue`. It generated the fully executable end-to-end query with `Group by`, `Order by`, and `Limit` to meet the requirements. To make it easy for analysts to run the queries, code assistance is integrated with the notebook. Just by clicking **Add to notebook**, the SQL code will be available in a notebook cell that users can run directly.

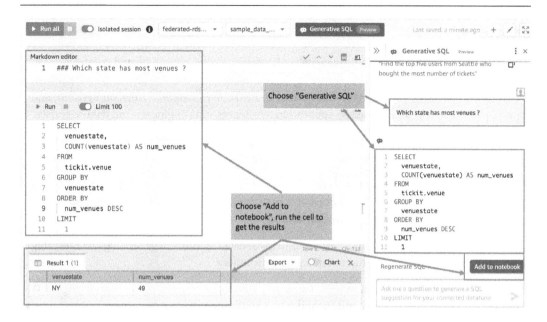

Figure 14.49 – Interact with code assistance from Amazon Redshift

Summary – Amazon Q with Amazon Redshift

As demonstrated, we can effortlessly generate end-to-end, error-free, and executable SQL by interacting with Amazon Q through a chat-style interface. Amazon Q seamlessly integrates with notebooks in the Amazon Redshift query editor v2. Users are not required to provide database and/or table details to the code assistant. It autonomously identifies the necessary tables and generates SQL code to fulfill the specified requirements in the prompt. Furthermore, to facilitate analysts in running queries, it is directly integrated with the notebook. Amazon Q, in conjunction with Amazon Redshift, proves to be a valuable asset for data analysts. In many cases, data analysts do not need to translate business requirements into technical steps. They can leverage the auto-generate SQL feature, bypassing the need to delve deep into database and table details.

Summary

In this chapter, we initially covered the integration of different AWS services with code companions to assist users in auto-code generation. Then, we explored the integration of Amazon Q Developer with some of the core services, such as AWS Glue, Amazon EMR, AWS Lambda, Amazon Redshift, and Amazon SageMaker, commonly used by application developers, data engineers, and data scientists.

We then discussed, in the prerequisites, the in-depth integration with sample common use cases and corresponding solution walk-throughs for various integrations.

AWS Glue integration with Amazon Q Developer, aiding data engineers in generating and executing ETL scripts using the AWS Glue Studio notebook environment. This includes a skeletal outline of a full end-to-end Glue ETL job using AWS Glue Studio.

AWS EMR integration with Amazon Q Developer to assist data engineers in generating and executing ETL scripts using the AWS EMR Studio notebook environment.

AWS Lambda console IDE integration with Amazon Q Developer, supporting application engineers in generating and executing end-to-end Python-based applications for file movement.

Amazon SageMaker studio notebook integration with Amazon Q Developer to help data scientists achieve major milestone steps in data collection, feature engineering, model training, and model deployment using different prompting techniques.

Amazon Redshift integration with Amazon Q to aid business analysts in generating SQL queries by simply providing business requirements. Users are not required to provide database and/or table details to the code assistant.

In the next chapter, we will look at how you can use Amazon Q Developer to get AWS-specific guidance and recommendations, either from the AWS console or from the documentation on a variety of topics such as architecture and best practices support.

References

- AWS Prescriptive Guidance - Data engineering: `https://docs.aws.amazon.com/prescriptive-guidance/latest/aws-caf-platform-perspective/data-eng.html`

- Jupyter kernel: `https://docs.jupyter.org/en/latest/projects/kernels.html`

- Amazon Q Developer with AWS Glue Studio: `https://docs.aws.amazon.com/amazonq/latest/qdeveloper-ug/glue-setup.html`

- TLC Trip Record Data: `https://www.nyc.gov/site/tlc/about/tlc-trip-record-data.page`

- Setting up Amazon Q data integration in AWS Glue: `https://docs.aws.amazon.com/glue/latest/dg/q-setting-up.html`

- Setting up Amazon Q Developer data integration in Amazon EMR: `https://docs.aws.amazon.com/amazonq/latest/qdeveloper-ug/emr-setup.html`

- Attach a compute to an EMR Studio Workspace: `https://docs.aws.amazon.com/emr/latest/ManagementGuide/emr-studio-create-use-clusters.html`

- Using Amazon Q Developer with AWS Lambda: `https://docs.aws.amazon.com/amazonq/latest/qdeveloper-ug/lambda-setup.html`

- Interacting with query editor v2 generative SQL: `https://docs.aws.amazon.com/redshift/latest/mgmt/query-editor-v2-generative-ai.html`

- Amazon Redshift "tickit" database: `https://docs.aws.amazon.com/redshift/latest/dg/c_sampledb.html`

- Direct marketing bank data: `https://sagemaker-sample-data-us-west-2.s3-us-west-2.amazonaws.com/autopilot/direct_marketing/bank-additional.zip`

- Amazon SageMaker Studio: `https://aws.amazon.com/sagemaker/studio/`

15
Accelerate Building Solutions on AWS

In this chapter, we will look at the following key topics:

- Key steps for building solutions on AWS
- Use case for leveraging Amazon Q features during the build process on AWS
- Awareness of AWS account resources

At this time, it's no secret that organizations are moving their IT infrastructure to the cloud at a rapid pace, with AWS being a market leader in offering cloud computing services for building their technology stack. When transitioning existing applications to AWS or creating new ones natively using many of the AWS services, there are certain steps that all builders need to take. A builder is a key persona in every organization, and those who leverage AWS services for their IT needs utilize AWS builders to address their business use cases.

An AWS builder is an individual who uses AWS to design, develop, deploy, and manage applications and solutions on the AWS cloud platform. These individuals may be developers, engineers, architects, data scientists, or other professionals working with AWS services to build innovative and scalable solutions for their organizations or clients. AWS builders leverage the wide array of tools and services provided by AWS to create robust and efficient applications that meet the specific needs and requirements of their projects. They are responsible for understanding AWS services, implementing best practices, and continuously improving their skills to drive innovation and success in their projects.

There is no denying that cloud computing has accelerated the IT build process compared to the on-premises world. However, large and complex projects still require significant time and effort before they can be brought into the production environment.

Let's quickly look at some of the key steps builders need to take to build projects on AWS.

Key steps for building solutions on AWS

The process builders typically go through to solve a use case using AWS services involves several steps:

1. **Understanding requirements**: Builders first need to understand the requirements of the use case, including the problem to be solved, desired outcomes, and any constraints or limitations.

2. **Selection of AWS services**: Based on the requirements, builders select the appropriate AWS services that best fit the use case. This may involve researching and evaluating various AWS services to determine which ones provide the required features and capabilities.

3. **Designing the solution architecture**: Builders design the architecture of the solution, including the overall system design, data flow, integration points, and scalability considerations. This involves creating diagrams, such as architecture diagrams or data flow diagrams, to visualize the solution.

4. **Setting up AWS resources**: Builders provision and configure the necessary AWS resources to implement the solution. This may include setting up compute instances (e.g., EC2), storage (e.g., S3), databases (e.g., RDS), networking (e.g., VPC), and other services as required.

5. **Developing application code**: Builders develop the application code or scripts needed to implement the solution. This may involve writing code in programming languages such as Python, Java, or Node.js, as well as creating configuration files or scripts for **infrastructure as code (IaC)** tools such as AWS CloudFormation or AWS **Cloud Development Kit (CDK)**.

6. **Testing and debugging**: Builders test the solution to ensure that it functions as expected and meets the requirements. This includes unit testing, integration testing, and end-to-end testing to validate the functionality and performance of the solution. Builders also debug any issues or errors that arise during testing.

7. **Deployment and rollout**: Once testing is complete, builders deploy the solution to the production environment. This may involve deploying application code to compute instances, configuring load balancers and auto-scaling groups, and configuring DNS settings. Builders carefully plan and execute the deployment to minimize downtime and disruptions to users.

8. **Monitoring and optimization**: After deployment, builders monitor the solution to ensure that it continues to operate smoothly and meets performance objectives. This involves monitoring metrics, logs, and alarms using AWS CloudWatch and other monitoring tools. Builders also optimize the solution for cost, performance, and scalability as needed.

9. **Documentation and knowledge sharing**: Builders document the solution architecture, configuration, and deployment process to facilitate knowledge sharing and future maintenance. This documentation helps other team members understand how the solution works and how to troubleshoot issues if they arise.

As you can see, there are many steps involved in building a solution for a given use case using many of the AWS purpose-built services. Each step presents its own challenges, and every builder has their own approach to navigating through them. The key point here is that, often, building complex solutions takes time, with much of it devoted to research. This research may involve looking into similar use cases solved previously, weighing the pros and cons of specific AWS services, exploring design patterns and their implementation, and likely troubleshooting issues and errors encountered during the build process.

Let's dive right into how you can accelerate the process of building solutions on AWS using Amazon Q, which focuses on the area where builders often spend a significant amount of time – research.

Use case for leveraging Amazon Q features during the build process on AWS

To better understand how Amazon Q can help accelerate the build process on AWS, we will introduce a use case and then go through the steps of solving it. Along the way, we will highlight the many features of Amazon Q that can reduce the time a builder would otherwise spend.

Keep in mind that, for simplicity, we are presenting a straightforward use case that can be solved in many different ways. However, once you go through the flow of the use case, it will help you understand the various features of Amazon Q when it comes to building solutions using AWS services. These features can be applied to any other complex use case an AWS builder might encounter in their organization. The key takeaway is to understand how these features save time and improve productivity, ultimately enhancing the experiences of AWS builders. Repeat after me – no AWS builder should be unhappy. Let's jump straight into the use case.

A company wants to build a web application where users can upload images for various purposes, such as profile pictures, product images, or content illustrations. To ensure optimal performance and user experience, the company aims to implement a scalable and efficient image processing pipeline to resize and optimize these images before storing them in the backend storage system.

Since AWS is their preferred cloud services provider, the business wants the IT team to create a simple and cost-effective solution in a short time using AWS services.

The use case may be simple here, but often, AWS builders in the organization receive complex requirements from the business that require methodical planning, research, and execution. The assumption is that AWS builders have a general sense of many of the AWS services, their features, and how they can help solve a use case. However, not every builder is the same, and not everyone has the same level of expertise. Some are entry-level builders who rely heavily on research and guidance from senior builders, while others know how to solve a use case but may not be familiar with all the best practices and alternative design patterns.

Important reminder: LLM-based solutions do not always return the same results

LLMs do not inherently produce idempotent results when deployed in applications. There are a few reasons for this:

Stochasticity: Most LLMs use some element of randomness when generating outputs, such as sampling from a probability distribution. This means each invocation will produce slightly different results.

Continuous learning: Many deployed LLMs continue training on new data, constantly updating their parameters. The model is essentially continuously changing, so results will evolve.

Interactivity: Conversational AI systems based on LLMs accumulate context and history, which impacts subsequent responses. The results depend on prior interactions.

External dependencies: Applications may pull in external data that influences the LLM's outputs. If that data changes, the outputs will as well.

Multimodal inputs: For LLMs that process images, audio, and so on, in addition to text, variations in those inputs lead to different outputs.

Since Amazon Q is also an LLM-based service, the results you see for this use case, as well as any use case you solve, may not always return the same output. Therefore, it is important to have a human in the loop to prompt it appropriately and also test and verify before deploying solutions in production.

Hope you didn't skip the preceding reminder note! Now, let's jump straight into how Amazon Q will help expedite solving this use case.

Conversational Q&A capability

The conversational ability of Amazon Q will help us through the initial steps to better understand the use case. You can ask Amazon Q different questions, and it will try to understand what you want to do. If you are an AWS builder with some knowledge of AWS services relevant to solving this use case, you can start with a broad question to Amazon Q and then narrow down options by prompting. For example, in the following screenshot, we ask for a solution to the use case without going into too much detail about what it should look like on AWS.

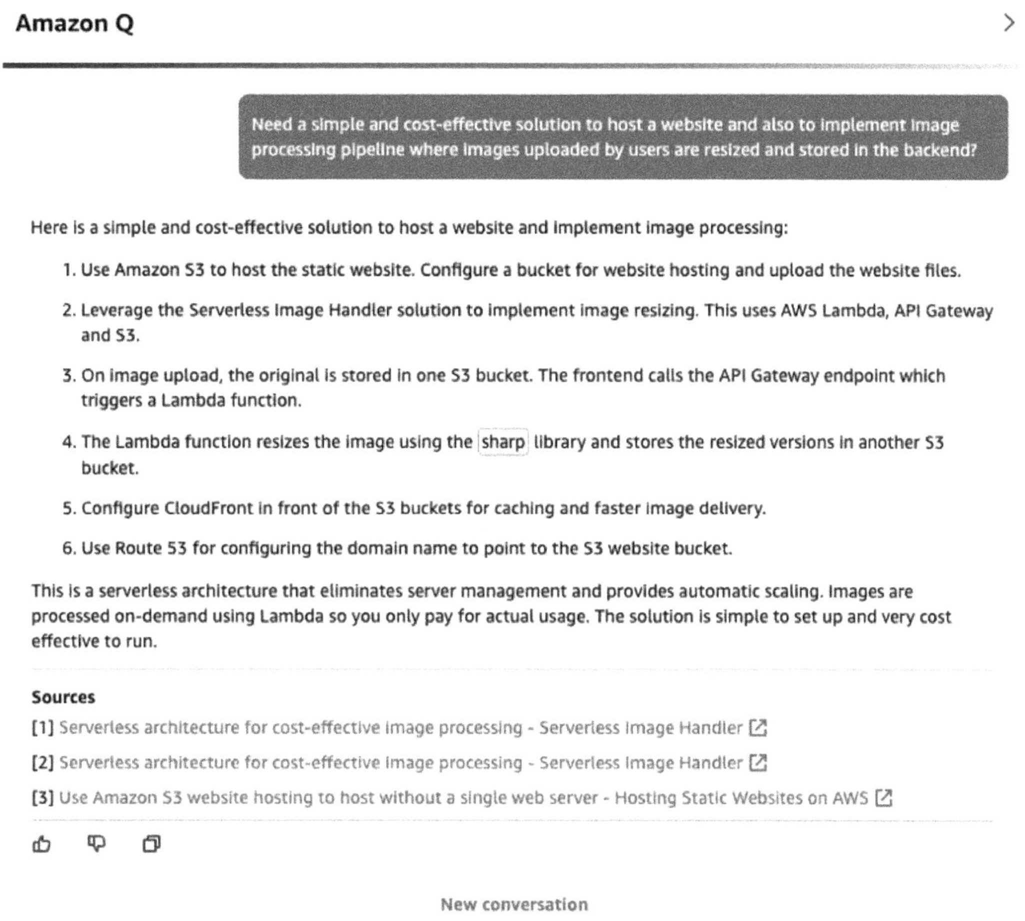

Amazon Q   >

Need a simple and cost-effective solution to host a website and also to implement image processing pipeline where images uploaded by users are resized and stored in the backend?

Here is a simple and cost-effective solution to host a website and implement image processing:

1. Use Amazon S3 to host the static website. Configure a bucket for website hosting and upload the website files.

2. Leverage the Serverless Image Handler solution to implement image resizing. This uses AWS Lambda, API Gateway and S3.

3. On image upload, the original is stored in one S3 bucket. The frontend calls the API Gateway endpoint which triggers a Lambda function.

4. The Lambda function resizes the image using the `sharp` library and stores the resized versions in another S3 bucket.

5. Configure CloudFront in front of the S3 buckets for caching and faster image delivery.

6. Use Route 53 for configuring the domain name to point to the S3 website bucket.

This is a serverless architecture that eliminates server management and provides automatic scaling. Images are processed on-demand using Lambda so you only pay for actual usage. The solution is simple to set up and very cost effective to run.

Sources

[1] Serverless architecture for cost-effective image processing - Serverless Image Handler ☑

[2] Serverless architecture for cost-effective image processing - Serverless Image Handler ☑

[3] Use Amazon S3 website hosting to host without a single web server - Hosting Static Websites on AWS ☑

New conversation

Figure 15.1 – Amazon Q conversational capability – general question

Since there weren't specific instructions in the question, Amazon Q provided a general outline of what the solution might look like. This is where you can further prompt the system to narrow down a possible solution. Since the use case requires us to build and host a simple website, we prompt Amazon Q to refine its recommendation further. The following screenshot shows this prompt and Q's response.

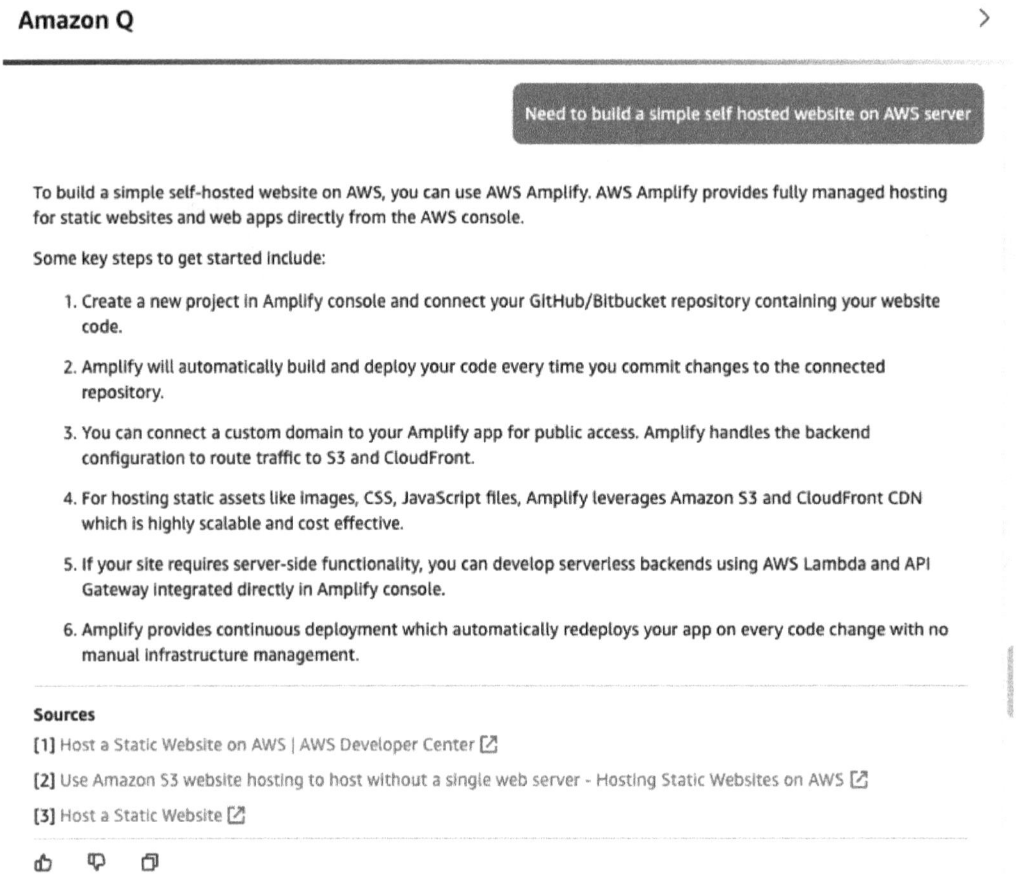

Amazon Q   〉

Need to build a simple self hosted website on AWS server

To build a simple self-hosted website on AWS, you can use AWS Amplify. AWS Amplify provides fully managed hosting for static websites and web apps directly from the AWS console.

Some key steps to get started include:

1. Create a new project in Amplify console and connect your GitHub/Bitbucket repository containing your website code.

2. Amplify will automatically build and deploy your code every time you commit changes to the connected repository.

3. You can connect a custom domain to your Amplify app for public access. Amplify handles the backend configuration to route traffic to S3 and CloudFront.

4. For hosting static assets like images, CSS, JavaScript files, Amplify leverages Amazon S3 and CloudFront CDN which is highly scalable and cost effective.

5. If your site requires server-side functionality, you can develop serverless backends using AWS Lambda and API Gateway integrated directly in Amplify console.

6. Amplify provides continuous deployment which automatically redeploys your app on every code change with no manual infrastructure management.

Sources

[1] Host a Static Website on AWS | AWS Developer Center ↗

[2] Use Amazon S3 website hosting to host without a single web server - Hosting Static Websites on AWS ↗

[3] Host a Static Website ↗

Figure 15.2 – Amazon Q conversational capability – prompt 1

Q understood the prompt, and it provided a recommendation to use the **AWS Amplify** service as a simple way to host the website. As you can see through this simple example, prompting the system is a crucial step in obtaining results that are customized to the use case you are trying to solve. Since the use case requires us to build a website and it will not just be a static web page, the prompt in the following screenshot further refines the output from Q.

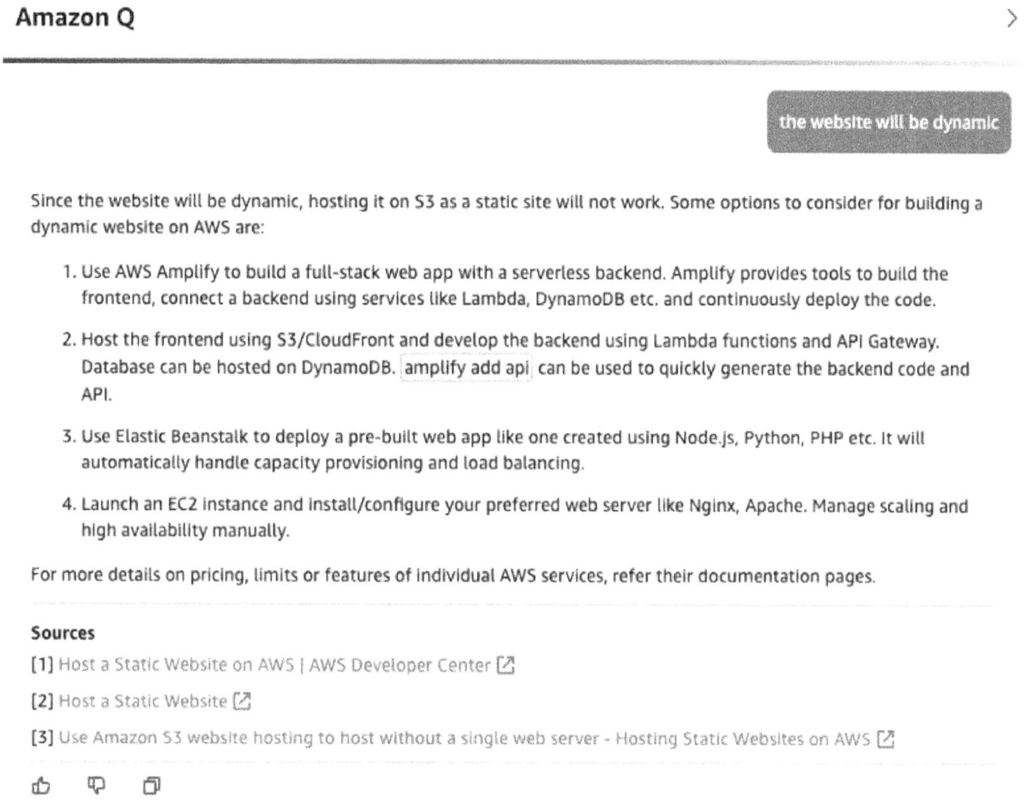

Amazon Q   >

the website will be dynamic

Since the website will be dynamic, hosting it on S3 as a static site will not work. Some options to consider for building a dynamic website on AWS are:

1. Use AWS Amplify to build a full-stack web app with a serverless backend. Amplify provides tools to build the frontend, connect a backend using services like Lambda, DynamoDB etc. and continuously deploy the code.

2. Host the frontend using S3/CloudFront and develop the backend using Lambda functions and API Gateway. Database can be hosted on DynamoDB. amplify add api can be used to quickly generate the backend code and API.

3. Use Elastic Beanstalk to deploy a pre-built web app like one created using Node.js, Python, PHP etc. It will automatically handle capacity provisioning and load balancing.

4. Launch an EC2 instance and install/configure your preferred web server like Nginx, Apache. Manage scaling and high availability manually.

For more details on pricing, limits or features of individual AWS services, refer their documentation pages.

Sources

[1] Host a Static Website on AWS | AWS Developer Center ↗

[2] Host a Static Website ↗

[3] Use Amazon S3 website hosting to host without a single web server - Hosting Static Websites on AWS ↗

👍   👎   🗇

Figure 15.3 – Amazon Q conversational capability – prompt 2

You can continue along this path of prompting until you receive enough recommendations to help solve a particular use case. Amazon Q also provides sources of information so that you can read more about it.

If you happen to have an idea of which AWS services you want to use for the use case, you can directly construct the very first question you ask Amazon Q with all the relevant considerations the system should make when coming up with a recommended solution. For example, in our use case, if we know that we could host the website on EC2 and then use S3 as a store to store the original and resized images, then we can prompt the system all at once, as seen in the following screenshot.

Amazon Q   >

> What are the steps to build a simple and cost-effective solution to host a dynamic website on EC2 server and also to implement image processing pipeline where images uploaded by users are resized and stored in the backend in S3?

To build a simple and cost-effective solution to host a dynamic website on EC2 server and implement an image processing pipeline where images uploaded by users are resized and stored in S3, here are the key steps:

1. Launch an EC2 instance and install/configure a web server like Nginx or Apache to host the website code.

2. Connect the EC2 instance to a database like RDS for MySQL or Aurora to store website content and user data.

3. Create an S3 bucket to store original and processed images.

4. Write code for your website to upload images to S3 on form submission.

5. Use AWS Lambda to process the uploaded images - resize them, generate thumbnails etc. This can be triggered using S3 event notifications.

6. The Lambda function should save the processed images back to the S3 bucket.

7. Integrate AWS SDKs or CLI in your website code to retrieve and display images from S3.

8. Configure routing and load balancing using CloudFront in front of the EC2/S3 architecture.

9. Use AWS services like IAM, VPC etc for security and network isolation.

Sources

[1] Is it cheaper to store resized images on s3 or is it cheaper to use lambda edge with cloudfront to resize the image on each request? ↗

[2] Use Amazon S3 website hosting to host without a single web server - Hosting Static Websites on AWS ↗

[3] What and How much will it cost me approximately and what do i need for the website and email and domain ↗

Figure 15.4 – Amazon Q conversational capability – well-defined requirements

As you can see from the recommendation, Amazon Q was able to come up with a solution using some of the AWS services you would like to use to solve the use case.

For the sake of a clear solution, let's assume that with the assistance of Amazon Q, we were able to nail down the flow of the solution as follows:

1. The user uploads an image through the web application interface.

2. The web application sends the uploaded image to an Amazon S3 bucket.

3. An S3 event notification triggers an AWS Lambda function.

4. The Lambda function retrieves the uploaded image from the S3 bucket and initiates the image processing directly.

5. The image processing tasks, such as resizing and compression, are performed within the Lambda function using libraries such as OpenCV or PIL.

6. Once the image processing is complete, the Lambda function saves the processed image back to the S3 bucket or a designated location.

7. Optionally, metadata about the processed image can be stored in a database or logging system for monitoring and analytics purposes.

8. The processed image is now ready for retrieval and display within the web application or any other downstream systems.

The following diagram shows the solution architecture for this particular use case.

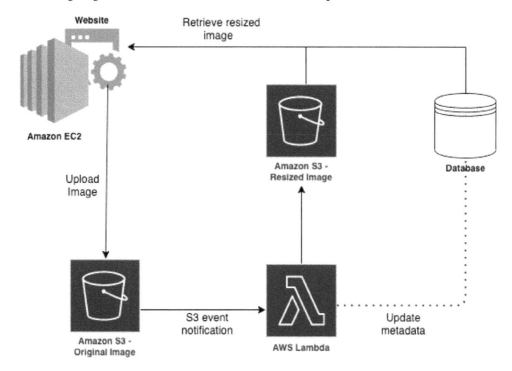

Figure 15.5 – Image processing pipeline – solution architecture

Since we are highlighting features of Amazon Q that would help fast-track building solutions on AWS, not all the steps will be covered in detail here. Also, the responses you receive from Q may be different from what we are getting, so we may not be able to show all the step-by-step instructions in this chapter for you to build this use case. The idea here is to show you how you can leverage these features of Q for your own use case, and at the end of the process, you will be able to quantify how much time Amazon Q saved you in the whole process based on all the steps you took, including design, development, building, testing, debugging, deployment, and documentation.

We now understand how the conversational feature of Q was able to help narrow the general flow of the solution along with its architecture. Let's understand the next feature of Amazon Q: optimally selecting the correct EC2 instance for the use case – something builders often have to decide based on the use case, taking into consideration the pros and cons of each such instance.

Selecting the optimal Amazon EC2 instance

To host the website, we need to get an Amazon EC2 instance up and running with an Apache HTTP server. The next question for the builder would be which EC2 instance should they select? Let's see how Amazon Q can help.

When you go to the EC2 console to launch an EC2 instance, there is now an advisor link in the **Instance type** section. The following screenshot shows the **Get advice** link.

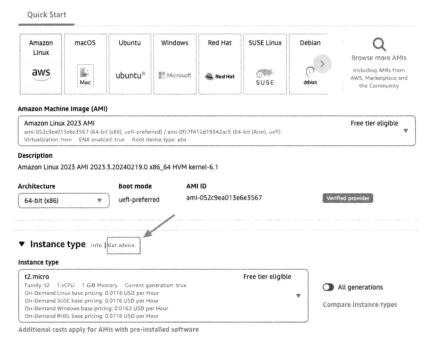

Figure 15.6 – Amazon Q – EC2 instance selection advisor

For our use case, we need a low-cost EC2 instance for hosting the website. We will pick the appropriate use case option, as shown in the following screenshot.

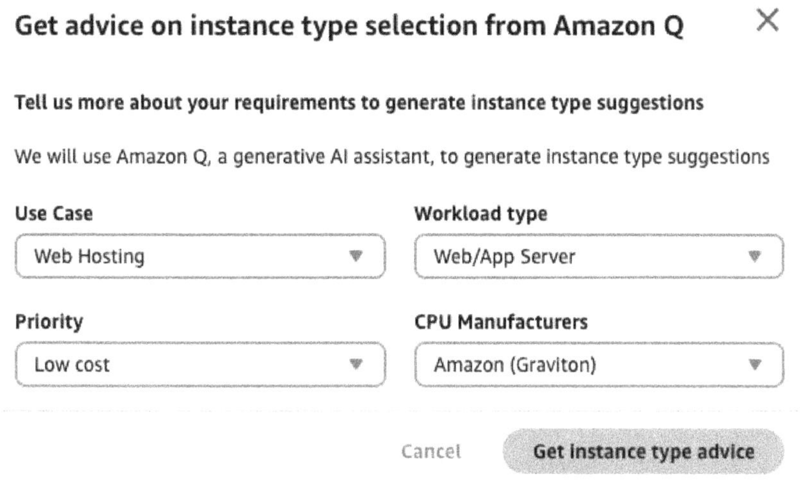

Get advice on instance type selection from Amazon Q ✕

Tell us more about your requirements to generate instance type suggestions

We will use Amazon Q, a generative AI assistant, to generate instance type suggestions

Use Case	Workload type
Web Hosting ▼	Web/App Server ▼

Priority	CPU Manufacturers
Low cost ▼	Amazon (Graviton) ▼

Cancel   **Get instance type advice**

Figure 15.7 – Amazon Q – EC2 instance selection for web hosting

Based on our criteria, Q responds with optimal EC2 instances that we can start with for our use case. Q just helps narrow down the choices; the actual benchmarking of cost/performance still needs to be done as part of the process.

The following screenshot shows the recommended EC2 instances for web hosting, keeping costs low.

Amazon Q   >

What are the EC2 instance families that are based on Graviton platform and are best suited for deploying Web/App Server workload type for Web Hosting use case? My highest preference is for lowest cost EC2 instance families. Use the format below to create the final output: EC2 Instance Selection Assistant Recommends Instance Families: Comma separated list of recommended instance Families only goes here. EC2 Instance Selection Assistant reasoning for recommendations: Bullet list of step-by-step reasoning for the EC2 Instance Selection Assistant recommendations goes here. Do not skip steps.

EC2 Instance Selection Assistant Recommends Instance Families: M7g, C7g, C6g, T4g, M6g

EC2 Instance Selection Assistant reasoning for recommendations:

1. The M7g and C7g instance families are based on the latest generation Graviton3 processors and provide the highest performance for a Web/App Server workload.

2. The T4g instance family provides lower performance than M7g/C7g but offers a lower cost option.

3. The C6g and M6g instance families are based on the previous generation Graviton2 processors and have lower performance than M7g/C7g but provide high availability.

4. Instance families like C7gd, M7gd, C6gd are optimized for high storage performance while C7gn, C6gn provide high network bandwidth. But none provide both high network and storage performance.

5. For a web hosting or application development/testing use case, the higher performance M7g or C7g instances would be best to optimize for high throughput and low latency. T4g provides a lower cost alternative with tradeoff in performance.

Sources

[1] Amazon EC2 FAQs – AWS | What are the specifications of the first-generation AWS Graviton Processors? ⤤

[2] AWS Elastic Beanstalk FAQs - Amazon Web Services (AWS) | Which platform branches are supported by Graviton on Elastic Beanstalk? ⤤

[3] The Amazon EC2 instances for your Elastic Beanstalk environment - AWS Elastic Beanstalk ⤤

Ask me anything about AWS   ⊳

Max 1000 characters

Figure 15.8 – Amazon Q – EC2 instance type recommendation for web hosting

In this section, we demonstrated how Amazon Q helped select the correct type of EC2 instance based on the use case. The remaining setup to enable a website on an EC2 instance can vary depending on how the HTTP server is set up. The web server can either be bootstrapped and started when the EC2 instance is being booted or, in many cases, an **Amazon Machine Image** (**AMI**) that has all the necessary setup is used to start the web server. Implementation details are out of the scope of this book as they may vary for individual use cases.

The next feature of Amazon Q addresses a topic that frustrates builders the most: networking issues. Builders are passionate about building applications. When they are ready to test or deploy something, networking issues can strike and derail the progress. The most frustrating part about networking is that, often, it's very difficult to know the exact issue just based on the error message. Getting to the root cause of the issue requires networking knowledge and troubleshooting skills. This process sometimes takes a lot of time.

To assist with this topic, Amazon Q provides a network troubleshooting feature that allows fast resolution of such issues in the AWS Management Console itself. Let's get into the details of this feature.

Network troubleshooting assistance

Most times, especially in production environments, all the infrastructure required for the project is configured and maintained via IaC. If the developer is tasked with setting up the networking themselves, preferably in a dev account, then they can start by asking Amazon Q about the steps required to set up networking for our use case.

The following screenshot shows the steps listed by Q for setting up networking for a public-facing website.

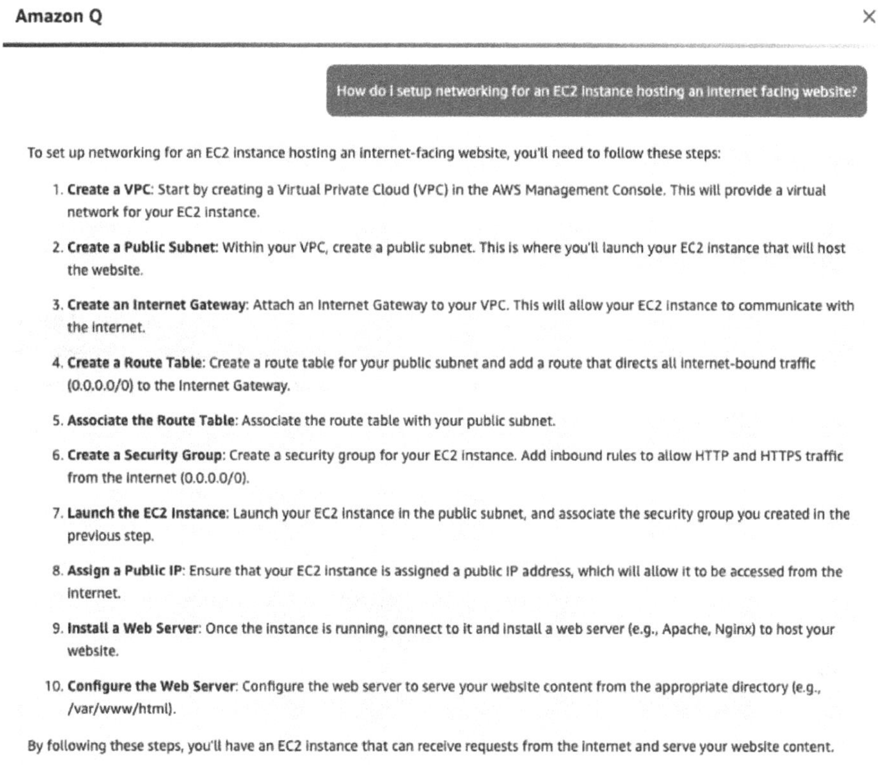

Amazon Q   ✕

How do I setup networking for an EC2 instance hosting an internet facing website?

To set up networking for an EC2 instance hosting an internet-facing website, you'll need to follow these steps:

1. **Create a VPC**: Start by creating a Virtual Private Cloud (VPC) in the AWS Management Console. This will provide a virtual network for your EC2 instance.

2. **Create a Public Subnet**: Within your VPC, create a public subnet. This is where you'll launch your EC2 instance that will host the website.

3. **Create an Internet Gateway**: Attach an Internet Gateway to your VPC. This will allow your EC2 instance to communicate with the internet.

4. **Create a Route Table**: Create a route table for your public subnet and add a route that directs all internet-bound traffic (0.0.0.0/0) to the Internet Gateway.

5. **Associate the Route Table**: Associate the route table with your public subnet.

6. **Create a Security Group**: Create a security group for your EC2 instance. Add inbound rules to allow HTTP and HTTPS traffic from the internet (0.0.0.0/0).

7. **Launch the EC2 Instance**: Launch your EC2 instance in the public subnet, and associate the security group you created in the previous step.

8. **Assign a Public IP**: Ensure that your EC2 instance is assigned a public IP address, which will allow it to be accessed from the internet.

9. **Install a Web Server**: Once the instance is running, connect to it and install a web server (e.g., Apache, Nginx) to host your website.

10. **Configure the Web Server**: Configure the web server to serve your website content from the appropriate directory (e.g., /var/www/html).

By following these steps, you'll have an EC2 instance that can receive requests from the internet and serve your website content.

Figure 15.9 – Amazon Q – networking recommendation for web hosting

Once you set up all the networking elements such as a VPC, a subnet, route tables, and so forth, you can then verify whether the setup is correct or not. For specific network analysis, Amazon Q relies on VPC Reachability Analyzer to provide explanations and detailed path analysis.

In the following screenshot, Amazon Q's network troubleshooting feature checked whether the VPC we set up can connect to the internet.

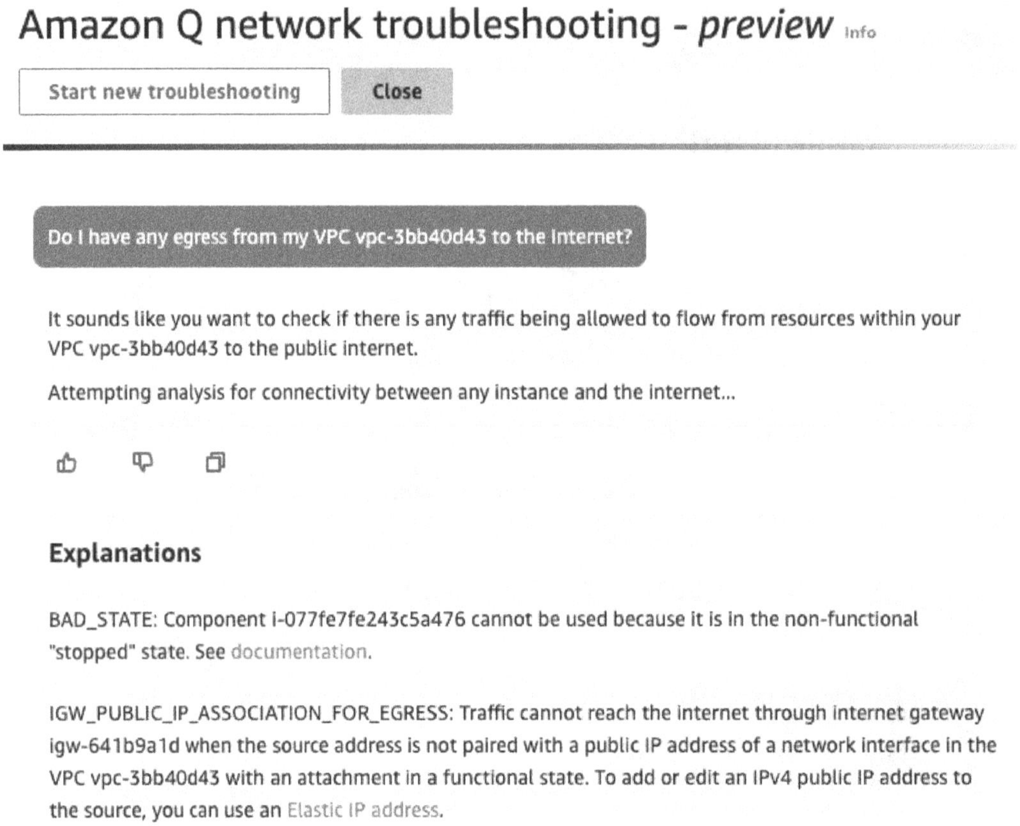

Amazon Q network troubleshooting - *preview* Info

Start new troubleshooting		Close

Do I have any egress from my VPC vpc-3bb40d43 to the Internet?

It sounds like you want to check if there is any traffic being allowed to flow from resources within your VPC vpc-3bb40d43 to the public internet.

Attempting analysis for connectivity between any instance and the internet...

👍   👎   🗗

Explanations

BAD_STATE: Component i-077fe7fe243c5a476 cannot be used because it is in the non-functional "stopped" state. See documentation.

IGW_PUBLIC_IP_ASSOCIATION_FOR_EGRESS: Traffic cannot reach the internet through internet gateway igw-641b9a1d when the source address is not paired with a public IP address of a network interface in the VPC vpc-3bb40d43 with an attachment in a functional state. To add or edit an IPv4 public IP address to the source, you can use an Elastic IP address.

Figure 15.10 – Amazon Q – networking recommendation for web hosting

The ability of Amazon Q to work in tandem with VPC Reachability Analyzer saves countless hours spent trying to troubleshoot what went wrong in the networking configuration. Once you fix the issues, you can also confirm whether the desired networking state is achieved by asking Amazon Q simple questions.

For instance, the following screenshot shows the builder checking whether the EC2 instance has connectivity to the internet, and Q will provide the result along with the networking path details.

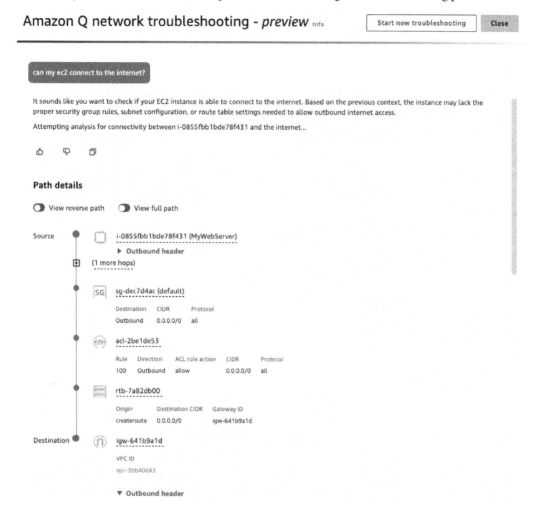

Figure 15.11 – Amazon Q – internet connection check for web hosting

We just showed a couple of questions to showcase the capability of Amazon Q; however, in real scenarios, you can ask it many more networking questions, such as the following:

- Can you help me figure out why I can't connect from the bastion host to my private instance in the application subnet?

- Which of my security groups are preventing access to my Amazon Aurora database?

- Why am I not able to SSH into EC2 instance i-xxxx?

Sometimes problems are not easy to solve or there might be some other issues with the service you are using. In such instances, you can also use Amazon Q to quickly raise a support case with AWS. Let's quickly look at this aspect.

Chatting with AWS Support

If at any point you are not able to troubleshoot an issue, you can ask Amazon Q to help you talk to someone in support, by asking Q to engage support directly. The following screenshot shows the options provided if you have just a basic support plan. When you have premium plans, it will allow you to create a support case right from the Q console. A link to different AWS Support plans is provided in the *References* section at the end of this chapter.

I am unable to connect to my ec2 instance. Can I talk to someone?

Based your issue, it looks like you have a technical query. You can engage our support engineers by posting to AWS re:Post, which is included in your Basic Support plan.

A direct line to our support engineers, who will help you troubleshoot technical issues, is available with a Premium Support plan.

If you'd like to access immediate support, you may explore premium support plans. ☑
Or, you may request to see your eligibility for a free 60 day trial for Business Support.

Go to AWS re:Post ☑

Request trial details ☑

Cancel

Figure 15.12 – Amazon Q – support case

The other option is to first see if Q can assist with the issue. If the response provided by Q does not seem to help, you can also click the thumbs-down icon in Q, and it will help you create a support case, as seen in the following screenshot.

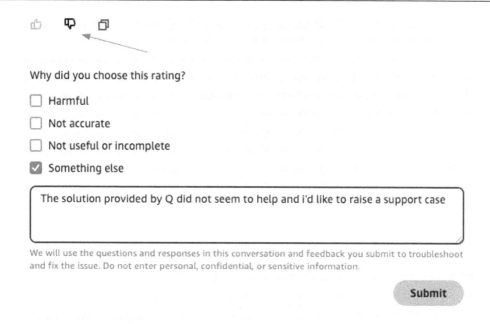

Figure 15.13 – Amazon Q –support case by clicking the thumbs-down icon

This way, it will also allow you to create a support case from the Q chat itself, with the issue context auto-populating support case fields. This saves time as you don't have to leave the screen to fill out a separate support case.

Our use case can now move forward with the next building block. Let's continue, and while solving the next portion of our use case, we will also introduce another key feature of Amazon Q.

Troubleshooting console errors with Amazon Q

To demonstrate the next feature of Amazon Q, let's move on to the next step of the solution, which involves creating a Lambda function. This function will be triggered when an image is uploaded to the S3 bucket from the website. The Lambda function will utilize an image processing library to resize the image and then place it in another S3 bucket. Additionally, the Lambda function will update the metadata for the resized image in a database, enabling the website to reference it and retrieve the correct image from the S3 bucket to display on certain web pages.

In the previous chapter, we demonstrated how Amazon Q can accelerate AWS Lambda code generation. Using the same techniques, you can auto-generate the Lambda function logic for this use case as well. The following is a sample Lambda function for our use case. It highlights the core image processing part without all the necessary additional features. The code is for your reference only and you can use Amazon Q to help generate the code on your behalf. The output may vary if we try to regenerate it, so care should be taken to validate its correctness:

```python
import boto3
from PIL import Image
import io

def lambda_handler(event, context):
    # Get the S3 bucket and key from the event
    s3 = boto3.client('s3')
    source_bucket = event['Records'][0]['s3']['bucket']['name']
    source_key = event['Records'][0]['s3']['object']['key']

    # Download the original image from S3
    response = s3.get_object(Bucket=source_bucket, Key=source_key)
    image_content = response['Body'].read()

    # Open the image using PIL
    image = Image.open(io.BytesIO(image_content))

    # Resize the image
    resized_image = image.resize((200, 200))  # Adjust the size as
needed

    # Upload the resized image to a new S3 bucket
    destination_bucket = 'your-destination-bucket-name'
    destination_key = f"resized/{source_key}"

    # Save the resized image to a BytesIO object
    resized_image_io = io.BytesIO()
    resized_image.save(resized_image_io, format='JPEG')  # Change the
format if needed
    resized_image_io.seek(0)

    # Upload the resized image to S3
    s3.put_object(Bucket=destination_bucket, Key=destination_key,
        Body=resized_image_io)

    return {
        'statusCode': 200,
```

```
            'body': 'Image resized and uploaded successfully!'
    }
```

We will set up this Lambda function and add a trigger for invocation whenever an object is created in the source S3 bucket. So, every time the website uploads an image to this bucket, the "create object" event in S3 will trigger the Lambda function.

The following screenshot shows the Lambda trigger that we added.

Figure 15.14 – AWS Lambda function – event trigger

During the build lifecycle, builders often spend a significant amount of time debugging and fixing issues encountered during testing. Before Amazon Q, builders had to do their own research to understand what an error was and how to fix it. But now, with Amazon Q, builders can troubleshoot errors directly from the AWS Management Console. Let's explore this feature and understand how we can use it to troubleshoot issues for our use case.

Troubleshooting Lambda errors with Amazon Q

In the AWS Management Console, Amazon Q can diagnose and address errors encountered while utilizing AWS services, including instances of insufficient permissions, misconfigurations, and exceeding service limits. This functionality is accessible for errors occurring within services such as Amazon EC2, Amazon ECS, Amazon S3, and AWS Lambda. While Amazon Q can resolve many common console errors, it does not handle basic validation errors, and it does not retain a record of previous troubleshooting sessions.

Let's understand how we can use this feature for our use case. Now, to test whether our Lambda function is working correctly, we will use the test option available in the AWS Lambda console. For the test, we will pre-upload a sample image in the source bucket and then execute the following test:

```
{
   "Records": [
      {
         "eventVersion": "2.0",
         "eventSource": "aws:s3",
         "awsRegion": "us-west-2",
         "eventTime": "2022-02-18T10:05:09.775Z",
         "eventName": "ObjectCreated:Put",
         "userIdentity": {
            "principalId": "EXAMPLE"
         },
         "requestParameters": {
            "sourceIPAddress": "127.0.0.1"
         },
         "responseElements": {
            "x-amz-request-id": "EXAMPLE123456789",
            "x-amz-id-2": "EXAMPLE123/5678abcdefghijklambdaisawesome/
mnopqrstuvwxyzABCDEFGH"
         },
         "s3": {
            "s3SchemaVersion": "1.0",
            "configurationId": "testConfigRule",
            "bucket": {
               "name": "image-processing-tvxgdfh",
               "ownerIdentity": {
                  "principalId": "EXAMPLE"
               },
               "arn": "arn:aws:s3:::image-processing-tvxgdfh"
            },
            "object": {
               "key": "example.jpg",
               "size": 1024,
               "eTag": "0123456789abcdef0123456789abcdef",
               "sequencer": "0A1B2C3D4E5F678901"
            }
         }
      }
   ]
}
```

This test simply simulates the event message the Lambda function would receive when an image is uploaded to the source bucket. The test feature in the Lambda console allows builders to verify whether the lambda is working as intended before the code is put into production.

When we click the **Test** button in the AWS Lambda console, we immediately encounter an error, as seen in the following screenshot.

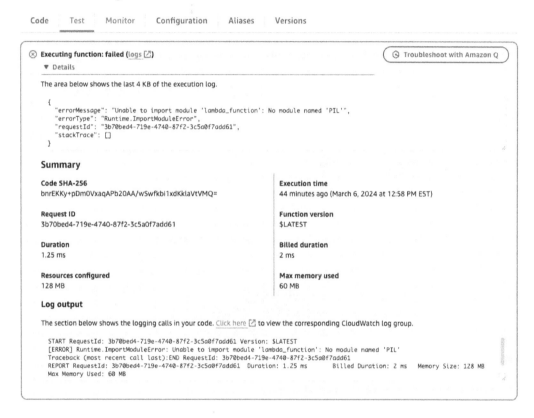

Figure 15.15 – AWS Lambda function – error during test

As seen from this error, it appears that the PIL module is not being recognized. The **Troubleshoot with Amazon Q** option appears next to the error. When you click that option, Amazon Q will first analyze the issue and provide an analysis of the issue, as seen in this case in the following screenshot.

Troubleshoot with Amazon Q   ✕

Amazon Q is helping you resolve the following error:

```
Executing function: failed (logs ) START RequestId: 3b70bed4-719e-4740-87f2-3c5a0f7add61 Version: $LATEST
[ERROR] Runtime.ImportModuleError: Unable to import module 'lambda_function': No module named 'PIL'...
```

Expand

Analysis

The error indicates that the Lambda function is unable to import the PIL module, which likely means it is not installed or available within the Lambda execution environment. As the function is attempting to import PIL during execution, this suggests a dependency is not being properly satisfied.

(Help me resolve)

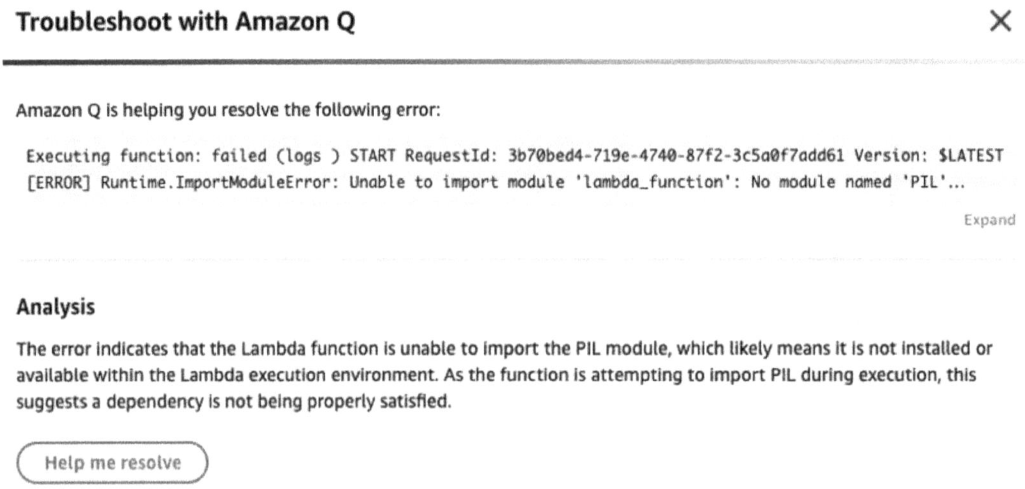

Figure 15.16 – Amazon Q – troubleshoot AWS Lambda error

When you select the **Help me resolve** option, Amazon Q will provide a possible resolution to the problem. This way, builders don't have to conduct external research on how to fix the error. The following screenshot shows the resolution steps for our Lambda function error.

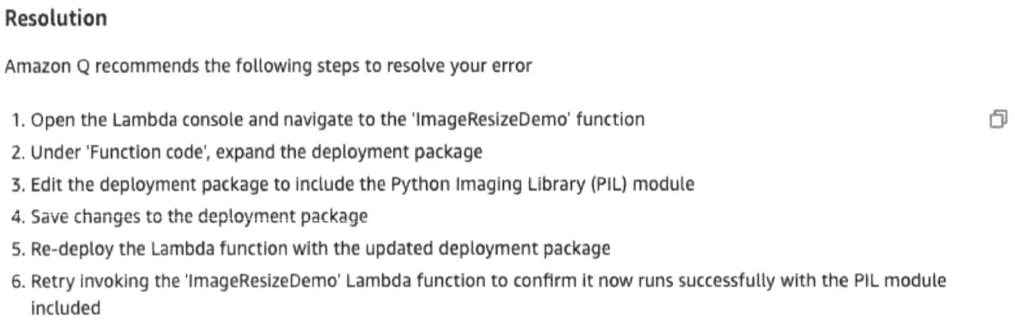

Resolution

Amazon Q recommends the following steps to resolve your error

1. Open the Lambda console and navigate to the 'ImageResizeDemo' function
2. Under 'Function code', expand the deployment package
3. Edit the deployment package to include the Python Imaging Library (PIL) module
4. Save changes to the deployment package
5. Re-deploy the Lambda function with the updated deployment package
6. Retry invoking the 'ImageResizeDemo' Lambda function to confirm it now runs successfully with the PIL module included

Figure 15.17 – Amazon Q – resolution for AWS Lambda error

So basically, the Python Imaging Library was missing, and Amazon Q suggested steps to add it to the Lambda function. At the beginning of this use case, I added an important note regarding solutions provided by LLM-based assistants such as Amazon Q. The solution offered may vary for different invocations of Amazon Q. To demonstrate this point, I closed the resolution window and requested that Q troubleshoot the same issue again. This time, Amazon Q came up with a different recommendation to solve this problem.

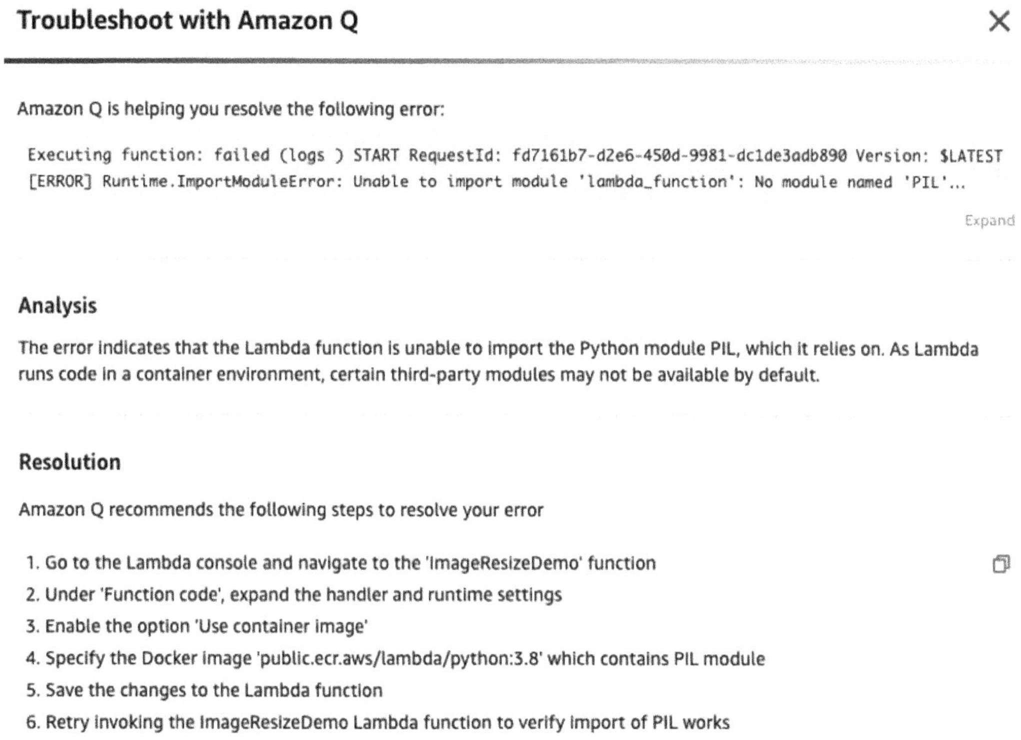

Troubleshoot with Amazon Q   ✕

Amazon Q is helping you resolve the following error:

```
Executing function: failed (logs ) START RequestId: fd7161b7-d2e6-450d-9981-dc1de3adb890 Version: $LATEST
[ERROR] Runtime.ImportModuleError: Unable to import module 'lambda_function': No module named 'PIL'...
```

Expand

Analysis

The error indicates that the Lambda function is unable to import the Python module PIL, which it relies on. As Lambda runs code in a container environment, certain third-party modules may not be available by default.

Resolution

Amazon Q recommends the following steps to resolve your error

1. Go to the Lambda console and navigate to the 'ImageResizeDemo' function
2. Under 'Function code', expand the handler and runtime settings
3. Enable the option 'Use container image'
4. Specify the Docker image 'public.ecr.aws/lambda/python:3.8' which contains PIL module
5. Save the changes to the Lambda function
6. Retry invoking the ImageResizeDemo Lambda function to verify import of PIL works

Figure 15.18 – Amazon Q – resolution for AWS Lambda error – variation 1

If I seek another instance of troubleshooting, this time, I receive another recommendation to solve the same issue, as seen in the following screenshot.

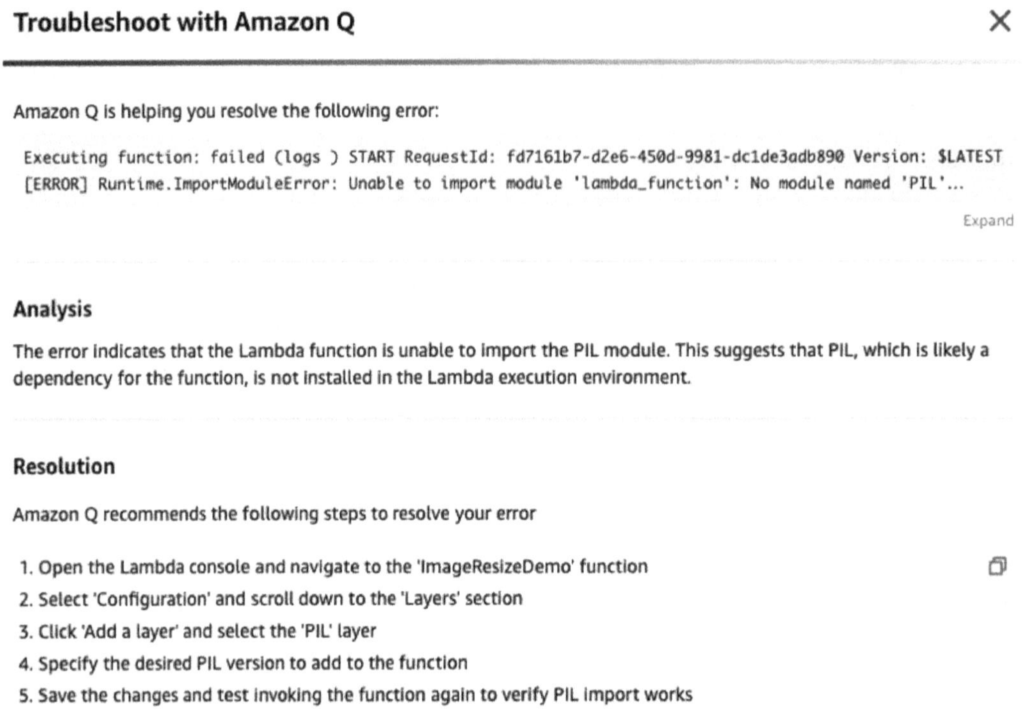

Troubleshoot with Amazon Q   ✕

Amazon Q is helping you resolve the following error:

```
Executing function: failed (logs ) START RequestId: fd7161b7-d2e6-450d-9981-dc1de3adb890 Version: $LATEST
[ERROR] Runtime.ImportModuleError: Unable to import module 'lambda_function': No module named 'PIL'...
```

Expand

Analysis

The error indicates that the Lambda function is unable to import the PIL module. This suggests that PIL, which is likely a dependency for the function, is not installed in the Lambda execution environment.

Resolution

Amazon Q recommends the following steps to resolve your error

1. Open the Lambda console and navigate to the 'ImageResizeDemo' function
2. Select 'Configuration' and scroll down to the 'Layers' section
3. Click 'Add a layer' and select the 'PIL' layer
4. Specify the desired PIL version to add to the function
5. Save the changes and test invoking the function again to verify PIL import works

Figure 15.19 – Amazon Q – resolution to AWS Lambda error – variation 2

Now, you may wonder why this happened in this case. Firstly, a problem may have multiple solutions, and each solution has its own steps. Which solution would be the best fit for our problem is up to the builders to try and decide. Not every solution may be the best fit, as Amazon Q may not be aware of any external dependencies your team or specific implementation of the project may have. Builders need to do their due diligence in how they implement the final solution. Amazon Q just assists you in narrowing down options that may have taken you longer to research on your own.

To keep moving, I adopted the recommendation provided by Amazon Q in *Figure 15.14*. To get a quick resolution to the library issue, I added an external ARN, `'arn:aws:lambda:us-west-2:770693421928:layer:Klayers-p312-Pillow:1'`, as a reference to the PIL library in my Lambda function layer section.

I reran the test to see if that resolved my issue. But, immediately, I encountered a different error, as seen in the following screenshot.

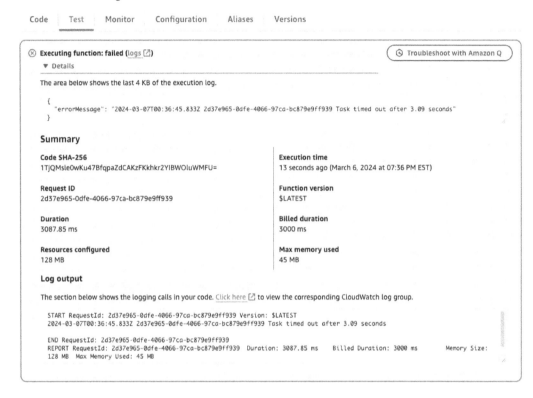

Figure 15.20 – Amazon Q – another AWS Lambda error

Looks like the library issue got resolved, but the function timed out. Instead of combing through the logs for the reason for this issue, I clicked the **Troubleshoot with Amazon Q** button again. The following screenshot shows the analysis of the issue along with the resolution provided by Amazon Q.

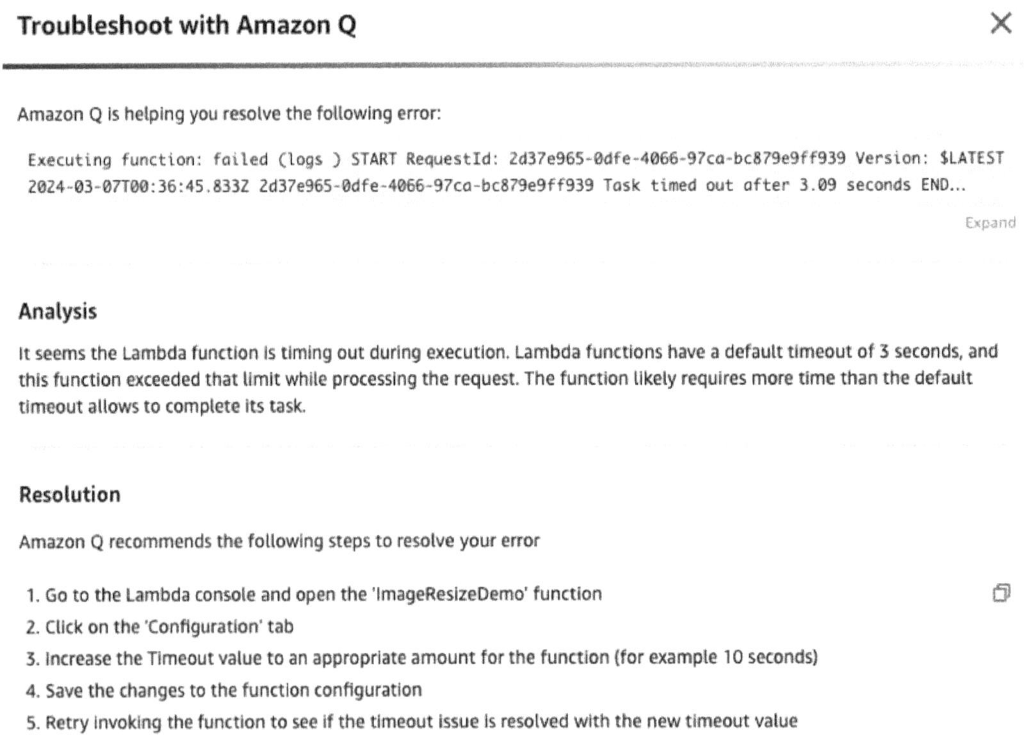

Troubleshoot with Amazon Q   ✕

Amazon Q is helping you resolve the following error:

```
Executing function: failed (logs ) START RequestId: 2d37e965-0dfe-4066-97ca-bc879e9ff939 Version: $LATEST
2024-03-07T00:36:45.833Z 2d37e965-0dfe-4066-97ca-bc879e9ff939 Task timed out after 3.09 seconds END...
```

Expand

Analysis

It seems the Lambda function is timing out during execution. Lambda functions have a default timeout of 3 seconds, and this function exceeded that limit while processing the request. The function likely requires more time than the default timeout allows to complete its task.

Resolution

Amazon Q recommends the following steps to resolve your error

1. Go to the Lambda console and open the 'ImageResizeDemo' function
2. Click on the 'Configuration' tab
3. Increase the Timeout value to an appropriate amount for the function (for example 10 seconds)
4. Save the changes to the function configuration
5. Retry invoking the function to see if the timeout issue is resolved with the new timeout value

Figure 15.21 – Amazon Q – analysis and resolution for the AWS Lambda timeout error

Sometimes, simple issues cause frustration and waste time trying to figure out the cause. Amazon Q understood the issue and provided exact steps to resolve the problem in seconds. This example may be simple but it highlights the productivity gains all AWS builders will experience once they start leveraging Amazon Q to troubleshoot issues in the AWS Management Console.

Finally, after increasing the timeout, the Lambda test was completed successfully, and the following screenshot confirms that the image was resized.

Figure 15.22 – Amazon Q – successful execution of the AWS Lambda function

You can check the destination S3 bucket to find a resized image in there.

This brings us to the end of our use case; however, another feature of Q can help you understand what resources exist in your account. In other words, Amazon Q is also aware of the AWS resources that are being used in your accounts.

Awareness of AWS account resources

So far in this chapter, we have seen how Amazon Q assists builders in creating solutions on AWS by generating architecture patterns, best practices, and also code. But what if the builders need assistance with AWS resources that are already built in their AWS account and want to know more details about them without having to find this information individually? Amazon Q is now not only aware of the AWS resources in your account but can also answer questions about costs, helping you better understand and create optimization strategies. Let's quickly look at these features.

Analyzing resources with Amazon Q

Before you ask Q to analyze your resources, ensure that IAM permissions are granted for the action you are asking Q to perform. For example, if you are asking Q to list all S3 buckets in your account, then the `s3:ListAllMyBuckets` IAM permission must be granted.

The following screenshot shows all the EC2 instances I used for my use case.

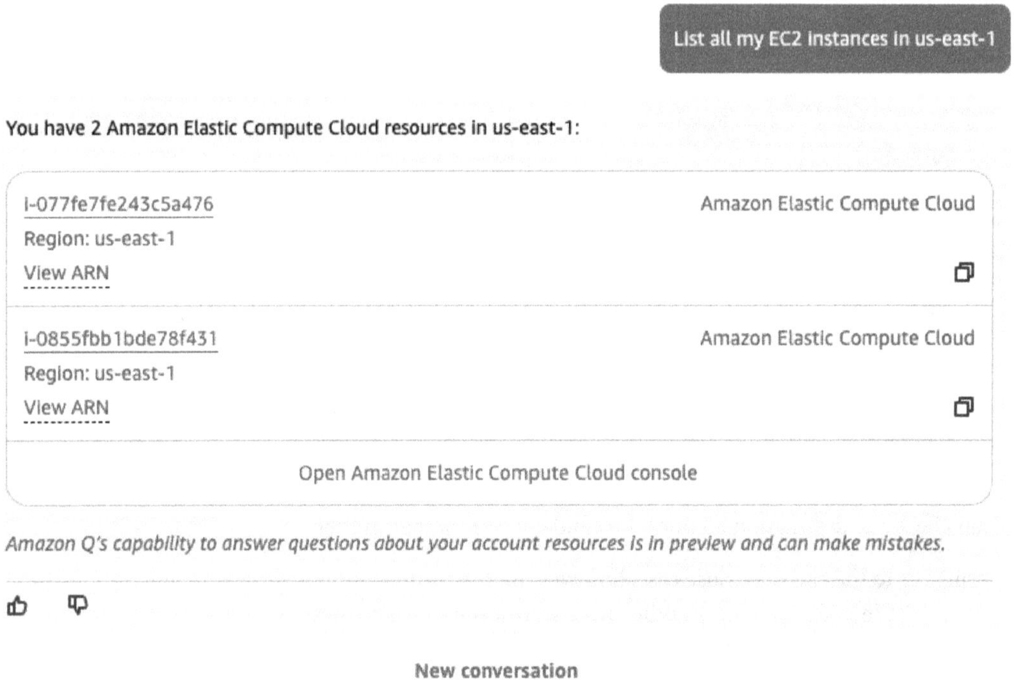

List all my EC2 instances in us-east-1

You have 2 Amazon Elastic Compute Cloud resources in us-east-1:

I-077fe7fe243c5a476 Region: us-east-1 View ARN	Amazon Elastic Compute Cloud
I-0855fbb1bde78f431 Region: us-east-1 View ARN	Amazon Elastic Compute Cloud

Open Amazon Elastic Compute Cloud console

Amazon Q's capability to answer questions about your account resources is in preview and can make mistakes.

New conversation

Figure 15.23 – Amazon Q – AWS resource analysis

You can ask refined questions to get many such insights into the AWS resources that you have provisioned. In the following screenshot, we are checking the encryption settings for the bucket we provisioned for our use case in this chapter.

Figure 15.24 – Amazon Q – AWS resource setting analysis

Before we conclude this chapter, we definitely want to know the costs associated with the use case. In fact, all teams using AWS for their workloads explore costs along with their management to ensure expenses do not proliferate due to unnecessary actions. Amazon Q can help with this aspect too.

Analyzing costs with Amazon Q

Amazon Q enables seamless cost management by retrieving, explaining, and analyzing your cost data, allowing you to understand your expenses.

When you inquire about your costs, Amazon Q provides detailed responses that include specific cost figures, associated metrics, time periods, and granular details, along with any applied groups or filters. Additionally, it offers a link to an AWS Cost Explorer view with the same specifications, enabling you to visualize or download the data for further analysis.

The following is a simple example of asking Q for costs for our EC2 instances for a particular month of the year.

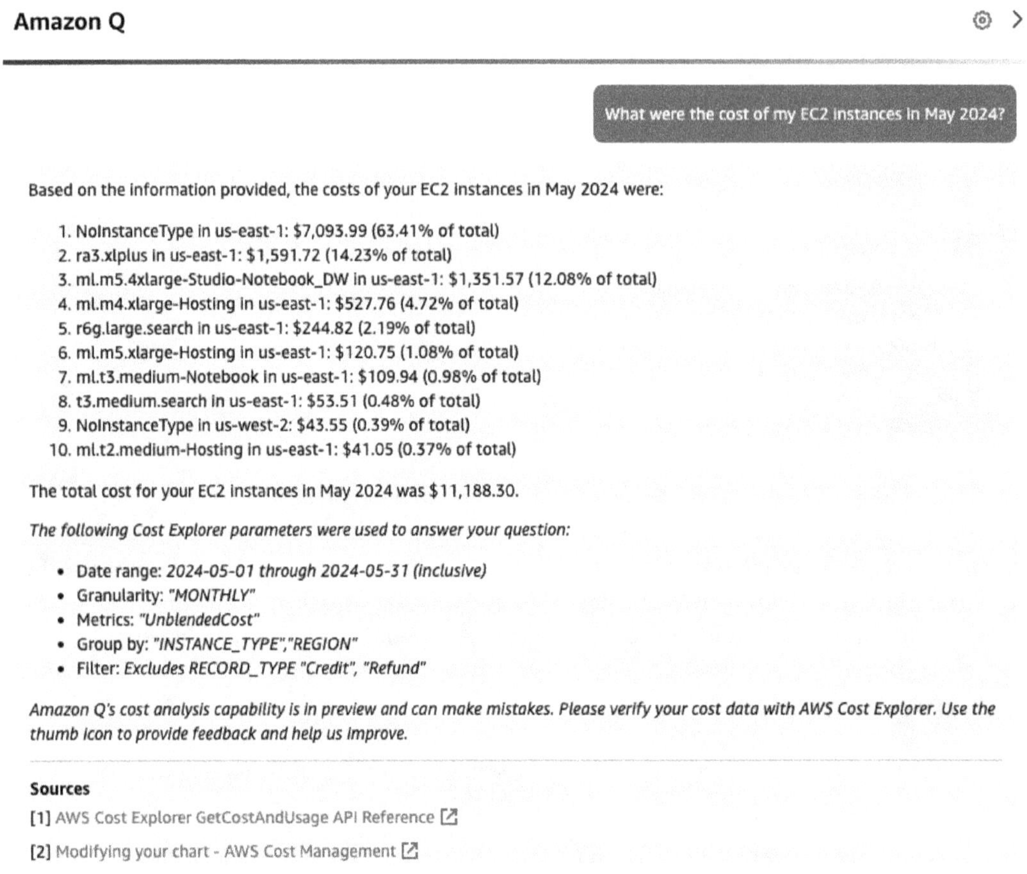

Amazon Q   ⚙ >

> What were the cost of my EC2 instances in May 2024?

Based on the information provided, the costs of your EC2 instances in May 2024 were:

1. NoInstanceType in us-east-1: $7,093.99 (63.41% of total)
2. ra3.xlplus in us-east-1: $1,591.72 (14.23% of total)
3. ml.m5.4xlarge-Studio-Notebook_DW in us-east-1: $1,351.57 (12.08% of total)
4. ml.m4.xlarge-Hosting in us-east-1: $527.76 (4.72% of total)
5. r6g.large.search in us-east-1: $244.82 (2.19% of total)
6. ml.m5.xlarge-Hosting in us-east-1: $120.75 (1.08% of total)
7. ml.t3.medium-Notebook in us-east-1: $109.94 (0.98% of total)
8. t3.medium.search in us-east-1: $53.51 (0.48% of total)
9. NoInstanceType in us-west-2: $43.55 (0.39% of total)
10. ml.t2.medium-Hosting in us-east-1: $41.05 (0.37% of total)

The total cost for your EC2 instances in May 2024 was $11,188.30.

The following Cost Explorer parameters were used to answer your question:

- Date range: *2024-05-01 through 2024-05-31 (inclusive)*
- Granularity: *"MONTHLY"*
- Metrics: *"UnblendedCost"*
- Group by: *"INSTANCE_TYPE","REGION"*
- Filter: *Excludes RECORD_TYPE "Credit", "Refund"*

Amazon Q's cost analysis capability is in preview and can make mistakes. Please verify your cost data with AWS Cost Explorer. Use the thumb icon to provide feedback and help us improve.

Sources

[1] AWS Cost Explorer GetCostAndUsage API Reference ↗

[2] Modifying your chart - AWS Cost Management ↗

Figure 15.25 – Amazon Q – AWS cost analysis

For Amazon Q to analyze the costs, AWS Cost Explorer needs to be enabled. After that, you can chat with Q and ask any variation of cost exploration questions, such as the following:

- How much did we spend on EC2 in 2024?
- What were my EC2 costs by instance types for the past 6 months?
- Which linked accounts spent the most on Redshift last month?

This brings us to the end of the chapter; however, we are just scratching the surface in terms of what Amazon Q can do. With generative AI rapidly evolving, expect many more new features and capabilities in Amazon Q pertaining to solving use cases in AWS.

Summary

In this chapter, we covered how Amazon Q can assist in building solutions on AWS. Amazon Q helps with many aspects of the build process, including providing architecture patterns and best practices, diagnosing AWS console errors, troubleshooting networking issues, creating support cases, and understanding all provisioned AWS resources along with their costs.

In the next and final chapter, we will look at another key aspect for DevOps builders and how Amazon Q can help fast-track many aspects of it.

References

AWS support plans: `https://aws.amazon.com/premiumsupport/plans/`

16

Accelerate the DevOps Process on AWS

In this chapter, we will look into the following key topics:

- Challenges during the DevOps process
- Introduction to Amazon CodeCatalyst
- Exploring Amazon Q's capabilities in Amazon CodeCatalyst
- Amazon Q's feature development capability in Amazon CodeCatalyst
- Amazon Q's summarizing capability in Amazon CodeCatalyst

In any organization, software development is not just about writing code. The software engineering process also involves something known as DevOps. It is a combination of "development" and "operations," involving a set of practices aimed at improving collaboration and communication between software **development** (**Dev**) and IT **operations** (**Ops**) teams. It focuses on automating processes, increasing efficiency, and delivering high-quality software products more quickly and reliably.

DevOps emphasizes a culture of collaboration, **continuous integration and continuous delivery** (**CI/ CD**), the automation of infrastructure and workflows, and monitoring and feedback loops to enable faster development cycles, improved deployment frequency, and more stable operating environments.

Many tools are available in the marketplace to serve various purposes in the DevOps life cycle, including version control, continuous integration, continuous delivery, configuration management, infrastructure as code, monitoring, and collaboration. AWS also has a service called Amazon CodeCatalyst that helps with this process. But first, we need to understand some of the challenges faced during the DevOps process.

Challenges during the DevOps process

Even though DevOps tools help in many aspects of the software development process, some challenges still persist during this process. Let's understand some of the challenges so that you will appreciate what Amazon Q brings to the DevOps process in CodeCatalyst:

- **Complexity**: Many DevOps tools can be complex to set up, configure, and maintain, requiring specialized knowledge and skills. For instance, setting up a CI/CD pipeline with Jenkins involves multiple steps: installing and configuring Jenkins on a server, integrating it with version control systems such as Git, setting up build scripts, configuring plugins for various stages of the pipeline, managing user permissions, and ensuring the server is secure and regularly updated. This process can be time-consuming and requires a deep understanding of both Jenkins and the underlying infrastructure.

- **Integration issues**: Integrating multiple DevOps tools into a cohesive pipeline can be challenging, leading to compatibility issues and data silos.

- **Limited automation**: While automation is a key principle of DevOps, not all tools offer robust automation capabilities, leading to manual workarounds and inefficiencies. For instance, consider a scenario where a team uses a deployment tool that lacks automated rollback features. If a deployment fails, team members must manually intervene to revert the system to a previous stable state. This manual process can be time-consuming and prone to errors and disrupt the streamlined workflow that DevOps aims to achieve. As a result, the lack of robust automation in the tool leads to inefficiencies and increased operational overhead.

- **Learning curve**: DevOps tools often have steep learning curves, requiring time and resources for teams to become proficient in their use, which can slow down adoption and implementation.

- **Lack of collaboration**: Siloed teams and departments can hinder collaboration and communication, leading to inefficiencies and bottlenecks in the DevOps pipeline.

- **Continuous testing**: Implementing comprehensive testing strategies, including unit tests, integration tests, and automated regression tests, can be challenging, particularly in complex environments with frequent code changes.

- **Skills gap**: Finding and retaining skilled DevOps professionals with expertise in automation, cloud computing, containerization, and other relevant technologies can be challenging in a competitive job market.

Let's quickly understand the basics of Amazon CodeCatalyst so that it's easy to understand how Amazon Q can accelerate the DevOps process.

Introduction to Amazon CodeCatalyst

Amazon CodeCatalyst streamlines software development for teams embracing continuous integration and continuous delivery practices. By consolidating essential tools, it simplifies work planning, code collaboration, and application development with built-in CI/CD capabilities. Seamlessly integrating AWS resources into projects is facilitated through direct connections with AWS accounts. This all-in-one tool manages every stage and facet of the application life cycle, enabling swift and assured software delivery.

The following figure highlights all the different aspects of software development that CodeCatalyst helps with.

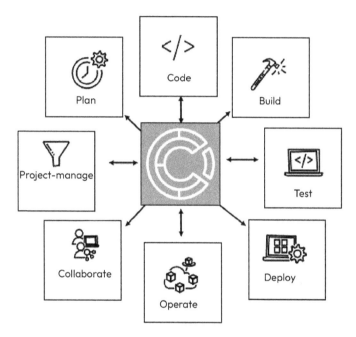

Figure 16.1 – Amazon CodeCatalyst capabilities

Let's walk through some of the key things that CodeCatalyst can do. These are also referenced in the AWS documentation for CodeCatalyst, a link to which is provided in the *References* section at the end of the chapter:

- **Code collaboration**: Collaborate seamlessly with your team on code through branches, merges, pull requests, and comments in your source code repositories. Quickly create development environments to work on code without the need to clone or set up connections to repositories.

- **Build, test, and deploy an application with workflows**: Define workflows with build, test, and deploy actions to manage the CI/CD of your applications. Initiate workflows manually or set them to start automatically based on events such as code pushes or the creation or closing of pull requests.

- **Prioritize work with issue tracking**: Use issues to establish backlogs and track the status of ongoing tasks with boards. Maintaining a healthy backlog of items for your team to address is integral to software development.

- **Monitoring and notifications**: Monitor team activity and resource status, and configure notifications to remain informed about important changes.

Amazon CodeCatalyst is a comprehensive service with multiple capabilities and features. Providing a detailed overview is beyond the scope of this book. However, if you are interested in exploring CodeCatalyst as a DevOps tool, feel free to go through the workshop that is listed in the *References* section at the end of this chapter.

To fast-track learning and see CodeCatalyst in action, multiple blueprints are provided within the service. These blueprints will create all the infrastructure needed to implement an end-to-end DevOps process for a specific use case. One such blueprint is for implementing a modern three-tier web application. You can find a step-by-step tutorial for it in the link provided in the *References* section.

With the introduction to DevOps done, let's jump straight into the theme of this chapter: how Amazon Q can help with the DevOps process inside Amazon CodeCatalyst.

Exploring Amazon Q's capabilities in Amazon CodeCatalyst

If your project resides in a source repository within Amazon CodeCatalyst, GitHub Cloud, or Bitbucket Cloud, and Amazon Q features are enabled, you can leverage some of Q's capabilities to expedite the DevOps process. Developers often face time constraints, leading to a backlog of tasks. Consequently, they may overlook providing detailed explanations for code changes in pull requests, assuming their peers will decipher them independently. Likewise, both pull request creators and reviewers may lack sufficient time to thoroughly analyze comments, especially in cases of multiple revisions.

In modern software development, engineers encounter significant challenges during the DevOps process, leading to bottlenecks and inefficiencies. One prominent issue is the time-consuming nature of code review and collaboration within teams. Software engineers often struggle to provide comprehensive explanations for their code changes in pull requests, while reviewers face difficulties in thoroughly analyzing and understanding the changes, particularly in cases with multiple revisions. These inefficiencies result in delays, reduced productivity, and missed opportunities for innovation. Additionally, adding new features to the code is a time-consuming process; at the same time, reviewers may face difficulties in comprehending complex changes, especially when dealing with large code bases.

With Amazon Q integration with Amazon CodeCatalyst, team members can streamline their workflows, allowing more time to focus on critical aspects of their work. We are in the infancy stage of incorporating Amazon Q into the DevOps process, but so far, here are some of the key features of Amazon Q that work with Amazon CodeCatalyst:

- **Feature development**: Enables developers to input ideas in an issue and get fully tested application code that is ready for merge in a pull request. With just a few clicks and inputs in natural language, the whole workflow is simplified.

- **Auto-generate a pull request summary**: Amazon Q can analyze all comments left on code changes within a pull request and generate a concise summary of the feedback.

- **Auto-generate a pull request description**: Amazon Q can analyze the code changes and generate a description of the changes in the pull request, making it easier to review and approve changes.

There are other areas where Q assists within CodeCatalyst, such as auto-selecting blueprints and providing task recommendations. We will go through each of these features in detail in this chapter. But first, as always, another reminder about the non-deterministic nature of LLM-generated output.

Reminder

Before we start showcasing Amazon Q's capabilities with CodeCatalyst, it is important to remember a few things. Responses received from LLMs are generally not idempotent. Idempotence refers to the property of an operation where applying it multiple times has the same effect as applying it once.

In the context of LLMs, each response generated is based on the specific input prompt provided at that moment. While the same prompt may produce similar or related responses across multiple iterations, there is no guarantee that the responses will be identical each time. Factors such as model initialization, randomization, and the specific context of the prompt can influence the variations in the generated responses.

Therefore, even in the example we will be using in the following sections, if you send the same prompt to Amazon Q, you may receive different responses each time, making the responses non-idempotent. Use your own judgment on what to accept as accurate and what to prompt the system again for to receive a revised approach, before you proceed to try out all the steps.

Now, let's start with the feature development capability in Amazon CodeCatalyst, our DevOps service, in which generative AI capabilities of Amazon Q are already enabled.

Amazon Q's feature development capability in Amazon CodeCatalyst

The feature development capability in Amazon Q enables users to customize and enhance Q's question-answering skills for specific domains. Concerning the DevOps process using Amazon CodeCatalyst, feature development could entail issuing a fix to a bug identified in the application, making improvements, or adding new functionality. For all these tasks, the code base requires alteration, testing, troubleshooting for any issues during this process, and then pushing the changes back into production for deployment. While it may seem straightforward, in large code bases with many components and insufficient comments for clarity, especially for newly onboarded developers, completing the end-to-end task quickly becomes challenging.

Let's walk through an example of how Amazon Q assists with feature development inside Amazon CodeCatalyst. Here, we assume that you have a basic understanding of the different components of CodeCatalyst. If this is your first time using the service, we encourage you to try out the workshop linked in the *References* section at the end of this chapter. Once you complete those, you will be able to relate better to some of the aspects we discuss in this section. Let's jump straight to the use case.

Use case for leveraging Amazon Q feature development in Amazon CodeCatalyst

To demonstrate the feature development capability, we will select the AWS Glue ETL blueprint available when you create a project. We've chosen this example to provide a comprehensive view of how developers can expedite their ETL code in Glue. Additional techniques for accelerating coding with Glue are covered in *Chapter 14*.

The following screenshot shows the AWS Glue ETL blueprint inside CodeCatalyst.

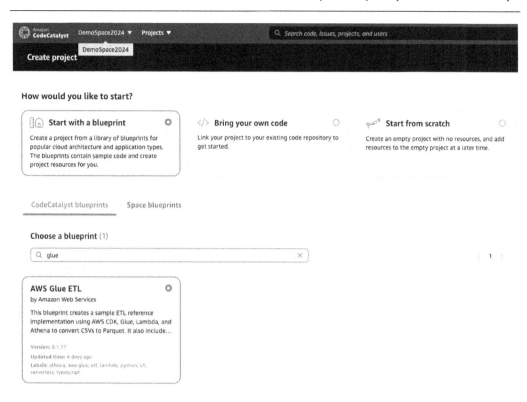

Figure 16.2 – Glue ETL blueprint in Amazon CodeCatalyst

In this example, since we already knew which blueprint we wanted, it was easy for us to read the description and select it for our project. However, some use cases may require us to read the descriptions of multiple available blueprints before selecting the most appropriate one. Amazon Q now assists you in creating a project by auto-selecting the best-matching blueprint based on the criteria you provide in natural language. Q will analyze the descriptions of all the blueprints and select the best one for you to create a project with, thus saving you valuable time.

The following example highlights the project creation process with Amazon Q suggesting the Glue blueprint based on the criteria we provided in the chat.

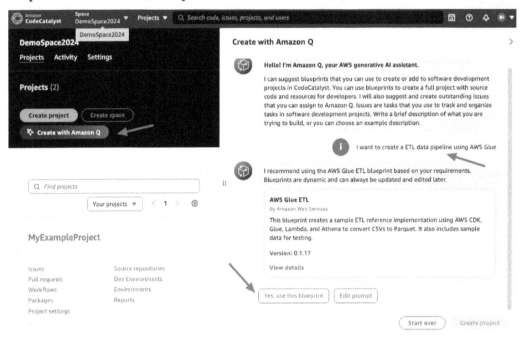

Figure 16.3 – Glue ETL blueprint suggestion by Amazon Q

Once you have completed the steps, the source code repository will be set up, and a CI/CD pipeline via a workflow will have been completed. The pipeline essentially initiates an AWS CloudFormation stack using the AWS **Cloud Development Kit** (**CDK**) code stored within the repository.

The following screenshot depicts the Glue ETL blueprint after its setup in CodeCatalyst. The architecture of this ETL project is also explained in the overview. It's a straightforward project where input data is processed using AWS Glue and stored in Amazon S3 in Parquet file format for querying by Amazon Athena. The workflow status at the bottom of the screen indicates completion, signifying that the CloudFormation stack has successfully deployed all the components of the architecture. You can navigate to the AWS Lambda and AWS Glue service consoles in AWS to inspect the artifacts.

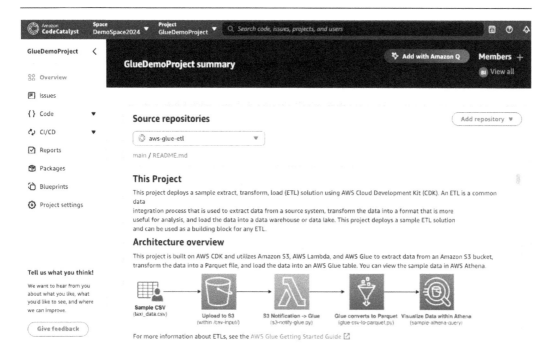

Figure 16.4 – Glue ETL blueprint deployed using Amazon CodeCatalyst

Notice the **Add with Amazon Q** option (at the top right of the preceding screenshot), which allows you to add new components by suggesting other blueprints based on your desired requirements. For this use case, let's assume we found everything we wanted in this single blueprint, so we will work with what we already have.

Now, let's assume that we have some enhancements to make in our Glue ETL pipeline. Constant enhancements and improvements to solutions are always encouraged, so we will also do something similar.

The Glue ETL blueprint stores the final output in the Parquet file format, which is widely used to set up data lakes. However, as the volume of processed and stored data increases, so does the storage cost. To enhance the storage efficiency of these Parquet files, compression algorithms can be applied to store them in a compressed format. However, determining the optimal compression algorithm that balances storage savings with performance can be challenging for developers. Without clear guidance on how to approach this task and test the compression-to-performance ratio, developers may spend considerable time on research before implementing changes. Alternatively, they can seek quick solutions by consulting Amazon Q's feature development capability, integrated into CodeCatalyst.

The following figure provides a good workflow of how Amazon Q helps with feature development.

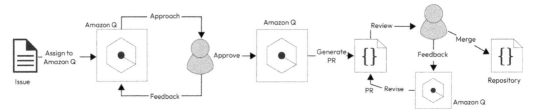

Figure 16.5 – Amazon Q feature development workflow

Let's understand all these steps in detail and see how Amazon Q can expedite this laborious DevOps process for the Glue project we have selected in this chapter.

Assigning issues to Amazon Q

Say an ETL developer knows that they need to apply some compression logic to the code but is unsure which algorithm to apply. Also, the developer is not sure what code would need to be changed throughout the code base. This is where Amazon Q can help. The developer will create an **issue** in CodeCatalyst and describe their issue in as much detail as possible.

The following screenshot shows an issue that we would create with a title and description of what our end goal is.

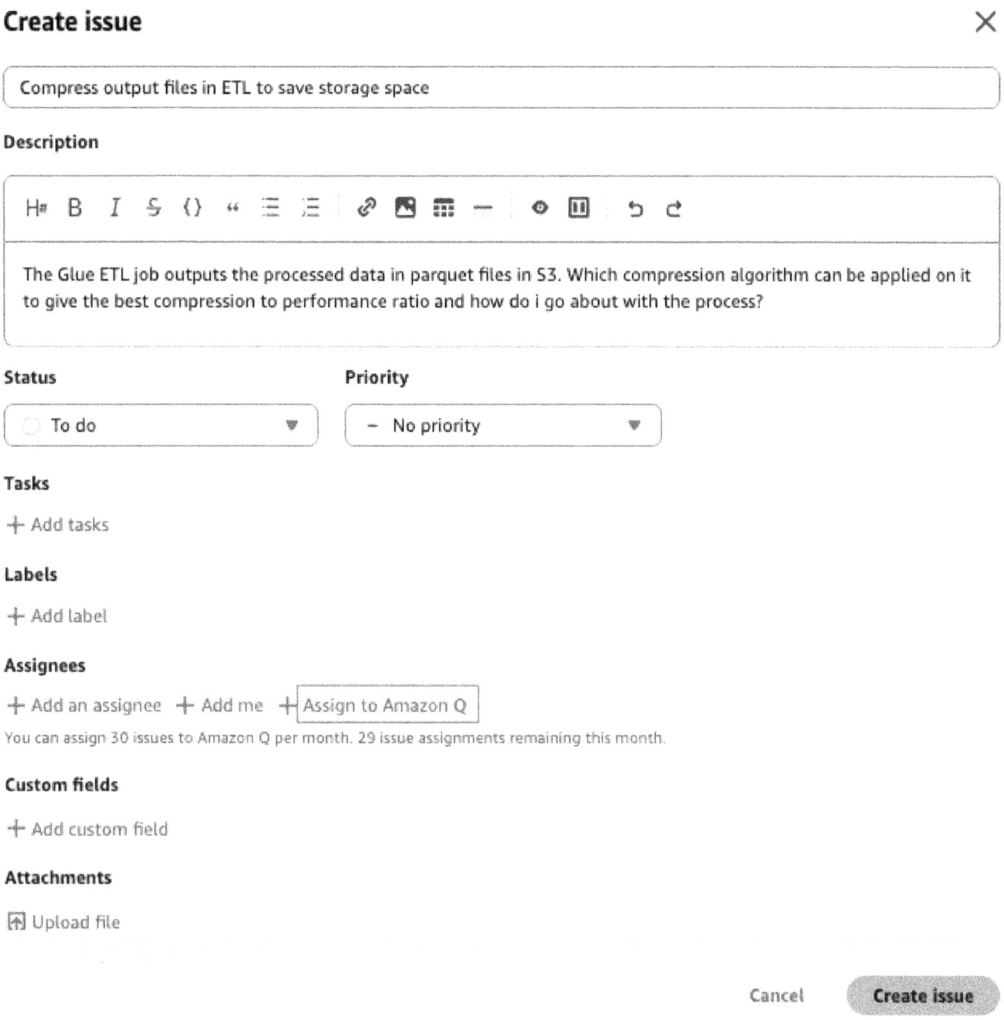

Create issue   ✕

Compress output files in ETL to save storage space

Description

| H# | B | *I* | S | {} | " | ☰ | ☷ | | 🔗 | 🖼 | ▦ | — | | ⊙ | ⏸ | | ↺ | ↻ |

The Glue ETL job outputs the processed data in parquet files in S3. Which compression algorithm can be applied on it to give the best compression to performance ratio and how do i go about with the process?

Status   **Priority**

◯   To do   ▼   —   No priority   ▼

Tasks

╋ Add tasks

Labels

╋ Add label

Assignees

╋ Add an assignee   ╋ Add me   ╋ Assign to Amazon Q

You can assign 30 issues to Amazon Q per month. 29 issue assignments remaining this month.

Custom fields

╋ Add custom field

Attachments

⤴ Upload file

Cancel   (**Create issue**)

Figure 16.6 – Create an issue in Amazon CodeCatalyst and assign it to Amazon Q

We can keep the status as **To do** and the priority as **No priority** as the default options for now. Note that instead of assigning this to a user, we have the option to assign this to Amazon Q instead.

Once it's assigned to Amazon Q, it wants us to specify how it should proceed with the flow. First, we provided the source repository that this issue should work against. Then, it's asking whether the user wants to review code change suggestions along the way. Based on user comments, Q can adjust its approach a certain number of times. For the vast majority of cases, it's always good to keep this option enabled as the suggestions may not always align with our desired code. Next, it's also asking whether Q can modify the workflow files on its own. In this case, we disabled this just to be on the safe side. Q can now also recommend tasks based on its analysis of the issue. For now, we'll keep it turned off but we will look at this option again during the next stage.

The following screenshot shows our selection when we assign the issue to Amazon Q.

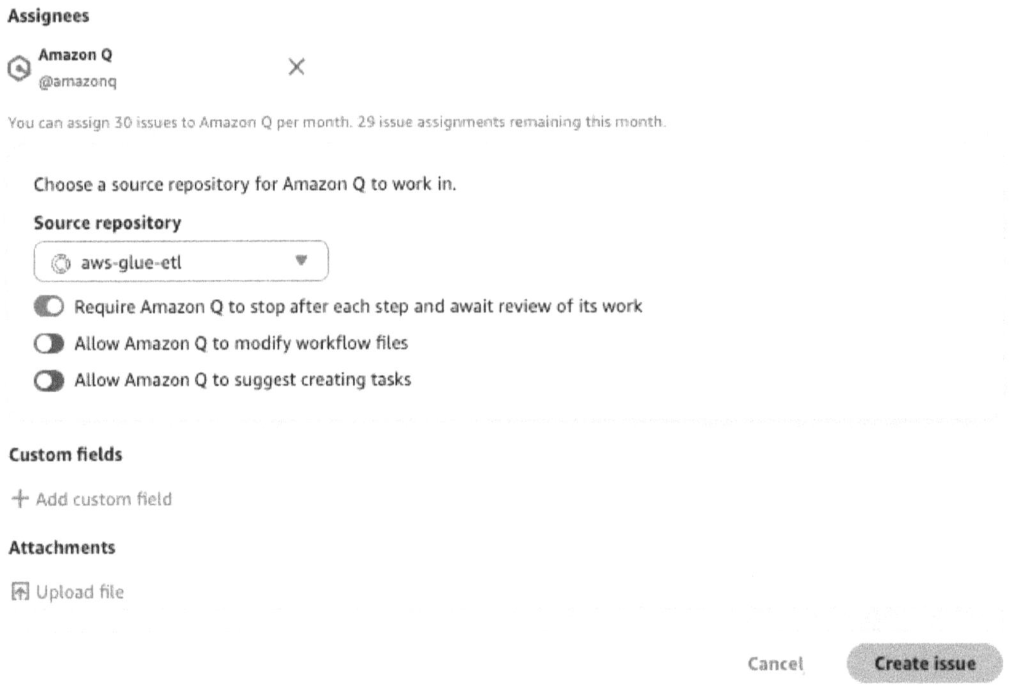

Assignees

Amazon Q
@amazonq   ✕

You can assign 30 issues to Amazon Q per month. 29 issue assignments remaining this month.

Choose a source repository for Amazon Q to work in.

Source repository

aws-glue-etl   ▾

Require Amazon Q to stop after each step and await review of its work

Allow Amazon Q to modify workflow files

Allow Amazon Q to suggest creating tasks

Custom fields

+ Add custom field

Attachments

⤒ Upload file

Cancel   **Create issue**

Figure 16.7 – Initial input required by Amazon to proceed with the issue

Once we confirm our selection and create the issue, Amazon Q moves the issue to the **In progress** lane, as highlighted in the following screenshot.

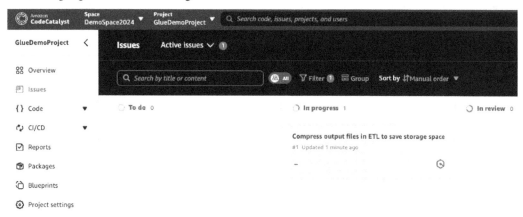

Figure 16.8 – Issue being analyzed by Amazon Q in the In progress lane

The next step is to generate an approach to solve the issue and provide the developer with the option to provide any feedback via prompts.

Generating an approach and providing feedback

This step is where all the real magic happens behind the scenes. Q will first evaluate any prerequisites and then proceed to read the code repository you provided in the previous step.

The following screenshot highlights the approach Q will take to reach the end goal. Observe that it waits for the user's response before it can proceed further.

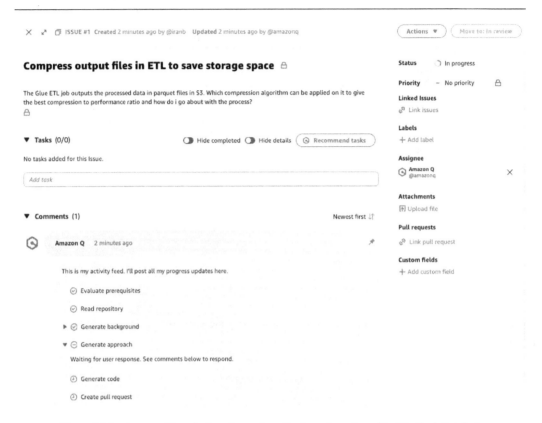

Figure 16.9 – Amazon Q analyzing steps when the issue is assigned to it in CodeCatalyst

The **Recommend tasks** button is also presented on this screen, where Q can analyze the issue and assign more manageable tasks to users for faster actions. In our use case here, we don't need to use this feature, but feel free to try it out for more complex issues you want to create in your workplace.

Based on what it learned, it will generate a summary background of what it understood about the project. The following screenshot highlights the background generated for the AWS Glue ETL project.

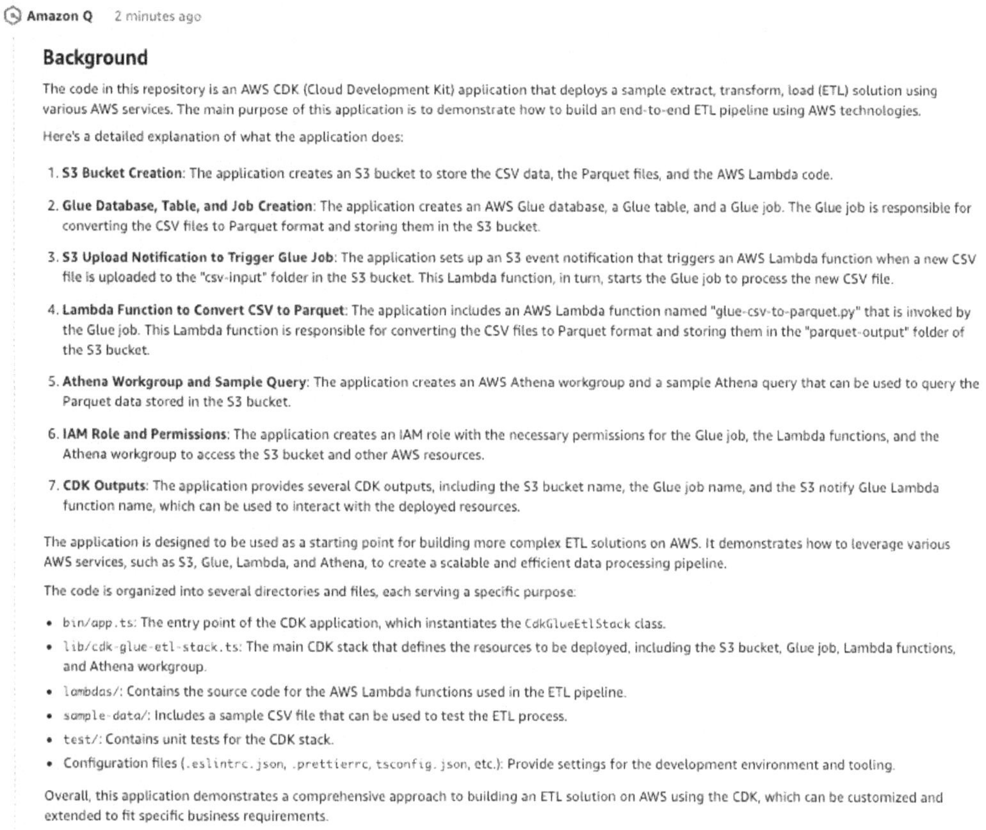

Amazon Q   2 minutes ago

Background

The code in this repository is an AWS CDK (Cloud Development Kit) application that deploys a sample extract, transform, load (ETL) solution using various AWS services. The main purpose of this application is to demonstrate how to build an end-to-end ETL pipeline using AWS technologies.

Here's a detailed explanation of what the application does:

1. **S3 Bucket Creation**: The application creates an S3 bucket to store the CSV data, the Parquet files, and the AWS Lambda code.

2. **Glue Database, Table, and Job Creation**: The application creates an AWS Glue database, a Glue table, and a Glue job. The Glue job is responsible for converting the CSV files to Parquet format and storing them in the S3 bucket.

3. **S3 Upload Notification to Trigger Glue Job**: The application sets up an S3 event notification that triggers an AWS Lambda function when a new CSV file is uploaded to the "csv-input" folder in the S3 bucket. This Lambda function, in turn, starts the Glue job to process the new CSV file.

4. **Lambda Function to Convert CSV to Parquet**: The application includes an AWS Lambda function named "glue-csv-to-parquet.py" that is invoked by the Glue job. This Lambda function is responsible for converting the CSV files to Parquet format and storing them in the "parquet-output" folder of the S3 bucket.

5. **Athena Workgroup and Sample Query**: The application creates an AWS Athena workgroup and a sample Athena query that can be used to query the Parquet data stored in the S3 bucket.

6. **IAM Role and Permissions**: The application creates an IAM role with the necessary permissions for the Glue job, the Lambda functions, and the Athena workgroup to access the S3 bucket and other AWS resources.

7. **CDK Outputs**: The application provides several CDK outputs, including the S3 bucket name, the Glue job name, and the S3 notify Glue Lambda function name, which can be used to interact with the deployed resources.

The application is designed to be used as a starting point for building more complex ETL solutions on AWS. It demonstrates how to leverage various AWS services, such as S3, Glue, Lambda, and Athena, to create a scalable and efficient data processing pipeline.

The code is organized into several directories and files, each serving a specific purpose:

- `bin/app.ts`: The entry point of the CDK application, which instantiates the `CdkGlueEtlStack` class.
- `lib/cdk-glue-etl-stack.ts`: The main CDK stack that defines the resources to be deployed, including the S3 bucket, Glue job, Lambda functions, and Athena workgroup.
- `lambdas/`: Contains the source code for the AWS Lambda functions used in the ETL pipeline.
- `sample-data/`: Includes a sample CSV file that can be used to test the ETL process.
- `test/`: Contains unit tests for the CDK stack.
- Configuration files (`.eslintrc.json`, `.prettierrc`, `tsconfig.json`, etc.): Provide settings for the development environment and tooling.

Overall, this application demonstrates a comprehensive approach to building an ETL solution on AWS using the CDK, which can be customized and extended to fit specific business requirements.

Figure 16.10 – Amazon Q background generation step in CodeCatalyst

The background is helpful in summarizing what the code base is all about. The next step is the main one, as it presents the approach it would take to solve the issue that was provided as input in the description of the issue.

The following screenshot highlights the approach for the compression use case we presented to Q.

⊙ **Amazon Q**   3 minutes ago

Approach

To accomplish the task of compressing the output files in the ETL process, the following steps can be taken:

1. Evaluate the compression options for Parquet files:

    1. Review the available compression codecs for Parquet files, such as Snappy, Gzip, Brotli, and LZO.
    2. Analyze the trade-offs between compression ratio and decompression performance for each codec.
    3. Determine the most suitable compression codec based on the requirements, such as storage savings and query performance.

2. Update the Glue job configuration to use the selected compression codec:

    1. Open the file `lib/cdk-glue-etl-stack.ts`.
    2. Locate the `glueTable` definition (around line 77).
    3. Add the `compressionType` property to the `glueTable` configuration, and set it to the selected compression codec:

    ```
    const glueTable = new glue.Table(this, 'GlueTable', {
        // ... other existing properties
        compressionType: glue.CompressionType.SNAPPY, // or any other selected codec
    });
    ```

3. Verify the compression in the ETL process:

    1. Deploy the updated CDK stack to your AWS environment.
    2. Upload a sample CSV file to the "csv-input" folder in the S3 bucket.
    3. Observe the Glue job execution and the resulting Parquet files in the "parquet-output" folder.
    4. Verify the file sizes and compression ratios of the Parquet files.

4. Optimize the compression settings (optional):

    1. If the initial compression settings do not meet the desired storage savings or performance requirements, consider the following:
        1. Experiment with different compression codecs and their configurations.
        2. Analyze the impact of compression on query performance by running sample Athena queries against the Parquet data.
        3. Adjust the compression settings in the `glueTable` configuration accordingly.

5. Update the documentation:

    1. Open the `README.md` file.
    2. Add a new section or update the existing "Project resources" section to describe the compression settings used for the Parquet files.
    3. Provide guidance on how to customize the compression settings in the future, if needed.

Figure 16.11 – Amazon Q-generated approach in CodeCatalyst – part 1

At the end of the step, once you understand the changes proposed by Q, you have two options. In some cases, Q will provide perfect choices and everything proposed may seem 100% accurate. In such cases, you can simply proceed. However, it's important to keep in mind, as we have been constantly reminded, that LLM-based solutions may not always be exactly what you want, but Q is willing to learn and provide better or alternative solutions. This is where prompt engineering comes in, and Amazon Q allows you to provide prompts via the comments section so that it can modify its approach and provide a different solution.

The following screenshot highlights the options Q presents for its generated approach. For our use case, we will accept the approach Q presented and proceed to the next step.

6. Add unit tests for the compression settings:

   1. Open the test/glue-etl-stack.test.ts file.

   2. Add a new test case to verify the compression settings of the Glue table:

```
test('Glue table uses the expected compression codec', () => {
  template.hasResourceProperties('AWS::Glue::Table', {
    TableInput: {
      StorageDescriptor: {
        Compressed: 'Snappy', // or the expected compression codec
      },
    },
  });
});
```

   3. Run the updated unit tests to ensure the compression settings are correctly applied.

By following these steps, you can update the ETL process to use the selected compression codec for the Parquet output files, optimize the compression settings, and ensure the changes are properly documented and tested.

I need to know if you agree with this approach before I can continue. If you want me to change my approach, reply to this comment. If you want me to start work on the code, choose Proceed.

(Proceed)   (Reply)

Behram Irani (you)

Ha B *I* S {} " ☰ ☰ ⊙ ▥	Comment
Add a comment	

Figure 16.12 – Amazon Q-generated approach in CodeCatalyst – part 2

Observe how Amazon Q systematically lays down possible next steps. Firstly, it understood our requirement and suggested using Snappy compression, which would provide a good balance between compression and speed. It then provided code snippets that will need to be added or changed in every source code file to make this work. For developers, researching possible solutions to a problem and understanding all the places where code changes are required can take a significant amount of time and effort. Amazon Q did all this on our behalf in just a few minutes.

To add a bit of intrigue, one of the code snippets provided in the approach seems to have a problem. I could have provided additional prompts and attempted to get Amazon Q to provide me with the exact code that would work. However, in real life, sometimes issues slip through the cracks. But don't worry; we will have another opportunity to fix this at a later stage. This also demonstrates how Q provides other features to move toward the perfect final solution within CodeCatalyst.

For now, we accept the approach and instruct it to proceed. The next step in the workflow is the generation of pull requests.

Generating pull requests

Once Q receives our marching orders to proceed with the approved approach, it will generate the necessary code and create a pull request. At this step, the workflow moves to the review stage. Developers will once again appreciate how much time this is going to save them.

The following screenshot shows Amazon Q doing all the work of branching and creating the necessary changes and presenting all this in a pull request for the approver to review.

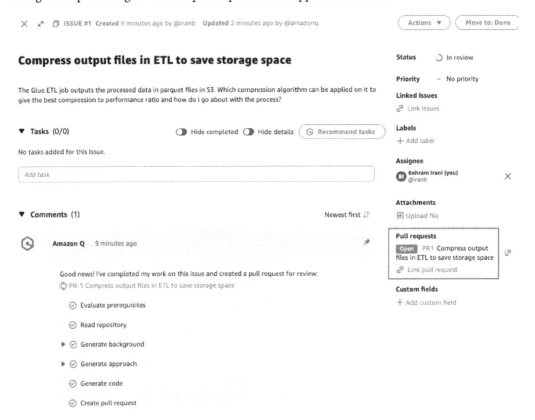

Figure 16.13 – Amazon Q completed the approach and generated a pull request

When you click on the pull request, the approach is laid out in the overview, making it easy for the reviewer to understand the changes that were made in the code base. The following screenshot shows the overview page of the pull request, which summarizes the changes.

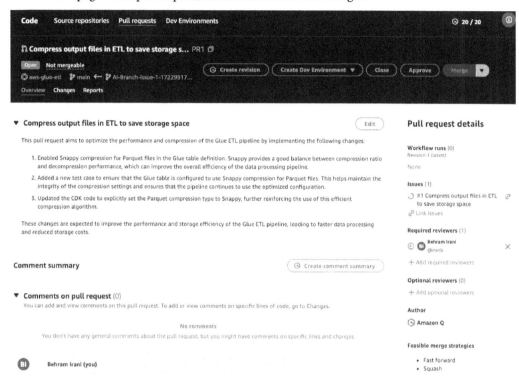

Figure 16.14 – Amazon Q-generated pull request overview in CodeCatalyst

The generation of a pull request, along with a detailed description of the changes, really helps with the next step of the workflow, which is to review and revise the suggested changes in the code base.

Reviewing and revising code changes

The approver would typically want to compare the proposed changes on a side-by-side comparison screen so that it's easy to understand all the changes in different code files. The following screenshot shows a comparison screen highlighting all the changes proposed by Amazon Q in green.

Figure 16.15 – Amazon Q-generated pull request code comparison in CodeCatalyst

Recall a while back we mentioned that there seems to be an issue in one of the code snippets suggested by the Amazon Q approach. Since we did not notice it during the previous step, it is prominent in the pull request when the actual code is presented for review. No team wants incorrect code to creep into the repository, hence it is usually a best practice to review and approve pull requests before the code is merged into the main branch.

Upon closer inspection, it is observed that `compression` is not a direct property of the Glue table in the CDK code file. Therefore, the code `compression:glue.Compression.SNAPPY` will fail to compile during the build stage. Instead of doing this all over again, CodeCatalyst will allow the reviewer to add comments right inside the pull request screen and make a revision request.

In the following screenshot, we added a comment next to the incorrect line of code, asking the system to use a different logic.

```
78              database: glueDatabase,
79              bucket: bucket,
80              s3Prefix: "parquet-output/",
81    +              // Use Snappy compression for Parquet files
82    +              // Snappy provides a good balance between
        compression ratio and decompression performance
83    +              compressionType: glue.CompressionType.SNAPPY,
```

H⍺ B I S {} " ≡ ≔ | 𝒫 🖼 ▦ — | Cancel   Save

⊙ ⊞ ↺ ↻

use storageParameters: [
    glue.StorageParameter.compressionType(glue.CompressionType.SNAPPY)
]|

Figure 16.16 – Add a comment in the pull request and request a revision in CodeCatalyst

When you hit the **Create Revision** button on the pull request screen, Amazon Q is invoked again. It takes the comment made in the code as a prompt and creates a new revision of the pull request with the correct logic this time. This is highlighted in the following screenshot, where revision v2 was able to come back with the correct code.

```
@ -78,7 +78,10 @@
78              database: glueDatabase,
79              bucket: bucket,
80              s3Prefix: "parquet-output/",
81    +          storageParameters: [
82    +            glue.StorageParameter.compressionType(glue.CompressionType.SNAPPY)
83    +          ],
84    +          tableName: "taxi_zone_lookup",
```

Figure 16.17 – New, updated code in the revised pull request in CodeCatalyst

Once the reviewer confirms everything looks correct, they can approve the pull request in CodeCatalyst as the next step. This is highlighted in the following screenshot.

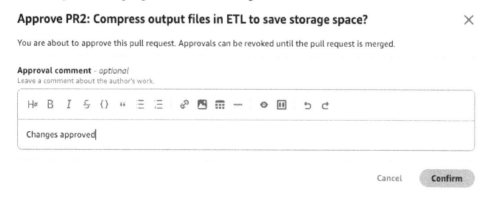

Approve PR2: Compress output files in ETL to save storage space?   ✕

You are about to approve this pull request. Approvals can be revoked until the pull request is merged.

Approval comment - *optional*
Leave a comment about the author's work.

H⍺ B I S {} " ≡ ≔ 𝒫 🖼 ▦ — ⊙ ⊞ ↺ ↻

Changes approved|

                                                    Cancel   Confirm

Figure 16.18 – New, updated code approved in the revised pull request in CodeCatalyst

This brings us to the final step in the workflow, which is to merge the approved code changes back into the repository.

Merging code changes in the repository

Once the review and approval are done, the code can be merged into the main branch, as seen in the following screenshot.

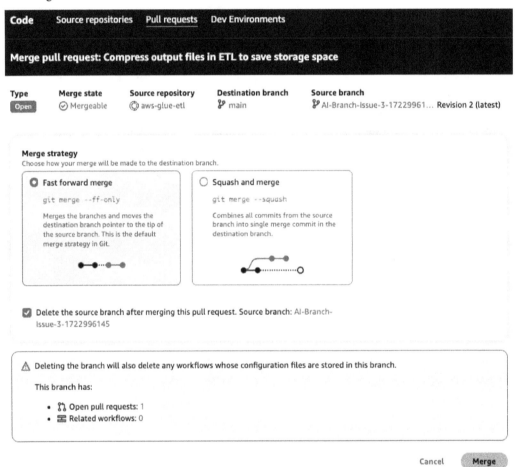

Figure 16.19 – Code merge process in CodeCatalyst

The merge request triggers the workflow to recompile, rebuild, and redeploy the code via the CI/CD process. You can always go to the workflows in the CI/CD section to confirm the successful completion of the build and deploy process. The following screenshot shows the successful build of the workflow after the pull request was merged.

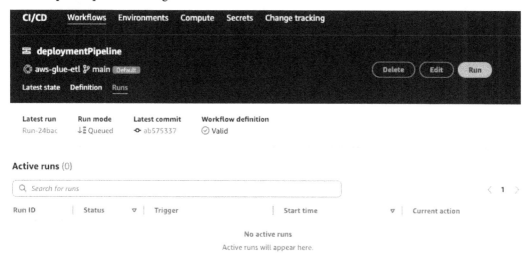

Figure 16.20 – Workflows in CodeCatalyst

For our Glue ETL use case, the workflow will compile the code and trigger the AWS CloudFormation stack via the CDK logic. The CloudFormation stack, in turn, will recreate/update the necessary infrastructure that was affected by the code changes. In our case, you can go to the Glue console and check the updated ETL job with the compression logic in it ready to be triggered for the next run.

Before we wind down the feature development capability in CodeCatalyst using Amazon Q, I also want to provide an alternate scenario of failure. In the world of software engineering, not everything goes as planned. What if during the code review process, the reviewer did not catch the code issue and did not fix it using the revision feature? Well, they would have approved the request, and the merge process would have triggered the workflow, eventually resulting in failure.

The following screenshot shows the failed workflow if the code issue had not been revised during the pull request process.

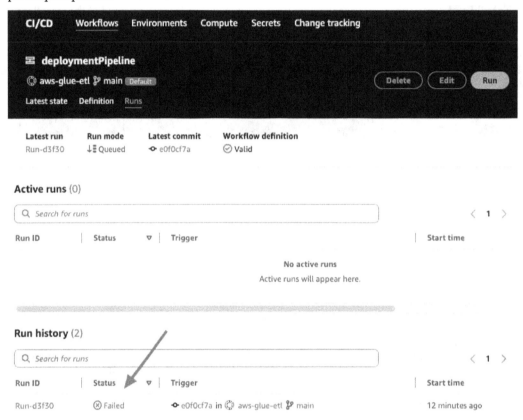

Figure 16.21 – Failed workflow in CodeCatalyst due to code issue

At this point, the development team enters firefighting mode and identifies the error causing the failure from the logs provided within the failed workflow. The following screenshot shows the error that caused the failure.

```
CDKBootstrapAction                                                                                          ×
⊗ Failed   Start time: 10 minutes ago      Duration: 2 minutes 37 seconds

Logs    Summary    Configuration    Variables

221 [01:11:18] output: cdk.out
222 [01:11:18] env: {
223 CDK_DEFAULT_REGION: 'us-west-2',
224 CDK_DEFAULT_ACCOUNT: '926211135825',
225 CDK_OUTDIR: 'cdk.out',
226 CDK_CLI_ASM_VERSION: '34.0.0',
227 CDK_CLI_VERSION: '2.99.1'
228 }
229 /codecatalyst/output/src3409/src/git.us-west-2.codecatalyst.aws/v1/DemoSpace2024/GlueDemoProject/aws-glue-etl/node_modules/ts-node/src/index.ts:
230 return new TSError(diagnosticText, diagnosticCodes, diagnostics);
231 ^
232 TSError: × Unable to compile TypeScript:
233 lib/cdk-glue-etl-stack.ts(83,9): error TS2345: Argument of type '{ database: glue.Database; bucket: cdk.aws_s3.Bucket; s3Prefix: string; compre:
234 Object literal may only specify known properties, and 'compressionType' does not exist in type 'S3TableProps'.
235 lib/cdk-glue-etl-stack.ts(105,34): error TS2339: Property 'ParquetCompression' does not exist on type 'typeof import("/codecatalyst/output/src3.
236
237 at createTSError (/codecatalyst/output/src3409/src/git.us-west-2.codecatalyst.aws/v1/DemoSpace2024/GlueDemoProject/aws-glue-etl/node_modules/ts-
238 at reportTSError (/codecatalyst/output/src3409/src/git.us-west-2.codecatalyst.aws/v1/DemoSpace2024/GlueDemoProject/aws-glue-etl/node_modules/ts-
239 at getOutput (/codecatalyst/output/src3409/src/git.us-west-2.codecatalyst.aws/v1/DemoSpace2024/GlueDemoProject/aws-glue-etl/node_modules/ts-nodo
240 at Object.compile (/codecatalyst/output/src3409/src/git.us-west-2.codecatalyst.aws/v1/DemoSpace2024/GlueDemoProject/aws-glue-etl/node_modules/ts
241 at Module.m._compile (/codecatalyst/output/src3409/src/git.us-west-2.codecatalyst.aws/v1/DemoSpace2024/GlueDemoProject/aws-glue-etl/node_module:
242 at Module._extensions..js (node:internal/modules/cjs/loader:1252:10)
243 at Object.require.extensions.<computed> [as .ts] (/codecatalyst/output/src3409/src/git.us-west-2.codecatalyst.aws/v1/DemoSpace2024/GlueDemoProj(
244 at Module.load (node:internal/modules/cjs/loader:1076:32)
245 at Function.Module._load (node:internal/modules/cjs/loader:911:12)
246 at Module.require (node:internal/modules/cjs/loader:1100:19) {
247 diagnosticCodes: [ 2345, 2339 ]
248 }
249 [01:11:25] Notices refreshed
250
251 Subprocess exited with error 1
252 [01:11:25] Error: Subprocess exited with error 1
```

Figure 16.22 – Failed workflow cause in the logs of the workflow in CodeCatalyst

One of the next steps after this would be to create another issue in CodeCatalyst, provide the error details in the description, and assign it back to Amazon Q for analysis and troubleshooting. You may have to engage in prompt engineering and provide multiple comments along the way for Amazon Q to arrive at the correct logic that would fix the issue. Once you go through the entire process, the new code will again be included in the pull request for approval and merging into the main branch.

Before we wind down this chapter, we will quickly look at a couple of other cool features in CodeCatalyst using Amazon Q that help developers save time and effort.

Amazon Q's summarizing capability in Amazon CodeCatalyst

Developers will understand the pain of trying to comprehend the changes made in a pull request submission. Typically, the people who make code changes are not the ones approving them. So, someone else needs to understand all the changes made in the code base, analyze their intended effect, and finally, approve them if they meet the standards.

Without proper comments, it's difficult to understand all the nuances of the changes made. Sometimes, even with comments, it gets overwhelming to grasp the full intention of the changes. This is where Amazon Q steps in to make life easier for developers and approvers.

Let's look at Amazon Q's ability to analyze code changes and generate a summary of the changes when creating a pull request, as well as create a summary based on all the comments made on the code changes in a pull request.

Pull request comment summary

When reviewing a pull request, users often leave numerous comments regarding the proposed changes. With multiple reviewers providing feedback, it can become challenging to identify common themes or ensure a thorough review of all comments across revisions. To streamline this process, the **Create comment summary** feature utilizes Amazon Q to analyze all comments left on code changes within a pull request and generate a concise summary of the feedback.

The following screenshot highlights the summary generated by Amazon Q for the compression changes we did in our pull request.

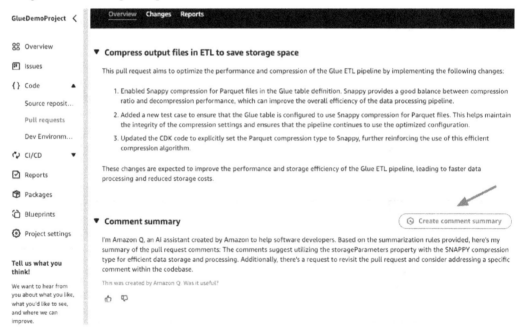

Figure 16.23 – Pull request content summary

Pull request description

Pull requests serve as a collaborative platform for reviewing code alterations, whether they're minor fixes, significant feature additions, or updates to released software versions. Including a summary of the code changes and their purpose in the pull request description aids reviewers in comprehending the alterations and fosters a historical perspective on the code evolution. However, developers frequently depend on the code itself to convey information, resulting in vague or insufficiently detailed descriptions that hinder reviewers' understanding of the changes or their underlying intent.

This is where the **Write description for me** feature comes in handy, as it is able to generate a description of the changes in the pull request, making it easier to review and approve changes. The following screen shows the **Write description for me** option on the pull request screen.

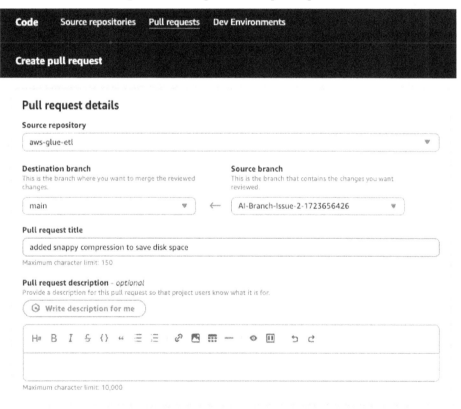

Figure 16.24 – Pull request description generation feature in CodeCatalyst

Amazon Q examines the disparities between the source branch containing the code modifications and the target branch where you intend to merge these alterations. Subsequently, it generates a summary outlining the nature of these modifications and offers its most accurate interpretation of their purpose and impact.

The following screenshot highlights the description generated by Amazon Q using all the information from the pull request.

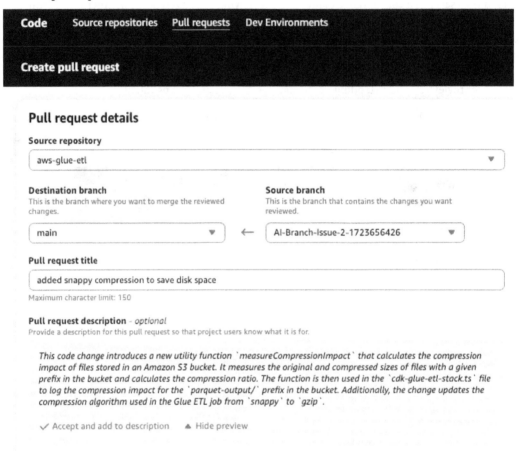

Figure 16.25 – Auto-generated description on the CodeCatalyst pull request screen

CodeCatalyst is a comprehensive service with numerous features of its own. We have only covered some of the next-generation features provided in the service via Amazon Q, which is a generative AI-powered assistant.

Summary

In this chapter, we covered some of the challenges faced by developers during the DevOps process. We then introduced Amazon CodeCatalyst, a comprehensive software development service designed for development teams to rapidly construct, deploy, and scale applications on AWS.

After that, we leveraged a use case to demonstrate the different generative AI capabilities of the service. Amazon Q allows developers to rapidly add new features to software. Developers can create issues and assign them to Amazon Q for comprehensive understanding and recommendations for different approaches needed to solve the particular task provided as input. Q can also be prompted to provide revised approaches based on the comments that users provide. It then generates the code and creates a pull request so that reviewers can examine the proposed changes and either approve or request revisions before merging them back into the main repository.

We also explored how Amazon Q provides pull request comment summaries and generates pull request descriptions. All these features boost developer productivity and save significant time during the entire software development process.

Final thoughts

Thank you for taking the time to go on the journey we took in this book. Even though we have reached the end of this book on Amazon Q Developer and covered many topics, we are still in the infancy stage of generative AI. Rapid innovation is happening across the board, so keep an open mind when it comes to seeing and exploring new features as they are rolled out. Many of the features we described may evolve for the better. The goal of writing this book was to give you a glimpse of the art possible with Amazon Q Developer.

Generative AI-powered assistants, such as Amazon Q Developer, are set to transform the landscape for both current software engineers and aspiring developers who may lack resources or skills. For established engineers, AI will significantly boost productivity by generating code snippets, providing real-time feedback, and automating tedious tasks such as debugging and documentation as we have seen in this book. This enables them to focus on more complex and creative aspects of development. AI can also facilitate better collaboration within teams by summarizing discussions and translating language barriers, fostering a more inclusive work environment.

For aspiring developers, particularly those with limited access to resources or formal education, AI offers personalized learning experiences tailored to their strengths and weaknesses. It provides instant feedback and learning support, making it easier to acquire new skills and understand complex concepts. AI-powered tools can democratize access to coding education by offering mentorship, career path recommendations, and resources that adapt to evolving industry trends. Additionally, AI can help break down barriers for individuals with physical or learning disabilities through voice-activated coding and other accessibility features.

So, to conclude, if you are feeling overwhelmed with learning anything new in the world of information technology, this is your moment to embrace generative AI-powered assistants such as Amazon Q Developer. The more you use them, the more you will empower yourself to achieve greater heights in your career.

Never stop learning; you'll be amazed by how many new doors it will open for you.

References

- Amazon CodeCatalyst documentation: `https://docs.aws.amazon.com/codecatalyst/latest/userguide/welcome.html`

- Amazon CodeCatalyst workshop: `https://catalog.workshops.aws/ccdevops/`

- Amazon CodeCatalyst – Modern three-tier web application tutorial: `https://docs.aws.amazon.com/codecatalyst/latest/userguide/getting-started-template-project.html`

- Managing generative AI features in Amazon CodeCatalyst: `https://docs.aws.amazon.com/codecatalyst/latest/adminguide/managing-generative-ai-features.html`

Index

`packtpub.com`

Subscribe to our online digital library for full access to over 7,000 books and videos, as well as industry leading tools to help you plan your personal development and advance your career. For more information, please visit our website.

Why subscribe?

- Spend less time learning and more time coding with practical eBooks and Videos from over 4,000 industry professionals

- Improve your learning with Skill Plans built especially for you

- Get a free eBook or video every month

- Fully searchable for easy access to vital information

- Copy and paste, print, and bookmark content

Did you know that Packt offers eBook versions of every book published, with PDF and ePub files available? You can upgrade to the eBook version at `packtpub.com` and as a print book customer, you are entitled to a discount on the eBook copy. Get in touch with us at `customercare@packtpub.com` for more details.

At `www.packtpub.com`, you can also read a collection of free technical articles, sign up for a range of free newsletters, and receive exclusive discounts and offers on Packt books and eBooks.

Other Books You May Enjoy

If you enjoyed this book, you may be interested in these other books by Packt:

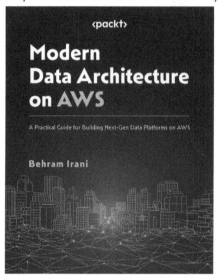

Modern Data Architecture on AWS

Behram Irani

ISBN: 978-1-80181-339-6

- Familiarize yourself with the building blocks of modern data architecture on AWS
- Discover how to create an end-to-end data platform on AWS
- Design data architectures for your own use cases using AWS services
- Ingest data from disparate sources into target data stores on AWS
- Build data pipelines, data sharing mechanisms, and data consumption patterns using AWS services
- Find out how to implement data governance using AWS services

Generative AI with Amazon Bedrock

Shikhar Kwatra, Bunny Kaushik

ISBN: 978-1-80324-728-1

- Explore the generative AI landscape and foundation models in Amazon Bedrock
- Fine-tune generative models to improve their performance
- Explore several architecture patterns for different business use cases
- Gain insights into ethical AI practices, model governance, and risk mitigation strategies
- Enhance your skills in employing agents to develop intelligence and orchestrate tasks
- Monitor and understand metrics and Amazon Bedrock model response
- Explore various industrial use cases and architectures to solve real-world business problems using RAG
- Stay on top of architectural best practices and industry standards

Packt is searching for authors like you

If you're interested in becoming an author for Packt, please visit authors.packtpub.com and apply today. We have worked with thousands of developers and tech professionals, just like you, to help them share their insight with the global tech community. You can make a general application, apply for a specific hot topic that we are recruiting an author for, or submit your own idea.

Share Your Thoughts

Now you've finished *Generative AI-Powered Assistant for Developers*, we'd love to hear your thoughts! Scan the QR code below to go straight to the Amazon review page for this book and share your feedback or leave a review on the site that you purchased it from.

https://packt.link/r/1-835-08914-3

Your review is important to us and the tech community and will help us make sure we're delivering excellent quality content.

Download a free PDF copy of this book

Thanks for purchasing this book!

Do you like to read on the go but are unable to carry your print books everywhere?

Is your eBook purchase not compatible with the device of your choice?

Don't worry, now with every Packt book you get a DRM-free PDF version of that book at no cost.

Read anywhere, any place, on any device. Search, copy, and paste code from your favorite technical books directly into your application.

The perks don't stop there, you can get exclusive access to discounts, newsletters, and great free content in your inbox daily

Follow these simple steps to get the benefits:

1. Scan the QR code or visit the link below

https://packt.link/free-ebook/978-1-83508-914-9

2. Submit your proof of purchase
3. That's it! We'll send your free PDF and other benefits to your email directly

www.ingramcontent.com/pod-product-compliance
Lightning Source LLC
Chambersburg PA
CBHW060650060326
40690CB00020B/4584